# LANDSCAPE GENETICS

# LANDSCAPE GENETICS

## CONCEPTS, METHODS, APPLICATIONS

Edited by

### Niko Balkenhol
Department of Wildlife Sciences, University of Göttingen, Germany

### Samuel A. Cushman
Forest and Woodlands Ecosystems Program, Rocky Mountain Research Station, United States Forest Service, USA

### Andrew T. Storfer
School of Biological Sciences, Washington State University, USA

### Lisette P. Waits
Fish and Wildlife Sciences, University of Idaho, USA

**WILEY** Blackwell

This edition first published 2016 © 2016 by John Wiley & Sons Ltd

*Registered office*: John Wiley & Sons, Ltd, The Atrium, Southern Gate, Chichester, West Sussex, PO19 8SQ, UK

*Editorial offices:* 9600 Garsington Road, Oxford, OX4 2DQ, UK
The Atrium, Southern Gate, Chichester, West Sussex, PO19 8SQ, UK
111 River Street, Hoboken, NJ 07030-5774, USA

For details of our global editorial offices, for customer services and for information about how to apply for permission to reuse the copyright material in this book please see our website at www.wiley.com/wiley-blackwell.

The right of the author to be identified as the author of this work has been asserted in accordance with the UK Copyright, Designs and Patents Act 1988.

*Library of Congress Cataloging-in-Publication Data*

Landscape genetics : concepts, methods, applications / edited by Niko Balkenhol, Samuel A. Cushman, Andrew T. Storfer, and Lisette P. Waits.
     pages cm
  Includes bibliographical references and index.
  ISBN 978-1-118-52528-9 (cloth) – ISBN 978-1-118-52529-6 (pbk.)
1.  Ecological genetics. 2.  Landscape ecology. 3.  Population genetics. I. Balkenhol, Niko.
  QH456.L36 2015
  576.5'8–dc23
                                                                          2015015467

A catalogue record for this book is available from the British Library.

Wiley also publishes its books in a variety of electronic formats. Some content that appears in print may not be available in electronic books.

Set in 9/11pt, PhotinaMTStd by Thomson Digital, Noida, India

1   2016

# CONTENTS

# LIST OF CONTRIBUTORS

**Michael F. Antolin**
Department of Biology, Colorado State University, USA

**Niko Balkenhol**
Department of Wildlife Sciences, University of Göttingen, Germany

**Samuel A. Cushman**
Forest and Woodlands Ecosystems Program, Rocky Mountain Research Station, United States Forest Service, USA

**Rodney J. Dyer**
Department of Biology, Virginia Commonwealth University, USA

**Bryan K. Epperson**
Department of Forestry, Michigan State University, USA

**Marie-Josée Fortin**
Department of Ecology and Evolutionary Biology, University of Toronto, Canada

**Olivier François**
Grenoble INP, Université Grenoble-Alpes, France

**Heather M. Galindo**
University of Washington Bothell, USA

**Erin Landguth**
Division of Biological Sciences, University of Montana, USA

**Stéphanie Manel**
Centre d'Ecologie Fonctionnelle et Evolutive (CEFE), France

**Kevin McGarigal**
Department of Natural Resources Conservation, University of Massachussetts, USA

**Brad H. McRae**
The Nature Conservancy, North America Region

**Melanie Murphy**
Department of Ecosystem Science and Management, Program in Ecology, University of Wyoming, USA

**Kim T. Scribner**
Department of Fisheries and Wildlife & Department of Zoology, Michigan State University, USA

**Kimberly A. Selkoe**
National Center for Ecological Analysis and Synthesis (NCEAS), University of California Santa Barbara, USA & Hawaii Institute of Marine Biology, University of Hawaii, USA

**Stephen F. Spear**
The Orianne Society, USA

**Andrew Storfer**
School of Biological Sciences, Washington State University, USA

**Helene H. Wagner**
Department of Ecology and Evolutionary Biology, University of Toronto, Canada

**Lisette P. Waits**
Fish and Wildlife Sciences, University of Idaho, USA

# WEBSITE

Please visit the website accompanying this book to learn about the newest developments in landscape genetics:

**www.landscapegenetics.info**

The website lists landscape genetic papers, provides links to analytical tools and research labs, and announces jobs, conferences, and training opportunities in landscape genetics.

# ACKNOWLEDGMENTS

This book is the result of a long and ongoing journey that started with the seemingly simple idea of teaching landscape genetics for graduate students and professionals. Since the first landscape genetics workshop and symposium were conducted at the 2007 International Association for Landscape Ecology (IALE) in Wageningen, Netherlands, we and several of the chapter authors have collaborated to teach multiple landscape genetics short courses, classes and workshops. We are grateful to the various funding agencies that allowed us to organize these training opportunities, including the Vokswagen Foundation (Germany), the National Center for Ecological Analysis and Synthesis (NCEAS, USA), the University of Idaho, American Genetics Association, and the Canadian Institute of Ecology and Evolution. In 2008, funding from NCEAS allowed Lisette Waits and Helene Wagner to initiate a distributed graduate seminar that has been taught to over 400 participants across the globe three times in the last six years by a community of landscape geneticists. We benefitted substantially from our experiences during these various courses while interacting with colleagues and students from many different countries, cultures, and disciplines. We thank you all for all your valuable feedback and encouragement over the last years and hope that the positive vibe and energizing momentum of landscape genetics is also reflected in this book.

We are also deeply grateful to the many reviewers of the chapters in this book whose comments greatly improved its quality.

*The Editors*
*Göttingen, Flagstaff, Pullman, Moscow*
*Spring 2015*

# GLOSSARY

**Adaptive locus**   See non-neutral locus.

**Adfluvial**   Species that migrate between freshwater lakes and streams or rivers.

**Advection**   Horizontal movement (e.g., of water), usually due to transport by currents.

**Agent-based models (ABMs)**   A class of computational models for simulating the actions and interactions of autonomous agents (both individual or collective individuals grouped into populations). Synonymous to individual-based models, which is the term more commonly used in ecology.

**Allele**   A unique genetic variant observed at a particular locus.

**Anadromous**   Aquatic species that migrate from salt water into fresh water to reproduce.

**Ancestry distribution models (ADM)**   Correlative models relating ancestry coefficients to environmental predictors.

**Anisotropy**   A situation where the autocorrelation in the data depends on direction as in directional asymmetry (e.g., when movement between two locations is more frequent or more probable in one direction than the opposite direction, e.g., due to the main wind direction across the study area).

**Assignment methods/assignment tests**   Methods that use genotypic information to evaluate population membership of sampled individuals.

**Autocorrelation**   A measure of the average similarity of any two observations depending on their spatial, temporal, or phylogenetic lag (i.e., distance). May be positive (nearby observations are more similar than distant ones), zero (independence, absence of significant autocorrelation), or negative (nearby observations are less similar than distant ones).

**Backward simulator**   Genetic signature is reconstructed typically using coalescent theory from time 0 to time $n$ where $n < 0$.

**Bandwidth**   Area associated with a graph edge (edge "width").

**Betweenness (Node/Edge)**   The number of times this element is traversed between all "shortest paths" in the graph.

**Biophysical model**   In the study of ocean and lake connectivity, biophysical models are simulations of larval dispersal and/or population dynamics based on representation of the geographic distribution of habitat, current flow, and life history parameters of the taxa of interest. More complex models incorporate the effects of seasonality, temperature, productivity, tides, and other environmental characteristics that impact demographic rates.

**Bottlenecks**   When a population goes through a period where its effective population size is extremely small, resulting in an increase in the effects of genetic drift and a consequent loss of genetic variation.

**Catadromous**   Aquatic species that migrate from fresh water in salt water to reproduce.

**Centrality (Node/Edge)**   A general term for measures of the relative position of a node or an edge in terms of direct or indirect connectivity or facilitation of flow through a network. There are four major types of centrality: degree, betweenness, closeness, and eigenvector.

**Chloroplast DNA (cpDNA)**   A circular DNA molecule found in the chloroplast of plants.

**Clique (Node)**   In general, a group of nodes more connected than expected by chance.

**Coalescent**   A body of theory that investigates time of divergence from a common ancestor. In population genetics, this theory can be applied to understanding differences in allele frequencies among populations.

**Conditional genetic distance (cGD)**   A measure of genetic dissimilarity based upon conditional genetic covariance. This measure differs from genetic distance measures in that it is not a pair-wise measure but the distance through a Population Graph constructed based on all of the data.

**Cycles (Edge)**   A sequence of paths that forms a closed loop.

**Degree**   The number of connections a node has to other nodes.

**Deme**   A deme is a group of individuals that is sufficiently genetically isolated from other groups of individuals and can also be considered a population.

**Dendritic**   Description of the geometric pattern of branching, consisting of a main stem and branches that decrease in size and increase in number hierarchically from downstream to upstream reaches of the network.

**Diameter (Graph)**   The shortest path through the graph with the longest length (e.g., the major axis of a graph).

**Dispersal kernel**   Probability distribution of the distance travelled by individuals or their propagules (e.g., larvae, seeds, pollen).

**Eddy**   A circular movement of water, counter to a main current, causing a small whirlpool.

**Edge (Graph)**   Connection between locations.

**Effective population size (Ne)**   The number of breeding individuals in an idealized population that would show the same amount of change in allele frequencies under random genetic drift or the same amount of inbreeding as the population under consideration.

**Essential parameter(s)**   Variables used to control the essential processes. In most simulations, some of these essential parameters are held constant, while others are varied, so that quasi-experimental studies can be conducted.

**Essential process(es)**   Processes that are included in a model because they are assumed to be vital for the functioning of the modelled system. They are often also the processes that are of interest to the study. In landscape genetic simulations, essential processes could, for example, be movement, mating, and reproductive fitness of individuals.

**Exploratory data analysis (EDA)**   A set of descriptive methods for summarizing data using visual representations from computer packages without statistical models.

**Extent**   The area within the landscape boundary and defines the population for the analysis.

**Factor analysis**   A statistical model that attempts to explain a set of observed variables in terms of combinations of unobserved variables called factors.

**Fitness surface**   In analysis of adaptive evolution in a spatially explicit context, a fitness surface represents the local fitness of a particular genotype at each location in the landscape and is used to model selection for different genotypes as a function of landscape conditions.

**Forward simulator**   Genetic divergence is simulated from time 0 to time $n$ where $n > 0$.

**Functional connectivity**   The degree to which landscape composition facilitates or impedes movement.

**Genetic clines**   Large-scale continuous variation of allele frequencies in geographic space.

**Genotype**   The combination of alleles observed at a particular locus for a specific individual.

**Grain**   In the context of landscape definition, grain refers to the dimension of the smallest resolved element in the landscape, and is typically corresponds to pixel size in a raster landscape and to minimum mapped patch size in a vector representation of a landscape.

**Graph**   Collection of nodes connected by edges representing similarity or connections. A graph representing flow (e.g., movement of individuals, information, genes) is often referred to as a network.

**Graph topology**   Overall structure of a graph.

**Gravity model**   Network flow model based on Newton's gravitational interactions formula.

**Groundwater**   Water located beneath the ground's surface in soil and in fractures of bedrock.

**Haploid**   An organism or a cell that contains only one set of chromosomes in its genome.

**Haplotype**   An allele or combination of alleles passed on from a single parent.

**Headwater**   Source of a river or stream.

**Heterozygote**   In a diploid organism, an individual that has two different alleles at each of its homologous chromosomes.

**Homozygote**   In a diploid organism, an individual that has two copies of the same allele on homologous chromosomes.

**Hydrodynamic connectivity**   Transfer of water, energy, matter, or organisms caused by the motion of fluids, commonly in reference to ocean currents.

**Hydrographic regime**   Aquatic setting defined by dominant physical features such as waves, tides, and current pattern, and associated depth profile (e.g., sandbars and channels).

**Hydrologic connectivity**   Transfer of water, energy, matter, or organisms within or between elements of the hydrologic cycle.

**Hyphoretic zone**   *A region mixing of groundwater and surface water beneath and alongside a stream bottom.*

**Inbreeding**   Mating between closely related individuals.

**Individual-based models (IBMs)**   a class of computational models for simulating the actions and interactions of autonomous agents (both individual or collective individuals grouped into populations). Equivalent to ABMs, but the term individual-based is more commonly used in ecology.

**Induced spatial dependence**   Spatial autocorrelation created by response to spatially structured landscape factors. If these factors are correctly included in the model, the autocorrelation will be removed from the residuals, but if some factors are missing or incorrectly specified, the residuals may show spatial autocorrelation.

**Influential points**   Unusual observations that may have a large influence on significance tests and parameter estimates. Such points may be difficult to detect in multivariate regression and in statistical methods based on distance matrices (Mantel tests, multiple regression of distance matrices (MRMs)).

**Inherent spatial autocorrelation**   Autocorrelation created by a biological process affecting the response (e.g., allele frequencies) directly, such as spatially restricted mating and dispersal.

**Isotropy**   When a spatial process, such as dispersal in a species, has the same intensity in all compass directions.

**Jet**   A stream of water moving in a rapid and organized fashion, relative to a main current.

**Landscape**   A landscape is a system of interacting ecological patterns and processes at any scale and can be considered to be an area that is spatially heterogeneous in at least one factor of interest at a scale relevant to the pattern-process relationships related to that factor of interest.

**Landscape definition**   The spatial definition of the landscape data utilized for analysis, including the thematic content, thematic resolution, grain and extent.

**Landscape genetics**   Research that combines population genetics, landscape ecology, and spatial analytical techniques to explicitly quantify the effects of landscape composition, configuration, and matrix quality on microevolutionary processes, such as gene flow, drift, and selection, using neutral and adaptive genetic data.

**Lentic**   Standing or still water; lentic aquatic systems include ponds, lakes, and wetlands.

**Likelihood**   A quantity that describes the probability of the data in a particular statistical model conditional on the model parameters.

**Linkage disequilibrium**   Non-random association of alleles at two or more loci; these alleles tend to be inherited together significantly more often than expected by random chance.

**Locus (Loci)**   A locus is a particular location in the genome of an organism. The term loci is the plural form of locus.

**Lotic**   Flowing water, such as rivers, springs, and streams.

**Mitochondrial DNA (mtDNA)**   A circular DNA molecule of the mitochondrion which is haploid and generally passed on only from mother to offspring.

**Model-based clustering (MBC)**   A set of approaches where the data are clustered using statistical mixture models.

**Model evaluation**   The process of evaluating whether a certain model structure is useful for addressing a specific research question.

**Moran eigenvector maps (MEM)**   A method for extracting spatial eigenvectors (see definition below) at multiple scales whose associated eigenvalues are proportional to their Moran's $I$.

**Moran's $I$**   A measure of spatial autocorrelation. Varies mostly between 1 (perfect positive autocorrelation) and $-1$ (perfect negative autocorrelation).

**Multivariate regression**   Linear model relating the variation in a matrix Y of multiple response variables (e.g., table of allele frequencies) to one (simple regression) or more than one (multiple regression) predictors X.

**Mutation**   A change in the DNA sequence of an organism. Mutations that occur in the DNA of germ cells can be passed on to the next generation.

**Neutral locus**   A locus where the frequency of alleles is not affected by selection.

**Node**   A point location (e.g., in a graph). In addition to location (a), nodes can have size (area), shape (polygon), and characteristics (e.g., habitat quality).

**Non-neutral or adaptive locus**   A locus where the frequency of alleles is affected by selection.

**Non-stationarity**   Refers to violations of the assumption of stationarity (i.e., that the process that generated the pattern has the same intensity and variability over the entire study area) and can be detected when either the correlation between genetic data Y and landscape predictors X, or the mean, variance, and spatial autocorrelation in the residuals U are not constant across the study area.

**Non-synonymous substitutions**  Mutations in coding DNA that result in changes in amino acid sequence.

**Nuclear DNA (nDNA)**  The DNA of the chromosomes found in the nucleus of a cell.

**ODD protocol**  Guidelines for describing a simulation model that enables other researchers to follow both the general model structure and the specifics of model parameterization and implementation. The three letters stand for Overview, Design concepts, and Details.

**Panmixia**  Random mating of individuals within a population or geographic area.

**Pattern-oriented modeling**  Synonymous to pattern-process modeling.

**Pattern-process modeling**  Evaluates whether an underlying process inferred through empirical induction can produce the patterns observed in the data, and how well it can do so.

**Population graph**  Graph theoretic representation of genetic relationships among sample locations.

**Principal component analysis (PCA)**  A method of data reduction that replaces a set of observed variables, such as allele frequencies, that may be correlated among themselves by a set of synthetic variables, the PCA axes, that are orthogonal and perfectly uncorrelated with each other. The first PCA axis is defined to capture a maximum of the variation in the original data set, the second PCA axes a maximum of the remaining variance, and so forth.

**Residual analysis methods**  A set of exploratory data analysis methods used to detect evidence of violations of regression assumptions.

**Resistance surface**  A representation of the landscape in which each location is assigned a cost or resistance which affects movement and gene flow through the landscape.

**Riparian**  At the interface between terrestrial and freshwater habitats, such as the edge of a river.

**Robustness analysis**  The process of evaluating whether simulation results are robust to changes in the actual model structure (i.e., by changing essential processes).

**Sensitivity analysis**  A technique used to determine how different values of an independent variable will impact a particular dependent variable under a given set of assumptions. This technique is used within specific boundaries that will depend on one or more input variables, such as the effect that changes in mutation rates will have on a landscape genetic inference.

**Simplifying assumption(s)**  Decisions about what processes and parameters not to include in the model. These assumptions basically represent hypotheses about how a system works.

**Spatial autocorrelation**  Measure of the degree to which individuals that are closer together in space are more genetically similar than those further apart.

**Spatial eigenvectors**  A set of completely uncorrelated, orthogonal synthetic variables that provides a spectral decomposition of the data based on the spatial coordinates of the sampling locations. Each spatial eigenvector, when plotted in space, will show a sine-type pattern with a specific period. On a regular transect, the largest-scale pattern has a period of the entire transect length, whereas the smallest-scale pattern has a period of twice the distance between adjacent sampling locations, so that high and low values alternate. For two-dimensional or irregular sampling designs, the patterns may be more complex.

**Spatially-explicit**  Spatial locations of individuals or groups of individuals (i.e., populations) are defined or monitored.

**Stream reach**  Segment of stream or river bounded by the confluence of another stream or river.

**Sweepstakes reproductive success**  When offspring of a small number of individuals dominate the production of new cohorts, resulting in a strong genetic drift that can enhance genetic differentiation of samples compared across space and time, often in an apparently random or unexpected way. The phenomenon, which has been theorized but rarely shown robustly, requires that individual fecundity is very high (e.g., oysters that spawn millions of larvae at a time) and that the conditions favorable for survival and settlement of larvae are patchy in space and time.

**Synonymous substitutions**  Mutations in coding DNA that do not result in changes in amino acid sequence. Often called "silent substitutions".

**Tessellation**  A pattern of non-overlapping polygons without gaps is used to partition a focal area. In landscape genetics, tessellation is generally used to create polygons around each point where genotypes are collected.

**Tidal bore**  A breaking wave caused by the leading edge of an incoming tide that may entrain material, such as larvae, that travel with it. The wave may be positioned at depth below the surface layer.

**Thematic content**  The factors included within the landscape definition, such as landcover categories, abiotic variables, linear features, and topographical landforms. Choice of which factors to include in the landscape definition should be guided by *a priori* hypotheses about how these features will affect the pattern-process relationship of interest, which in landscape genetics may be gene flow, genetic diversity or adaptive evolutionary processes.

**Thematic resolution**  The resolution or functional form at which the variables included in the landscape definition (thematic content) are represented in the landscape definition. For categorical variables such as land cover this will involve choice of how many map classes to represent. For continuous variables such as elevation or canopy cover thematic resolution will be the functional form of the relationship between the variable and the response, such as linear, quadratic, exponential, Gaussian, and power functions. Choices about thematic resolution should be governed by *a priori* hypotheses or prior knowledge about the relationship between the landscape variable and the pattern-process relationship of interest.

**Tributary**  A river or stream flowing into a larger river or lake.

**Turbulence**  Disorganized or chaotic water movement. Turbulence promotes mixing and can transport larvae in unpredictable, circuitous routes.

**Uncertainty analysis**  A technique used to assess the confidence in model variables or results.

**Upwelling/relaxation cycles**  Upwelling is the rising of deep water to the surface, often associated with strong wind pushing coastal surface waters offshore. When winds cease, prevailing currents push the same water back to shore during the relaxation phase. Release of coastal marine larvae is sometimes associated with upwelling/relaxing cycles, allowing larvae to complete development offshore where predation is less and then return to their natal habitat with little down coast movement.

**Validation**  Establishing evidence that provides a high degree of assurance that simulation programs accomplish intended requirements.

**Verification**  Ensuring that simulation programs are correctly working (i.e., that no programming errors occur).

**Waterscape features**  Features of an aquatic environment that may structure the genetic variation of populations and individuals by effecting movement through water and the entry and exit of habitat patches. Examples include wind and water current speeds and directions, water clarity, temperature and nutrient gradients, bathymetry and bottom type, and ecological factors controlling variation in population size or migration rates.

# Chapter 1

# INTRODUCTION TO LANDSCAPE GENETICS – CONCEPTS, METHODS, APPLICATIONS

*Niko Balkenhol,[1] Samuel A. Cushman,[2] Andrew Storfer,[3] and Lisette P. Waits[4]*

[1]*Department of Wildlife Sciences, University of Göttingen, Germany*
[2]*Forest and Woodlands Ecosystems Program, Rocky Mountain Research Station, United States Forest Service, USA*
[3]*School of Biological Sciences, Washington State University, USA*
[4]*Fish and Wildlife Sciences, University of Idaho, USA*

## 1.1 INTRODUCTION

Genetic variation is considered the most basic level of biological diversity and a prerequisite for the variability of species, populations, and ecosystems (Primack 2014). Diversity at the genetic level is also crucial for the fitness and survival of individuals, the viability of populations, and the ability of species to adapt to environmental change (Allendorf et al. 2012; Frankham et al. 2010). Thus, conserving genetic diversity is important in itself, and researchers in many disciplines, including ecology, evolution, and conservation, are interested in understanding the factors that shape patterns of genetic variation in nature. The foundations for understanding genetic diversity were laid more than 100 years ago (e.g., Hardy 1908; Weinberg 1908; Wright 1917), at

which time time, laboratory techniques did not yet allow the actual quantification of genes or DNA (deoxyribonucleic acid, see Chapter 3). Consequently, much of the early work of population geneticists was theoretical and conceptual. This changed after the discovery of the structure of DNA in 1953 by Francis Crick, James Watson, and Maurice Wilkins, and even more so after the development of PCR (polymerase chain reaction) by Kary Mullis in 1983. PCR made it possible to obtain large quantities of DNA even from minuscule samples, and the technique revolutionized many research disciplines, including medicine, forensics, genetic engineering, and population genetics.

Due to these technological advancements, genetic data also became more readily available to ecologists and conservationists, who increasingly realized the tremendous

*Landscape Genetics: Concepts, Methods, Applications,* First Edition. Edited by Niko Balkenhol, Samuel A. Cushman, Andrew T. Storfer, and Lisette P. Waits.

impact of human activities on biological diversity. In the 1970s and 1980s, genetic factors were recognized to be of fundamental importance for successful conservation strategies (e.g., Frankel 1970, 1974) and genetic diversity was explicitly considered in two of the earliest books on conservation biology (Soulé & Wilcox 1980; Frankel & Soulé 1981). Furthermore, human-caused loss and fragmentation of habitats were determined to be major drivers (e.g., Wilcove et al. 1986) and the ability to move among remaining habitat patches was identified as a key for the long-term conservation of populations and species in fragmented landscapes (e.g., Levins 1969; Hanski 1998). The consequences of changing environments also became a central topic of landscape ecology, which emerged as a scientific discipline in the 1980s (e.g., Naveh & Lieberman 1984; Forman & Godron 1986). Given these almost simultaneous developments in several research areas, it is not surprising that scientists began to combine concepts and methods from population genetics and landscape ecology to assess the influence of environmental heterogeneity on gene flow and genetic diversity (e.g., Pamilo 1988; Merriam et al. 1989; Manicacci et al. 1992; Gaines et al. 1997). Nevertheless, "landscape genetics" did not exist as a research area until it was formally defined in a seminal paper by Manel et al. (2003). This paper stimulated a tremendous interest in the scientific community, so that many novel methods for analyzing landscape genetic data were introduced (e.g., Guillot et al. 2005; Murphy et al. 2008) and the number of published landscape genetic studies grew quickly (reviewed in Holderegger & Wagner 2006; Storfer et al. 2010). Just ten years after its first formal

definition, landscape genetics had already contributed substantially to research in ecology, evolution, and conservation (see Manel & Holderegger 2013). Currently, landscape genetics still presents itself as a highly dynamic and rapidly advancing field. New methods are frequently suggested and novel research questions are identified as a result of both conceptual and technological improvements. The rapid growth of landscape genetics is both exciting and motivating, but it is also accompanied by tremendous challenges.

In this introductory chapter, we highlight some of these challenges and explain the rationale for this book and its particular structure. Before doing so, we provide a definition of what we feel constitutes landscape genetics. Furthermore, we provide a simple conceptual framework for landscape genetic analyses, which can be particularly useful for the novice landscape geneticist.

## 1.2  DEFINING LANDSCAPE GENETICS

Most readers of this book will already know that landscape genetics combines landscape ecology and population genetics. This is certainly correct, but is also not very specific or precise. To better understand landscape genetics, it is worthwhile to define the field more clearly. Three commonly used definitions of landscape genetics are shown in Table 1.1.

In the original definition of Manel et al. (2003) the focus was on "microevolutionary processes", which can be measured using genetic data. Thus, the

**Table 1.1** Overview of definitions of landscape genetics.

| Reference | Definition of landscape genetics* | Analytical consequence |
|---|---|---|
| Manel et al. (2003), page 189 | [Landscape genetics . . . ] aims to provide information about the interaction between landscape features and **microevolutionary processes, such as gene flow, genetic drift and selection.** | Need to quantify mircoevolutionary processes |
| Holderegger and Wagner (2006), page 793 | [ . . . ] landscape genetics endorses those studies that combine population genetic data, adaptive or neutral, with data on **landscape composition and configuration, including matrix quality.** | Need to quantify landscape heterogeneity |
| Storfer et al. (2007), page 131 | [ . . . ] research that **explicitly quantifies the effects** of landscape composition, configuration and [or] matrix quality on gene flow and [or] spatial variation. | Need to explicitly test for landscape-genetic relationships |

*Bold emphases are ours.

emphasis of this definition lies on the population genetic aspects of landscape genetics, but was not very specific about the 'landscape features' to be included in the analyses. The definition was extended by Holderegger and Wagner (2006) who clarified that landscape heterogeneity can be measured in terms of landscape composition, configuration and/or matrix quality (see Chapter 2 for explanations of these terms). Holderegger and Wagner (2006) also noted that landscape genetics can be conducted using different types of genetic data and that appropriate analyses and correct inferences depend strongly on whether the data is adaptive (i.e., under selection) or not (i.e., neutral; see also Holderegger et al. 2006). Finally, Storfer et al. (2007) highlighted that landscape genetics needs to quantitatively link landscape and genetic data to explicitly test for landscape-genetic relationships. This aspect is particularly important, because it allows landscape genetics to move beyond descriptive studies that visually assess spatially coinciding patterns in genetic and landscape data, towards quantitative models that make it possible to predict the genetic consequences of environmental change (e.g., Jay et al. 2012, Wasserman et al. 2012).

Putting these three definitions together, we can define landscape genetics as *research that combines population genetics, landscape ecology, and spatial analytical techniques to explicitly quantify the effects of landscape composition, configuration, and matrix quality on microevolutionary processes, such as gene flow, drift, and selection, using neutral and adaptive genetic data.*

## 1.3 THE THREE ANALYTICAL STEPS OF LANDSCAPE GENETICS

The definitions provided above lead to a simple conceptual framework for landscape genetic data analysis. Specifically, three general steps are necessary to reach the goals of landscape genetics (see last column in Table 1.1). First, we have to measure genetic variation so that we can quantify the miroevolutionary processes we are interested in. This step relies heavily on population genetic approaches and involves the description of the genetic composition of individuals or populations sampled across space – see Chapters 3, 7, and 9 for details.

Second, we have to quantify landscape heterogeneity so that we can capture the composition, configuration, and/or matrix quality of the study landscape – see Chapters 2 and 8. Third, we have to statistically link

landscape heterogeneity and genetic variation, so that we can explicitly and quantitatively test for landscape–genetic relationships (Chapters 5 and 10).

Note that the order of steps one and two is not crucial and could be reversed. For example, in this book the chapter on landscape ecology (Chapter 2) precedes the chapter on population genetics (Chapter 3) because we felt that it is often more sensible to first think about the landscape and its characteristics and next think about the genetic processes occurring in that landscape for a particular study species. In reality, the two steps will ideally be considered simultaneously, as only this will lead to optimal study design and strong inferences (Chapter 4).

Obviously, this three-step framework simplifies actual landscape genetics studies, because many decisions have to be made during all steps, because analytical choices in one step will affect options for another step, and because some methods actually combine multiple steps within a single analysis. Thus, finding optimal combinations of methods for all three steps and for the specific research questions, study landscape and species is not trivial and is unlikely to be covered by a single, cookbook-style recipe. Nevertheless, viewing landscape genetics in terms of the three basic analytical steps can help tremendously when designing a landscape genetic study and when trying to navigate through the thick jungle of landscape genetic methods. Thus, we encourage readers to keep this simple framework in mind when working through the other chapters of this book.

## 1.4 THE INTERDISCIPLINARY CHALLENGE OF LANDSCAPE GENETICS

Regardless of what definition of landscape genetics is used, they all highlight the fact that the field combines multiple, usually autonomous disciplines. Consequently, landscape genetics is often described as "interdisciplinary". However, the simple combination of various research approaches, concepts, and theories does not necessarily constitute true interdisciplinarity. Specifically, the level of integration across the various disciplines involved determines whether a scientific field is multidisciplinary, interdisciplinary, or transdisciplinary (e.g., Morse et al. 2007). A multidisciplinary field involves various disciplines that address a research topic collaboratively, but still rely on their traditional disciplinary approaches and paradigms.

Thus, answers to the research question are often found within involved disciplines and overall conclusions are drawn by comparing and combining results obtained from the different research approaches.

In an interdisciplinary field, the research should be much more coordinated among the different disciplines. This coordination involves the standardization of vocabulary, mutually defined research questions and study design, and the synchronization of conceptual frameworks used in each discipline. Addressing a topic through interdisciplinary research should lead to knowledge that impacts all of the involved disciplines. Thus, research that creates new disciplinary knowledge by simply addressing a question through an unusual analytical approach "borrowed" from some other discipline is not really interdisciplinary.

Finally, in a transdisciplinary field, disciplinary boundaries no longer exist, as research approaches from formerly distinct disciplines are fully integrated into a single conceptual framework. This framework involves all aspects of research, from problem definition and study design to actual data analysis and interpretation of results. Importantly, transdisciplinary research should lead to new ways of thinking about a problem and thus to the development of novel theories and research areas.

In our opinion, not all current landscape genetic research is truly interdisciplinary. While we are beginning to develop analytical and conceptual frameworks specifically for landscape genetics (e.g., Wagner & Fortin 2013; Chapter 5), many landscape genetics methods are still borrowed from other disciplines and usually focus on a single analytical step at a time. Various barriers to truly interdisciplinary research exist, with two of the most substantial ones being (a) the difficulties of effectively communicating across different disciplines and (b) the lack of experts that have received enough training across involved disciplines to close communicative gaps and overcome disciplinary boundaries. This is definitely also the case for landscape genetics.

Currently, very few researchers possess the background knowledge and skills to be experts in all of the subjects involved in landscape genetics. Most scientists to date have received disciplinary training and are either experts in landscape ecology, or population genetics, or spatial data analysis, but not in all three areas. This kind of complementary expertise may not even be available within a single university, and very few academic curricula include a comprehensive combination of population genetics, landscape ecology, and spatial quantitative data analysis (Wagner et al. 2012). Thus, for many students, the only options thus far for learning about the different components of landscape genetics is either to attend a landscape genetics seminar (see Wagner et al. 2012), or to read the many, often very brief and rather technical, published papers on landscape genetics. In our experience, the latter is often not very efficient, because of the rapid developments in the field and because papers are usually targeted towards a specific audience for which a certain level of preexisting knowledge on the topic can safely be assumed. For example, it is not necessary to explain the term "genotype" in a genetics journal or to explain the term "spatial grain" in a landscape ecological journal. However, for the beginning landscape geneticist, unknown technical terms, methods, and concepts used in landscape genetic publications often add up to a substantial mass of ambiguity or even confusion. Furthermore, readers without adequate background knowledge might be able to redo the analysis presented in a certain paper, but it will be unlikely that they will be able to critically evaluate the study and make significant contributions to the advancement of the field. Hence, short-term progress in landscape genetics depends on collaborations among disciplinary experts that know enough about all aspects of landscape genetics to effectively communicate with each other, evaluate published studies, and identify existing limitations and possible improvements (Cushman 2014). Similarly, the long-term future of the field depends on providing sufficient training opportunities for the next generation of landscape geneticists.

### 1.4.1  The two scopes of landscape genetic research

In addition to these challenges, which are quite typical for any interdisciplinary field, landscape genetics is also based on approaches that can be used for quite different purposes. Specifically, we currently see at least two major research avenues that are followed in landscape genetics. On the one hand, there are those studies that are interested in understanding how landscape characteristics affect microevolutionary processes. These studies follow the original idea of landscape genetics as defined by Manel et al. (2003) and are interested in genetic variation itself. Researchers following this avenue often have a background in genetics or evolutionary biology and are currently especially interested in using genomic approaches in landscape genetics (see Chapter 9).

On the other hand, there are those studies that are not interested in genetic variation or microevolution in itself, but rather use genetic data to infer underlying ecological processes, such as dispersal or disease transmission. Researchers following this research avenue are usually trained in ecology, and increasingly try to combine landscape genetics with other field methods, such as mark-recapture or telemetry (e.g., Cushman & Lewis 2010).

These different scopes of research further complicate the interdisciplinary nature of landscape genetics, simply because someone interested in evolutionary questions will emphasize different data types and methods compared to someone investigating ecological questions. Clearly, ecological and evolutionary processes and resulting population dynamics and biodiversity patterns are often strongly intertwined (e.g., Hairston et al. 2005, Palkovas & Hendry 2010), and landscape genetics has tremendous potential for untangling the relative roles of ecology and evolution in shaping biological patterns (e.g., Wang et al. 2013). However, it will be difficult to realize this potential if researchers interested in evolution neglect the data and methods provided by ecologists or if ecologists shy away from using the novel data and tools developed by geneticists. Thus, to realize the potential of landscape genetics for eco-evolutionary research, we need to maintain and strengthen the communication and collaboration among different disciplines, and ideally provide a reference baseline as a starting point for future developments in the field.

Overall, learning about, applying, and improving landscape genetics remains a challenging task because of the multiple, highly diverse disciplines involved in the field and because of the different research foci in the field.

## 1.5 STRUCTURE OF THIS BOOK – CONCEPTS, METHODS, APPLICATIONS

With this book, we aim to facilitate the first steps of learning about landscape genetics. We envisage the book as a guide for anybody wanting to learn about the field, as a tool for facilitating interdisciplinary communication and collaboration, and as a primer for disciplinary experts wanting to teach classes on landscape genetics at their home institutions. Ultimately, we also

hope that the book with serve as a baseline for critical discussions about landscape genetics and become a starting point for future advancements.

To reach these goals, we structured chapters in this book into three interrelated parts, each reflecting a slightly different purpose. Chapters in the first part deal with *Concepts* in a broad sense. These chapters serve as an introduction to the three major steps involved in landscape genetics research and are intended for readers with little or no experience in landscape genetics. These textbook-style chapters obviously include introductions to landscape ecology (Chapter 2) and population genetics (Chapter 3), as the two most fundamental disciplines involved in landscape genetics. In addition, this first section includes a chapter on the basics of landscape genetic study design (Chapter 4) and on the spatial analytical approaches for statistically linking landscape and genetic data (Chapter 5). While these latter two chapters are again quite fundamental, they include aspects we hope will be interesting and novel even for more experienced landscape geneticists.

Chapters in the second part of the book deal with *Methods* and are more in-depth treatments of certain topics that we currently deem particularly important. These chapters are intended for the advanced reader who is interested in the technical details of specific analytical approaches. This second part includes a chapter on landscape genetic simulation modeling (Chapter 6), because simulations hold tremendous potential for advancing landscape genetic methods and theory development. Genetic assignment and clustering methods are among the most commonly used approaches for quantifying genetic structure (analytical step 1) and are therefore covered separately in Chapter 7. Similarly, resistance surfaces are often used to quantify landscape heterogeneity in landscape genetic studies (analytical step 2), so they are treated in detail in Chapter 8. Chapter 9 then deals with genomic approaches that are increasingly used to generate large quantities of genetic data. These approaches are highly valuable for (adaptive) landscape genetics, especially because they can separte the patterns and processes of neutral markers genetic data from markers under selection. Chapter 10 introduces graph theory and network approaches, as these are increasingly used in both landscape genetics and genomics.

The third part of the book contains chapters that summarize *Applications* of landscape genetics in different systems. This section includes a chapter on landscape influences on plant population genetics (Chapter 11), a chapter on landscape genetic applications for understanding connectivity in terrestrial animals (Chapter 12), and a chapter on landscape genetic approaches for aquatic systems (Chapter 13).

Finally, in Chapter 14, we highlight some of the conclusions contained in the preceding chapters, identify emerging challenges and offer suggestions for future research and development needs in landscape genetics. Throughout the book, bold italics indicate terms that are described in the Glossary.

### 1.5.1 Limitations and potential of this book

The structure of the book, with the combination of introductory-level chapters and quite advanced discussions of certain topics, is rather unusual. However, we believe that this structure reflects the actual situation of contemporary landscape genetics. On the one hand, we still lack a single source of information that covers all of the required basics for interested beginners. On the other hand, certain – mostly analytical – aspects of landscape genetics have advanced rapidlly and are already too complex to be dealt with in a basic manner. Closing the void between the basics and advanced applications in a single book is obviously very challenging. We are aware of the limitations of this book and have no doubt that for many disciplinary experts, the basic chapters will not be as detailed or as all-embracing as they could be. For example, much more could be said about landscape ecology and about population genetics, respectively. As pointed out above, the intention of the book is not to replace the excellent textbooks on these disciplines (e.g., Allendorf et al. 2012; Turner et al. 2010) but rather to provide beginners with first introductions to the respective topics. These introductions will obviously not convert readers into experts, but they should enable beginners to (a) gain a sufficient overview of the disciplines involved and provide a starting point for further explorations; (b) better understand and critically evaluate published landscape genetic studies; and (c) more effectively communicate about landscape genetic research with experts from different disciplines.

Similarly, the advanced chapters provide more details on certain methods and are mostly intended for readers interested in actually applying these methods. The goal of these chapters is to provide guidance to identify common pitfalls and the most crucial assumptions, advantages, and limitations of different approaches. Finally, the three application chapters illustrate that landscape genetic methods can be used in a variety of circumstances and for different research questions. However, these three chapters are not intended to provide syntheses or even comprehensive reviews of the current state-of-the-art in landscape genetics. The field is already too diverse to allow for full reviews of all studies falling under the grand umbrella of landscape genetics. At the same time, the field may be too nascent for extensive syntheses. Nevertheless, the three application chapters should generate many ideas and motivate readers to apply or improve landscape genetic approaches for their own research.

Overall, we realize that the book is limited. Readers who desire a single source that covers all of the aspects involved in landscape genetics in a basic, yet detailed and application-relevant manner, will likely be disappointed. We hope that the book will nevertheless provide a useful overview and first starting point for learning about landscape genetics.

How successful this book will be in reaching its goals will largely depend on its readers, on their willingness to move towards truly interdisciplinary research, and on their motivation for advancing landscape genetics beyond its current state. Hence, we look forward to the feedback and discussion this book will generate and its impact on the future development of landscape genetics toward an inter- or even transdisciplinary field.

## REFERENCES

Allendorf, F. W., Luikart, G.H., & Aitken, S.N. (2012) *Conservation and the Genetics of Populations*. Wiley-Blackwell, Oxford.

Cushman, S. A. (2014). Grand challenges in evolutionary and population genetics: the importance of integrating epigenetics, genomics, modeling, and experimentation. *Frontiers in Genetics*, **5**, 197.

Cushman, S. A. & Lewis, J. S. (2010) Movement behavior explains genetic differentiation in American black bears. *Landscape Ecology* **25**, 1613–25.

Forman, R.T.T. & Godron, M. (1986) *Landscape Ecology*. John Wiley and Sons, Inc., New York.

Frankel, O.H. (1970) Sir William Macleay memorial lecture 1970: variation – the essence of life. *Proceedings of the Linnean Society of New South Wales* **95**, 158–69.

Frankel, O.H. (1974) Genetic conservation: our evolutionary responsibility. *Genetics* **78**, 53–65.

Frankel, O.H. & Soulé, M.E (1981) *Conservation and Evolution*. New York: Cambridge University Press, New York.

Frankham, R., Ballou, J.D., & Briscoe, D.A. (2010) *Introduction to Conservation Genetics*. Cambridge University Press, New York.

Gaines, M.S., Diffendorfer, J.E., Tamarin, R.H., & Whittam, T.S. (1997) The effects of habitat fragmentation on the genetic structure of small mammal population. *Journal of Heredity* **88**, 294–304.

Guillot, G., Mortier, F., & Estoup, A. (2005) Geneland: a program for landscape genetics. *Molecular Ecology Notes* **5**, 712–15.

Hairston, N.G., Ellner, S.P., Geber, M.A., Yoshida, T., & Fox, J.A., (2005) Rapid evolution and the convergence of ecological and evolutionary time. *Ecology Letters* **8**, 1114–27.

Hanski, I. (1998) Metapopulation dynamics. *Nature* **396**, 41–9.

Hardy, G.H. (1908) Mendelian proportions in a mixed population. *Science* **28**, 49–50.

Holderegger, R. & Wagner, H.H. (2006) A brief guide to landscape genetics. *Landscape Ecology* **21**, 793–6.

Holderegger, R., Kamm, U., & Gugerli, F. (2006) Adaptive vs. neutral genetic diversity: implications for landscape genetics. *Landscape Ecology* **21**, 797–807.

Jay F., Manel S., Alvarez N., Durand, E.Y., Thuiller, W., Holderegger, R., Taberlet, P., & Francois, O. (2012) Forecasting changes in population genetic structure of Alpine plants in response to global warming. *Molecular Ecology* **21**, 2353–68.

Levins, R. (1969) Some demographic and genetic consequences of environmental heterogeneity for biological control. *Bulletin of the Entomological Society of America* **15**, 237–40.

Manel, S. & Holderegger, R. (2013) Ten years of landscape genetics. *Trends in Ecology and Evolution* **28**, 614–21.

Manel, S., Schwartz, M.K., Luikart, G., & Taberlet, P. (2003) Landscape genetics: combining landscape ecology and population genetics. *Trends in Ecology and Evolution* **18**, 189–97.

Manicacci, D., Olivieri, I., Perrot, V., Atlan, A., Gouyon, P.H., Prosperi, J.M., & Couvet, D. (1992) Landscape ecology: population genetics at the metapopulation level. *Landscape Ecology* **6**, 147–59.

Merriam, G., Kozakiewicz, M., Tsuchiya, E., & Hawley, K. (1989) Barriers as boundaries for metapopulations and demes of *Peromyscus leucopus* in farm landscapes. *Landscape Ecology* **2**, 227–35.

Morse, W.C., Nielsen-Pincus, M., Force, J., & Wulfhorst, J. (2007) Bridges and barriers to developing and conducting interdisciplinary graduate-student team research. *Ecology and Society* **12**, 8. [online] URL: http://www.ecologyandsociety.org/vol12/iss2/art8/.

Murphy, M.A., Evans, J.S., Cushman, S.A., & Storfer, A. (2008) Representing genetic variation as continuous surfaces: an approach for identifying spatial dependency in landscape genetic studies. *Ecography* **31**, 685–97.

Naveh, Z. & Lieberman, A.S. (1984) *Landscape Ecology – Theory and Application*. Springer, New York.

Palkovacs, E.P. & A.P. Hendry (2010) Eco-evolutionary dynamics: intertwining ecological and evolutionary processes in contemporary time. *F1000 Biology Reports* **2**, 1.

Pamilo, P. (1988) Genetic variation in heterogeneous environments. *Annales Zoologici Fennici* **25**, 99–106.

Primack, R.B. (2014) *Essentials of Conservation Biology*. Sinauer Associates, Inc., Sunderland.

Soulé, M.E. & Wilcox, B. (1980) *Conservation Biology: An Evolutionary-Ecological Perspective*. Sinauer Associates, Inc., Sunderland.

Storfer, A., Murphy, M.A., Evans, J.S., Goldberg, C.S., Robinson, S., Spear, S.F., Dezzani, R., Delmelle, E., Vierling, L., & Waits, L.P. (2007) Putting the "landscape" in landscape genetics. *Heredity* **98**, 128–42.

Storfer, A., Murphy, M.A., Spear, S. F., Holderegger, R., & Waits, L.P. (2010) Landscape genetics: Where are we now? *Molecular Ecology* **19**, 3496–514.

Turner, M.G., Gardner, R.H., & O'Neill, R.V. (2010) *Landscape Ecology in Theory and Practice: Pattern and Process*. Springer, New York.

Wagner, H.H. & Fortin, M.-J. (2013) A conceptual framework for the spatial analysis of landscape genetic data. *Conservation Genetics* **14**, 253–61.

Wagner, H., Murphy, M., Holderegger, R., & Waits, L. (2012) Developing an interdisciplinary, distributed graduate course for 21st century scientists. *BioScience* **62**, 182–8.

Wang, I. J., Glor, R.E., & Losos, J.B. (2013) Quantifying the roles of ecology and geography in spatial genetic divergence. *Ecology Letters* **16**, 175–82.

Wasserman, T.N., Cushman, S.A., Shirk, A.S., Landguth, E.L., & Littell, J.S. (2012) Simulating the effects of climate change on population connectivity of American marten (*Martes americana*) in the northern Rocky Mountains, USA. *Landscape Ecology* **27**, 211–25.

Weinberg, W. (1908) Über den Nachweis der Vererbung beim Menschen. *Jahreshauptversammlung des Vereins für Vaterländische Naturkunde in Württemberg* **64**, 369–82.

Wilcove, D.S., McLellan, C.H., & Dobson, A.P. (1986) Habitat fragmentation in the temperate zone. In: Soulé, M.E. (ed.), *Conservation Biology: Science of Scarcity and Diversity*. Sinauer Associates, Inc., Sunderland.

Wright, S. (1917) Color inheritance in mammals. *Journal of Heredity* **8**, 224–35.

*PART 1*

# CONCEPTS

# Chapter 2

# BASICS OF LANDSCAPE ECOLOGY: AN INTRODUCTION TO LANDSCAPES AND POPULATION PROCESSES FOR LANDSCAPE GENETICISTS

*Samuel A. Cushman,*[1] *Brad H. McRae,*[2] *and Kevin McGarigal*[3]

[1]*Forest and Woodlands Ecosystems Program, Rocky Mountain Research Station, United States Forest Service, USA*
[2]*The Nature Conservancy, North America Region*
[3]*Department of Natural Resources Conservation, University of Massachussetts, USA*

## 2.1 INTRODUCTION

Ecology is defined as the interaction of organisms and their environment (Tansley 1935) and the environments in which organisms live are spatially structured at a range of scales (Levin 1992). Therefore, the interactions that connect organisms with their environment are affected by the composition and structure of the landscape in which they live (Watt 1947). A foundational idea in ecology is that the strength of interaction varies with location in the environment and with distance between individuals (Turner et al. 2001). Understanding the effects of location and distance on the interactions between organisms and between organisms and their environment is a key focus of **landscape** ecology (Wiens 1992). Landscape ecology focuses on the interactions of spatial patterns and ecological processes (Turner 1989). Given this emphasis on pattern–process relationships, much of the theory and methodology of landscape ecology focuses on how

*Landscape Genetics: Concepts, Methods, Applications*, First Edition. Edited by Niko Balkenhol, Samuel A. Cushman, Andrew T. Storfer, and Lisette P. Waits.

to characterize landscape heterogeneity, where it comes from, how it changes over time, and why it matters to ecological processes.

The emergence of landscape ecology was enabled by the confluence of several parallel developments over the past 40 years. The first is the recognition of the importance of broad-scale environmental issues. For example, habitat loss and fragmentation are widely seen as global drivers of a burgeoning extinction crisis (McKinney & Lockwood 1999), climate change is leading to dramatic reorganization of biological communities at broad scales (Thomas et al. 2004), and invasive species are having profound effects on ecosystems across the world (Mooney & Hobbs 2000). These broad-scale environmental challenges require solutions that are also at broad scales, which has motivated much landscape-level research and conservation planning. A second key in facilitating the development of landscape ecology is the emergence of scale and hierarchy concepts (Levin 1992). Landscape ecology focuses on the interaction of patterns and processes, and one of the key ideas is that both patterns and processes are scale-dependent (Wiens 1989). That is, each process acts at a characteristic scale in space and time and is affected by the structure and composition of the landscape at those characteristic scales (Fig. 2.1). A third key idea contributing to the emergence of landscape ecology is a transition from a view of ecosystems as internally homogeneous, discretely bounded, and equilibrial systems to one in which ecosystems are seen as spatially variable across scales and temporally dynamic (e.g., Watt 1947; Turner 1987). Many classical ideas in ecology and population genetics are based on simplified models of ecological systems in which spatial and temporal variation are ignored (Kareiva and Wennergren 1995). Landscape ecology seeks to inject space and time into the investigation explicitly and evaluate how the ecological processes and structure of the ecological system change as a result of spatial heterogeneity and temporal dynamics (Pickett & Cadenasso 1995). A critical fourth component in the emergence of landscape ecology is technological advances, particularly in remote sensing, computing, geographical information systems, and spatial statistical methods (Haines-Young et al. 2003; Sklar & Costanza 1991). These advances have enabled explicit consideration of spatial and temporal variability across scale in analysis that simply was not possible until very recently.

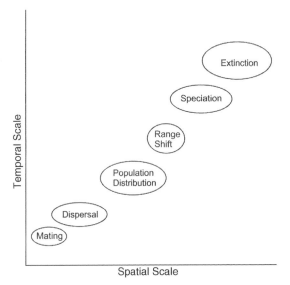

**Fig. 2.1** Different population processes occur at different scales in space and time. Processes of mating among members of a population occur at short time scales and fine spatial scales relative to dispersal, which typically occurs over greater distances. The population distribution is typically larger in extent than mating and dispersal processes (spatial scale) and typically changes more slowly in time. The population extent may shift its range at longer time scales. Speciation and extinction events typically occur at longer time scales and coarser spatial scales.

## 2.2  HOW LANDSCAPES AFFECT POPULATION GENETIC PROCESSES

The habitats in which organisms live vary spatially across scales, and these patterns interact with organism perception and behavior to drive the higher-level processes of population dynamics, gene flow, and adaptive evolution (Johnson et al. 1992; Wiens et al. 1993). The structure and composition of the landscape are key drivers of population distribution and gene flow, and interact with individual genetic characteristics to affect fitness. In particular, a disruption in habitat continuity may interfere with ecological processes necessary for population persistence (Fahrig 2003). For example, habitat loss and fragmentation may create discontinuities (i.e., patchiness) in the distribution of critical resources (e.g., food, cover, water) and environmental conditions (e.g., microclimates). These discontinuities may inhibit movement and gene flow across the population or the changes may result in reduced fitness, leading to population declines or adaptive evolution under directional selection. Although there are many ways that landscape structure and composition may affect population processes, they are dominated by: (1) area effects, (2) edge effects, and (3) isolation effects.

## 2.2.1   Area effects

One of the most important landscape-level influences on population processes is the effect that habitat extent has on gene flow and fitness. Habitat loss has consistently negative effects on biodiversity (Fahrig 2003), including reductions in species richness (Findlay & Houlahan 1997; Schmiegelow & Mönkkönen 2002; Steffan-Dewenter et al. 2002), declines in populations, and changes in species distributions (Bender et al. 1998; Sánchez-Zapata & Calvo 1999; Donovan & Flather 2002). Habitat loss changes the distribution of resources and can affect individual behavior and spatial activity patterns, changing the ability of the organism to acquire the resources needed to survive and reproduce (Mangel & Clark 1986; Wiens et al. 1993). For example, from an energetics perspective, if food resources become more patchily distributed, it may be more costly to acquire them (Mahan & Yahner 2000). In addition, moving between disjunct resource patches to acquire food resources may involve moving through suboptimal habitats that require higher energetic expenditures and expose individuals to higher rates of predation (Bergin et al. 2000) and reduce breeding (Kurki et al. 2000) and dispersal success (Belisle et al. 2001; With & Crist 1995; King & With 2002).

Most species require at least a minimum area of habitat in order to meet all life history requirements (e.g., Robbins et al. 1989). Theoretical studies predict a threshold habitat level below which the population cannot sustain itself (Fahrig 2002; Flather & Bevers 2002; Hill & Caswell 1999). The amount of habitat required for species persistence depends on species-specific behavioral and life-history characteristics (With & King 1999; Vance et al. 2003), and the effects of habitat loss on each species will depend on the interaction of its ecological requirements and capabilities with the degree of habitat loss in the surrounding landscape (McGarigal & Cushman 2002; Schmiegelow & Mönkkönen 2002; Fahrig 2003). For example, large-bodied, high-trophic-level species appear to be

particularly vulnerable to local extinction due to habitat loss (Gibbs & Stanton 2001). As habitat is lost in a landscape, the most area-sensitive species will be lost first. As the habitat loss continues, other species will drop out according to their minimum area requirements (e.g., Robbins et al. 1989; Bender et al. 1998; Flather & Bevers 2002). Thus, smaller patches generally contain fewer species than larger patches (Debinski & Holt 2000), and the set of species remaining in small patches is often a predictable subset of those found in large patches in the same region (Ganzhorn & Eisenbeiß 2001; Kolozsvary & Swihart 1999; Vallan 2000; Fahrig 2003). These area effects influence landscape genetic processes in several ways. First, as habitat is lost populations initially decline linearly with decreases in habitat extent, which results in reduction in effective population sizes, acceleration of genetic drift, and lower equilibrium heterozygosity and allelic richness. As habitat loss approaches the extinction threshold in some areas of the landscape, local populations may rapidly decline to extinction, resulting in gaps in distribution, which lead to attenuated gene flow, which creates spatial genetic structure and further reductions in heterozygosity and allelic richness. As the habitat area is further reduced beyond the extinction threshold the entire regional population will become extinct.

## 2.2.2   Edge effects

One of the strongest influences on population genetic processes is the effect of edges on organism movement and fitness. Edges are produced by natural discontinuities in geophysical factors (sometimes referred to as "inherent" edges), or by natural or anthropogenic disturbances (sometimes referred to as "induced" edges). These edges may be relatively permanent features of the landscape, for example, if they are produced by discontinuities in underlying abiotic factors (e.g., land–water interface), or transient features of a landscape, for example, if they are induced by disturbances (e.g., timber harvesting).

Depending on the ecology and life history of the species in question, edges can either inhibit or enhance movement and increase or decrease fitness (Kremsater & Bunnell 1999; Carlson & Hartman 2001; Laurance et al. 2001). Early wildlife management efforts were focused on maximizing edge habitat because it was believed that most species favored habitat conditions created by edges and that the juxtaposition of different habitats would increase species diversity (Leopold 1933). Indeed, this concept of edge as a positive influence guided land management practices for most of the 20th century. Recent studies, however, have suggested that changes in microclimate, vegetation, invertebrate populations, predation, brood parasitism, and competition along forest edges (i.e., edge effects) has resulted in the population declines of several vertebrate species dependent upon forest interior conditions (e.g., Strelke & Dickson 1980; Kroodsma 1982; Brittingham & Temple 1983). In fact, many of the adverse effects of forest fragmentation on organisms seem to be directly or indirectly related to these so-called negative edge effects.

One of the primary edge effects is the alteration of microclimate within habitat patches due to changes in the physical fluxes of radiation, wind, and water (Franklin and Forman 1987; Saunders et al. 1991; Baker & Dillon 2000). Following habitat loss, changes in these fluxes across the newly created edges can influence the microclimate of the remnant habitat patches (Saunders et al. 1991). Air temperatures at the edge of a forest remnant, for example, can be significantly higher than those found in either the interior of the remnant or the surrounding agricultural land (Geiger 1965; Kapos 1989). Similarly, with the conversion of natural vegetation to developed land uses, the entire pattern of momentum transfer over the landscape may be altered (Saunders et al. 1991). The wind profile does not fully equilibrate with the new land cover for some distance, perhaps for a distance as much as 100–200 times the height of the vegetation (Monteith 1975; Grace 1983). In addition, edges may allow below-canopy winds to penetrate the patch and modify relative humidity near the edge. Conversion of natural vegetation to developed land uses alters the rates of rainfall interception and evapotranspiration, and hence changes soil moisture levels (Kapos 1989). Altered surface and subsurface flows affect the timing and magnitude of peak flows (Hornbeck 1973) and the transport of soil and nutrients (Likens et al. 1970; Bormann et al. 1974). These watershed hydrological impacts influence the local moisture regimes along habitat edges.

One of the more obvious edge effects is increased rates of disturbance along edges, primarily as a result of increased exposure to wind (Franklin & Forman 1987; Saunders et al. 1991). Increased wind exposure at edges may result in damage to the vegetation, either through direct physical damage from pruning or windthrow (Moen 1974; Grace 1977) or by increasing evapotranspiration with reduced humidity and increased dessication (Tranquillini 1979; Lovejoy et al. 1986). Several authors have noted that edges may have more stressed, dead, and downed trees than do adjacent forests (Geiger 1965; Chen et al. 1992). This condition is conducive to insect infestations, which can cause additional disturbance.

The combination of climatic and disturbance effects along edges often produces marked changes in vegetation structure and composition and structure. Changes in light, temperature, wind, and moisture regimes affect seedling establishment, growth, and survival (Wales 1972; Gates & Mosher 1981; Ranney et al. 1981; de Casenave et al. 1995). Some species benefit from the modified microclimate near edges, others do not (e.g., Chen et al. 1992; Zen 1995). Similarly, increased disturbance rates at edges favor certain species. Overall, the altered physical environment can exert considerable influence on the composition and structure of vegetation near edges. Plant species common at successional edges (in contrast to permanent edges caused by inherent differences between adjacent natural communities, e.g., forest–water edge) include species that benefit from disturbance, as well as shade-intolerant, mid- and early-succession vegetation and non-native species (Ranney et al. 1981; Lovejoy et al. 1986; Alverson et al. 1988). Vegetation structure near edges reflects these compositional changes, and is further modified by the high rates of physical disturbance. Consequently, vegetation near edges usually consists of a diverse mixture of species and structures and is often characterized by high foliage height diversity and abundant dead wood (both snags and logs).

The alteration of the abiotic and biotic environment near habitat edges can reduce the quality of the edge habitat for some animal species and increase it for others. This reduction in habitat quality may be due to a less favorable microclimate (i.e., resulting in higher energetic costs), less favorable physical structure, fewer available food resources, adverse interspecific interactions (e.g., increased competition, predation, or parasitism), or a combination of these. For example, Mills (1995) documented that voles in southwest Oregon were almost nonexistent near forest edges compared to forest interiors and attributed the difference to the lack of truffles (the vole's preferred food) near forest edges due to the drier microclimate. Unfortunately, despite the many studies documenting trends in species abundances and distributions near edges, few studies have attempted to determine causes of the observed patterns (Kremsater & Bunnell 1999).

The several edge effects discussed above can influence spatial genetic processes in a number of ways. First, if edges act as barriers or filters reducing movement of organisms across the edge, this can create genetic substructure by blocking or reducing gene flow. This also will tend to increase the rate of genetic drift in the patches isolated by the edges, resulting in loss of genetic diversity across the subdivided population. In contrast, proximity to edges may facilitate movement of some species that are associated with those conditions, increasing gene flow across their populations. Second, increasing edge density will likely result in loss of habitat potential for species associated with interiors of large extensive habitat patches, leading to reduced local population sizes and loss of genetic diversity. Conversely, habitat capability for many species increases with edge density in the landscape, at least up to a point. For these species, increasing edge in the landscape would likely result in higher population densities, higher gene flow, and higher genetic diversity. Thus, the nature of the edge effect on population genetic structure will depend on whether the observed edges act as barriers or conduits for movement, or increase or decrease habitat capability for the organism in question.

### 2.2.3   Isolation effects

One of the ultimate drivers of spatial genetic structure in populations is the effect of landscape heterogeneity on movement patterns and resulting isolation of individuals and local populations. As heterogeneity increases in a landscape, movement of organisms will be affected, resulting in differing degrees of isolation depending on the abundance, distribution, and dispersal abilities of the species. When the landscape change results in a loss or fragmentation of habitat, and therefore supports fewer individuals (than the original contiguous habitat), there will be fewer local (within-patch) opportunities for intraspecific interactions. This may not present a problem for individuals (and the persistence of the population) if movement among patches is largely unimpeded by intervening habitats in the surrounding landscape and connectivity across the landscape can be maintained. However, if movement among habitat patches is significantly impeded, then individuals (and local populations) in remnant

habitat patches may become functionally isolated (McCoy & Mushinsky 1999; Rukke 2000; Virgos 2001; Bender et al. 2003; Tischendorf et al. 2003). The degree of isolation for any fragmented habitat distribution will vary among species depending on how they perceive and interact with landscape patterns (Dale et al. 1994; With & Crist 1995; Pearson et al. 1996; With et al. 1997); less vagile species with very restrictive habitat requirements and limited gap-crossing ability will likely be most sensitive to isolation effects (e.g. Marsh & Trenham 2001; Rothermel & Semlitsch 2002).

Local populations can become functionally isolated in several ways. First, the edge of the occupied patch may act as a filter or barrier that impedes or prevents movement, thereby disrupting emigration and dispersal from the patch (Wiens et al. 1985). Some evidence for this exists for small mammals (e.g., Wegner & Merriam 1979; Chasko & Gates 1982; Yahner 1986). Second, the distance from remnant habitat patches to other neighboring habitat patches may influence the likelihood of successful movement of individuals among habitat patches. Again, the distance at which movement rates significantly decline will vary among species depending on how they scale the environment. Therefore, a 100 m-wide agricultural field may be a complete barrier to dispersal for small organisms such as invertebrates (e.g., Kareiva 1987) or amphibians (Rothermel & Semlitsch 2002; Marsh et al. 2004), yet be quite permeable for larger and more vagile organisms such as birds. Lastly, the composition and structure of the intervening landscape mosaic may determine the permeability of the landscape to movements (known as landscape "resistance"; see Chapter 8). A landscape may be composed of a variety of continuous gradients, discrete patches, or networks of linear elements (such as roads or hydrological networks). Each of these elements may differ in its resistance to movement, facilitating movement through certain elements of the landscape and impeding it in others (e.g., Adriansen et al. 2003; Cushman et al. 2006; Spear et al. 2010). Regardless of how local populations within a landscape become functionally isolated, whether it is due to properties of the edges themselves, the distance between patches, or properties of the intervening landscape, the end result is the same – fewer individual movements across the landscape. This, in turn, can lead to genetic differentiation as a function of cumulative movement cost (isolation-by-resistance; IBR) and lead to loss of genetic diversity through drift in small isolated subpopulations (Charlesworth & Charlesworth 1987).

## 2.3 DEFINING THE LANDSCAPE FOR LANDSCAPE GENETIC RESEARCH

### 2.3.1 What is a landscape?

From an ecological perspective a landscape is a system of interacting ecological patterns and processes at any scale. Spatially, a landscape can be considered to be an area that is spatially heterogeneous in at least one factor of interest (Turner 2005). From an organism's perspective a landscape can be considered to be a heterogeneous distribution of resources and conditions, such as those that define its ecological niche, at a scale relevant to its ecology. A key point is that a landscape is not necessarily defined by its size. Rather it is defined by being a spatially heterogeneous area at an **extent** and resolution relevant to the phenomenon under consideration. In landscape genetics the phenomenon under consideration is usually the genetic structure of a population and the processes that govern it, such as gene flow or adaptive evolution (e.g., Manel et al. 2003; Segelbacher et al. 2010). Thus in landscape genetics the extent of the landscape may be defined by the extent occupied by the focal population and its structure may be defined by the ecological factors that drive the population process of interest (such as gene flow or selection). Given that the appropriate definition of a landscape will vary depending on the ecological system

and objectives of research, there are several key steps in defining the landscape appropriately for any given research project. These include defining a meaningful spatial extent for the landscape, choosing an appropriate conceptual model of landscape structure, selecting proper **thematic content** and **thematic resolution**, and selecting a proper spatial **grain**.

There are a range of conceptual models of landscape structure that are widely adopted in landscape ecological research (Fig. 2.2). For some questions landscape structure can be represented by a point pattern of element occurrences (Fig. 2.2a), which indicate the location of entities of interest, such as organisms or particular environmental features, in the landscape. Alternatively, when the extent and pattern of linear features, such as hydrological or road networks, are relevant, a landscape may be represented as a linear network (Fig. 2.2b). Commonly in past landscape ecology research landscapes have been represented as mosaics of different patch types, each representing a distinct ecological condition (Fig. 2.2c). Alternatively, in many cases landscape conditions are best represented as gradients that continuously vary across the study area, such as elevation, climate, or density of vegetation (Fig. 2d). In landscape genetics research the distribution of genetic samples (representing individuals or populations) is frequently represented as a point pattern or a graph matrix (Fig. 2.2e). The landscape in landscape genetic analysis may be represented by linear networks, patch mosaics, gradients, or a combination of these, depending on the nature of the system and the question at hand. The focus of landscape genetic

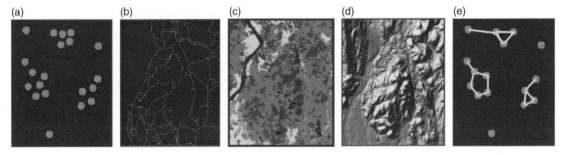

(a)    (b)    (c)    (d)    (e)

**Fig. 2.2** Conceptual models of landscape structure. The structure of a landscape can be represented in various ways. For example, the distribution of point elements (a) might be a suitable landscape model for a system in which the location of entities is the only factor that is important. Conversely, a linear network model (b) might be appropriate when the question involves connectivity of a hydrological network or the influences of a road network on fragmenting terrestrial habitats. A landscape mosaic model (c) could be chosen when the research goal is to assess the effects of different categorical land cover types on gene flow or selection. A gradient model (d) would be appropriate when gene flow or selection processes are affected by continuously varying attributes of a landscape such as elevation, density of vegetation or human population density. Landscape genetics studies sometimes represent spatial locations of genetic samples, with or without graph edges connecting them (e). (For a color version of this figure, please refer to the color plates section.)

research is associating differences in the genetic characteristics of individuals or populations at point locations (Fig. 2.2a) with the structure and connectivity of the landscape in which they reside (Fig. 2.2a to d).

### 2.3.2 Thematic content

It is a considerable challenge to determine how to represent the spatial structure of the environment in a way that is relevant to the population process under study. The choice of what attributes of environmental variation to represent (thematic content) can have immense implications for landscape genetic analysis. Consider a mountainous landscape in the US Rocky Mountains. There are many landscape features that one might choose to characterize, including elevation, forest cover, soil depth, geological parent material, climate, human landuse, roads, water sources, and other factors (e.g., Fig. 2.3). How does one choose which factors to represent in a model of landscape structure to use in landscape genetic studies? Ideally, the research team has selected a system and study organism based on prior knowledge that enables *a priori* development of research hypotheses. For example, if the researcher hypothesizes that gene flow of a particular species is related to elevation, forest cover, and roads then these factors would be reasonable choices to include as thematic content in the *landscape definition*. The apparent structure of the landscape is fundamentally dependent on the attributes selected to be represented thematically, and finding meaningful relationships between genetic processes and landscape patterns depends on correctly including the landscape features that most strongly affect the genetic process of interest.

### 2.3.3 Thematic resolution

Once factors to include have been selected, one must then choose how to represent them in terms of thematic resolution. Thematic resolution refers to the functional form (for continuous variables) or classification scheme (for categorical variables) used to define the values of a variable. For example, if one selected elevation as a factor of interest, would one choose to represent its effect on gene flow as linearly increasing with increasing elevation (Fig. 2.4a) or nonlinearly increasing, perhaps as a Gaussian function (Fig. 2.4c). Gene flow is usually most related to *functional connectivity* between locations.

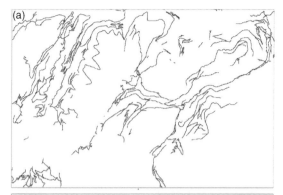

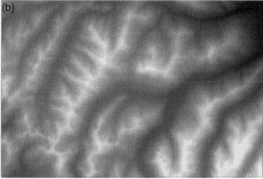

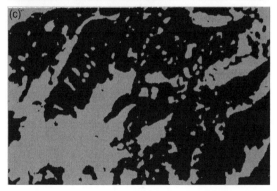

**Fig. 2.3** The thematic content of a landscape is the factors or variables represented spatially. For example, a landscape genetics study might hypothesize that roads fragment the population (a), or that gene flow is higher at low elevations than at high elevations (b), or that fitness is high in closed canopy forest but low in non-forest (c). It is critical to carefully select the thematic content of the landscape model to match the *a priori* hypotheses chosen to meet the analysis objective.

Predictions of functional connectivity can be made by calculating least-cost paths across a map representing hypothesized resistance to movement (e.g., red lines in

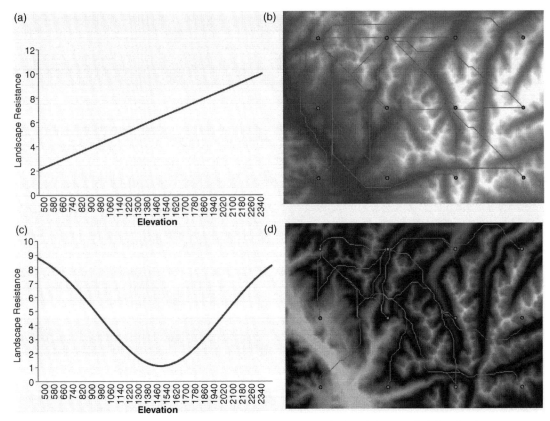

**Fig. 2.4** The thematic resolution is the resolution and functional form at which each factor included in the thematic content is represented. For example, the effect of elevation on a population process could be represented in various ways. For example, resistance to gene flow could increase linearly with elevation (a). In this case landscape resistance would be low in the valley areas (dark) and high on ridges (white) (b). The least-cost routes (lines) among a network of individuals (dots) would minimize cumulative cost by preferentially following low-elevation paths (b). Conversely, resistance to gene flow might be lowest at middle elevations (c). In this case resistance to gene flow would be lowest at intermediate elevations and higher in the deepest valleys and on the highest ridges (c), and least-cost routes connecting a network of individuals would preferentially follow paths that avoid low and high elevations (paths in d). Notably, even though (b) and (d) both have the same thematic content (elevation) and address the same question (gene flow), they produce very different predictions of the pattern and degree of connectivity among individuals. (For a color version of this figure, please refer to the color plates section.)

Fig. 2.4). If gene flow were related to elevation as a Gaussian function (as in Fig. 2.4d), but one used a linear function (as in Fig. 4b) to represent its effect, it is likely that the results would be misleading given that predicted connectivity among points is very different in these two representations of elevation effects. Similarly, suppose that one selected forest cover as a variable of interest; how would it be mapped for analysis? One could represent the effects of forest cover as four classes of different forest types (Fig. 2.2a), or as three classes

of different successional stages (Fig. 2.5b), or as a combination of the four classes and three seral stages (12 cover-seral classes as in Fig. 2.5c). If gene flow were governed by seral stage (as in Fig. 2.5b) but forest cover were represented as cover types (as in Fig. 2.5a) it is likely that the analysis would not find a relationship (or produce a misleading relationship) as these maps have very different spatial patterns and produce substantially different predictions of the connectivity among locations (lines in Fig. 2.5a to c).

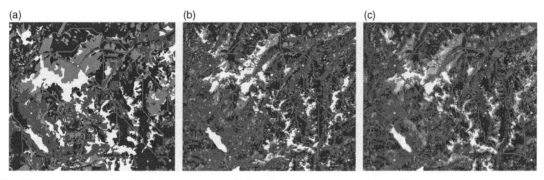

**Fig. 2.5** In a mosaic model of landscape structure and when the chosen thematic content is forest cover, there are a number of ways to represent the thematic resolution of forest in a landscape. For example, one could represent the effect of forest cover on a population process as four classes of different forest types (a), or as three classes of different successional stages (b), or as a combination of the four cover types and three seral stages (c). A network of individuals is shown as yellow dots and the least-cost paths connecting them are shown as red lines. The location of these least-cost routes and the relative cost of movement among pairs of points differ notably between these three thematic resolutions of forest cover. (For a color version of this figure, please refer to the color plates section.)

### 2.3.4  Spatial extent and grain

Given a choice of a particular variable to represent and its thematic resolution, the next critical question is over what extent and at what spatial grain should it be mapped for analysis? *Extent* is the area within the landscape boundary and defines the population for the analysis. *Grain* is the size of the individual units of observation. For example, a fine-grained map might represent information in 0.1 ha units, whereas a coarse-grained map might resample this information to 1 ha or 10 h a size units. Extent and grain define the upper and lower limits of resolution of a study and inferences about scale-dependency are constrained by the extent and grain of the data (Wiens 1989). One cannot reliably extrapolate beyond the extent of the sampled population, nor can one infer a pattern–process relationship finer than the grain of the data (Fig. 2.6). In practice grain and extent in landscape genetic studies are usually dictated by the available spatial data for a study area and the distribution of collected genetic samples. However, just as a mismatch of choice of what variables to measure or the thematic resolution at which to measure them can distort or mask the true ecological relationships, incorrect specification of grain and extent can strongly affect land-scape genetic analyses (e.g., Anderson et al. 2010; Cushman & Landguth 2010; Short Bull et al. 2011), although improperly specified grain and extent may often have less influence in landscape genetic analysis than misspecification of thematic content and

resolution (McRae et al. 2008; Cushman & Landguth 2010; Koen et al. 2010). For example, Fig. 2.7 shows how predicted connectivity among a set of source points changes in a landscape as a function of changing grain of the map, holding extent, thematic content, and thematic resolution constant.

### 2.3.5  *A priori* hypotheses should guide landscape definition

Ideally, one would *a priori* develop a suite of hypotheses regarding the potential relationships between thematic content, thematic resolution, grain and extent, and the process under study. These topics are explored in more detail in Chapters 5 and 8 of this book. The relevant point here is that choices regarding thematic content and thematic resolution have a large influence on the strength and nature of observed relationships in land-scape genetics research (e.g. Cushman & Landguth 2010; Shirk et al. 2010; Wasserman et al. 2010) and that great care must be taken to make appropriate choices as to what landscape features to include in an analysis and how to represent them in terms of the-matic resolution and spatial scale. These choices should be guided by *a priori* knowledge of the process and targeted hypotheses about potential pattern–process relationships. One does not need to have perfect knowl-edge of the process in advance and sophisticated meth-ods have been developed to explore large hypothesis spaces and seek optimized results (e.g., Shirk et al.

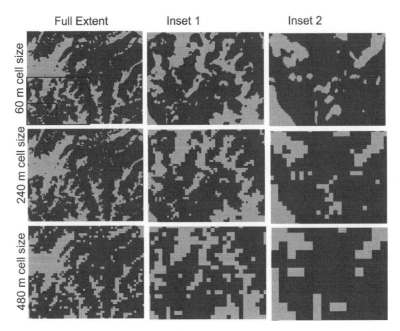

**Fig. 2.6** Illustration of the effects of changing grain and extent on landscape composition. This figure shows a two-class mosaic of forest and non-forest (e.g., mosaic model, forest cover as thematic content, binary mapping of forest (black) and non-forest (gray) as thematic resolution). The effects of different grains are shown in the rows of the nine-panel figure, with 60 m cell size in the top row, 240 m cell size in the middle panel, and 480 m cell size in the bottom row. The effects of different extents are shown in columns of the figure, with a "full extent" shown in the first column and two lesser window extents shown in columns 2 ("inset 1") and 3 ("inset 2"). The extents of these inset landscapes are shown in numbered boxes in the first panel. The changing patterns and extent of forest and non-forest across the panels show that landscape structure is highly sensitive to both the grain and extent of the analysis.

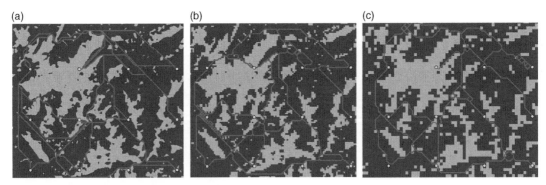

**Fig. 2.7** Effect of changing the grain on predicted connectivity patterns. This figure shows how changing the grain of a two-class mosaic of forest and non-forest affects predicted connectivity among a network of individuals. Individuals are shown as points and the least-cost paths connecting all combinations of individuals are shown as lines. The routes and relative cost among pairs of individuals change as the grain of the landscape map changes. (For a color version of this figure, please refer to the color plates section.)

2010; Wasserman et al. 2010). However, care must be exercised in proposing what variables, what thematic resolution, and what scales to analyze.

## 2.4 DEFINING POPULATIONS AND CHARACTERIZING DISPERSAL PROCESSES

Equally important to correctly defining the thematic content, thematic resolution, and spatial scale of the landscape for analysis is delineating the populations under study and characterizing their dispersal processes. Population connectivity is the net result of three factors: (1) the structure and resistance of the landscape, (2) the distribution and density of the population, and (3) the dispersal of the organism under study (Cushman & Landguth 2012). Correctly defining the landscape enables the study to correctly address issue (1) above, but issues (2) and (3) are equally important. Distribution and density are important because if a population is uniformly spread and occurs at a high density then there will likely be high genetic connectivity and high internal rates of movement regardless of the structure of the landscape. Conversely, if the population is very fragmented in its distribution and occurs at low density then gene flow will be primarily driven by patterns of occurrence in the landscape. Similarly, if a species has very high dispersal ability and can move through a wide range of land cover types then landscape heterogeneity will have relatively less influence on movement and gene flow than for a species with limited dispersal ability (e.g., Bohanak 1999; Govindaraju 1988; Landguth et al. 2010a). In the sections below we will describe three important models of population structure and discuss their implications for gene flow processes and landscape genetic analyses.

### 2.4.1 Panmictic populations

Much of classical population genetics theory is based on an idealized model in which populations are assumed to be discretely bounded and internally panmictic. When there is no non-random mating, mutation, selection, random genetic drift, gene flow, or meiotic drive both allele and genotype frequencies in a population remain constant from generation to generation (Hartl & Clark 1997). All real populations are likely to violate one or more of these assumptions. The effects of this will depend on which assumptions are violated. Non-random mating will result in deviations from Hardy–Weinberg proportions. A common cause of non-random mating is inbreeding, which causes an increase in homozygosity for all genes. Another cause is lack of panmixia driven by higher probability of mating with proximal individuals as a function of isolation-by-distance (IBD) or isolation-by-resistance. This is very commonly observed in landscape genetic studies. In addition, genetic migration provides genetic connectivity among two or more populations through long-distance dispersal. In general, genetic migration results in more homogeneous allele frequencies among the populations. In the presence of migration among populations Hardy–Weinberg proportions will normally not be observed. Third, genetic drift in small populations can cause random change in allele frequencies due to sampling effects. Genetic drift is not just an issue in small populations. In populations that are under strong isolation-by-distance or isolation-by-resistance, drift will tend to lead to relatively rapid loss of genetic diversity within local genetic neighborhoods (e.g., Wright 1943; Rousett 1997; Landguth et al. 2010a). In most real populations non-random mating, migration and drift have large influences on the genetic structure of the population and the discretely bounded, panmictic model of populations therefore is useful only as an ideal model with which to compare the real structure of populations. Several alternative models of population structure have been developed to more accurately reflect the effects of spatial structure of the landscape on non-random mating, migration, and drift.

### 2.4.2 Metapopulations

Metapopulations are literally populations of populations connected by dispersal (Levin 1974; Hanski & Gilpin 1991). Metapopulation theory has been formalized in a number of alternative models (Harrison 1991, 1994; Harrison & Taylor 1997), which largely differ in the rate and direction of individual movement among habitat patches. At its simplest, the theory holds that within each habitat patch the population has a finite probability of extinction, and likewise each patch has a particular colonization rate based on the number of occupied patches in the metapopulation (Fig. 2.8a). More sophisticated versions of the metapopulation concept involve spatially-explicit representation of the size, shape, and location of habitat patches, which allows for agent-based

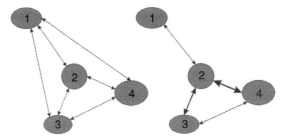

**Fig. 2.8** Two depictions of metapopulation representation of a population structure. (a) A simple, spatially-unstructured metapopulation in which all populations are equally connected and exchange migrants at equal rates regardless of size and proximity. (b) A spatially-explicit metapopulation in which the expected rate of exchange among pairs of populations is a function of both population size and proximity. Arrows represent the rate of migration among pairs of populations. In (a) the arrows connect all pairs of populations and are all of equal "weight", indicating an equal rate of exchange among all populations. In (b) arrows do not connect some populations that are farther apart than the maximum dispersal ability of the species (e.g. populations 1–3, 1–4), and the rate of exchange is highest between populations 2 and 4 because they are closest in proximity and because 4 is the largest population. Conversely, the rate of exchange between 1 and 2 is lowest because they are the pair of populations that is farthest apart and both populations are relatively small.

analysis of individual movement among patches to account for the effects of spatial structure of the landscape mosaic (Fig. 2.8b). In both versions of the metapopulation concept, populations in particular patches continually go extinct, but the metapopulation as a whole persists so long as the colonization rate is equal to the extinction rate. Metapopulations subject to high local extinction rates, but with correspondingly high rates of recolonization, have a high population turnover, but persist as long as the opposing rates are equal. Metapopulation dynamics reflect the rates of local extinctions and recolonizations as determined by inter-patch movement and factors affecting these processes. Individual movement between patches is perhaps the most important defining feature of a metapopulation. The theory predicts that subdivision and isolation of populations caused by fragmentation can lead to reduced dispersal success and patch colonization rates, which may result in a decline in the persistence of the local populations and an enhanced probability of regional extinction for the entire metapopulation (e.g., Lande 1987; With & King 1999). Specifically, increased population isolation increases extinction risk by reducing demographic and genetic input from immigrants and reducing the chance of recolonization after extinction (Lande 1988; Schoener & Spiller 1992; Gulve 1994). In population and landscape genetics, metapopulation concepts of population structure are often represented by the island and stepping-stone models, in which local populations are assumed to be ideal, panmictic Wright–Fisher systems that are connected by dispersal among them.

### 2.4.3  Gradient populations

The metapopulation model essentially preserves the core characteristics of the ideal panmictic population model. Specifically, each of the subpopulations is generally considered to have no internal structure. That is, within a subpopulation mating is assumed to be random. The model essentially represents a network of ideal populations linked by migration. In many situations this model fails to reflect real population structure. For example, populations of many plants and animals cannot readily be delineated into "patches" that approximate ideal panmictic subpopulations. Often populations are continuously distributed across extensive areas, spanning distances far larger than the dispersal ability of the individuals within it. In such situations it is often impossible to define boundaries to delineate "subpopulations" and the extent of the population is such that the assumption of non-random mating will be grossly violated. In such cases it is better to adopt a "gradient" concept of population structure (e.g., Wright 1943; Endler 1973; Templeton 1981; Cushman et al. 2010; Wasserman et al. 2010).

Very often the probability of mating with any particular individual or dispersing to any particular destination is a function of the distance that individual or destination is from the subject. This is the essential characteristic of the concept of isolation-by-distance, that mating choices or dispersal destinations are related to distance (Wright 1943). In landscape genetics mating and dispersal are often influenced by the spatial structure of the landscape, such that there is non-uniform probability of movement through different land cover types. In this case mating and dispersal probabilities will be affected by the intervening landscape between the origin and destination as embedded in the concept of landscape resistance (see Chapter 8).

Isolation-by-distance or isolation-by-resistance processes can create strong spatial genetic population structure as a result of non-random neighborhood mating, non-random dispersal, and elevated genetic drift in local genetic neighborhoods (e.g., Wright 1943; Endler 1973; Landguth et al. 2010b). One signature of this spatial genetic structure is the range of the genetic correlogram. A correlogram plots the correlation of one variable with itself as a function of the lag between the two observations. For example, in landscape genetics one might calculate the Mantel correlogram, showing the correlation of genetic distance among individuals as a function of the cost–distance between them (Borcard & Legendre 2012; Fig. 2.9).

The correlogram in Fig. 2.9 shows that the genetic distance between individuals increases as the cost distance between them increases, reflecting local non-random mating, and local dispersal. At short distances there is positive correlation between individuals, indicating that individuals that are separated by short cost distances are likely to be genetically similar, as a result of sharing a high proportion of alleles because

they reside in the same genetic neighborhood (Wright 1943). The correlogram shows that at longer cost distances the genetic correlation decreases to zero and then becomes negative. This negative correlation indicates that individuals separated by long cost distances (e.g., greater than ~60,000 in the correlogram shown in Fig. 2.9) are likely to be more genetically dissimilar than average across the full population, reflecting a low proportion of shared alleles because they reside in different genetic neighborhoods that do not directly exchange dispersers or mates.

These distance and cost-based processes result in populations that differ in many critical respects from either the classical panmictic model or the metapopulation model. In the panmictic model, any individual drawn from the population has the same probabilities of having any particular combination of alleles and degree of heterozygosity, as the population is assumed to be "well-mixed". In contrast, in a gradient population the alleles an individual will carry and its heterozygosity are highly dependent on its location within the population (e.g. Wright 1943). The movement ability, behavior, and population density of the organism in interaction with the structure of landscape resistance will determine the strength and pattern of internal genetic structure. Importantly, this internal genetic structure is non-transitive. The transitive law of logic states that if object 1 has some relationship to 2 and 2

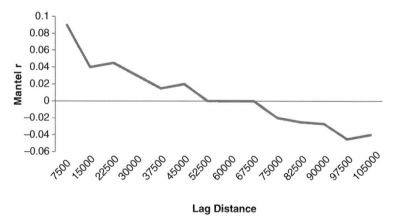

**Fig. 2.9** Mantel correlogram showing the change in correlation between the genetic distance and cost distance. The $y$ axis is the Mantel correlation among individuals and the $x$ axis is the cost distance between them. The correlogram shows that at short cost distances (up to 45,000 cost units) there is significantly positive correlation, indicating that individuals tend to be genetically similar when they are separated by relatively small cost distances. Conversely, at cost distances beyond ~60,000 cost units the correlation is negative, indicating that individuals separated by large cost distances tend to be genetically dissimilar.

(a)    (b)    (c)

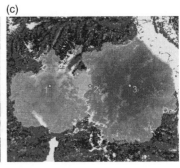

**Fig. 2.10** Depiction of genetic neighborhoods in a continuously distributed population on a resistance surface. The gray-scale background is a landscape resistance model in which gene flow is facilitated by a mature closed canopy forest; the resistance increases as the forest cover becomes open and is highest in non-forest cover types. The three numbered yellow dots represent the locations of three individuals taken from a continuous population that is distributed across the map. The blue patch surrounding point 2 in (b) is a dispersal kernel that originates on the location of individual 2 and extends outward to a maximum of 20,000 cost units. If this species had a genetic neighborhood extent of 20,000 cost units, this blue patch would indicate the extent of the genetic neighborhood centered on the location of individual 2. Both individuals 1 and 3 are within this genetic neighborhood (e.g., they are both covered by the blue dispersal kernel originating at individual 2). However, given that the dispersal kernel value is higher (darker blue) at the location of individual 3 than at the location of individual 1 (which is right on the edge of the kernel), one would predict higher genetic similarity between individuals 2 and 3 than between individuals 2 and 1, if gene flow is governed by the landscape features depicted in the resistance surface. In (c) the genetic neighborhood extent (20,000 cost units) is shown surrounding each of individuals 1 and 3. These two neighborhood kernels both overlap individual 2, but they do not overlap each other. That is, the genetic neighborhood extent surrounding individual 1 does not include individual 3 and vice versa. This shows the non-transitive nature of continuously distributed populations governed by isolation-by-distance or resistance. The extent of a local population is a function of each location and there are often no discrete boundaries between genetic neighborhoods. (For a color version of this figure, please refer to the color plates section.)

has the same relationship to 3, then 1 and 3 share that relationship as well. This applies to panmictic populations, such that if individual 1 and 2 are in the same population and individual 2 and 3 are in the same population, then we can deduce that 1 and 3 are in the same population and all share the same probability profile for allelic composition and heterozygosity (e.g., Fig. 2.10a). However, in a gradient population this is not the case. One may choose to define local populations in the gradient concept by the extent of the local neighborhood of genetically correlated individuals (e.g., Epperson 1993). Figure 2.10b shows that individuals 1 and 3 are in the same local population as individual 2 based on this definition, as their local genetic neighborhood kernels overlap the location of individual 2. However, individuals 1 and 3 are not in the same genetic neighborhood and so by this definition they are not in the same local population (Fig. 2.10c). All three individuals could be expected to have non-random genetic differences from one another as a result of local non-random mating, dispersal, and drift.

## 2.5  PUTTING IT TOGETHER: COMBINATIONS OF LANDSCAPE AND POPULATION MODELS

Depending on the ecological system, study organism, and objectives of analysis, there are a large combination of different landscape models and population types that could be appropriate for a given landscape genetic study (Table 2.1). It is important to think carefully about designing landscape analysis *a priori* in ways that are appropriate given the study species, the nature of its population, and the type of population process under study. Table 2.1 delineates some of the common combinations of landscape model, spatial population model and population process model for landscape genetic studies of gene flow. For example, row 1 in Table 2.1 represents a landscape model of a classic, discrete, isolated, panmictic population in a homogeneous landscape. The second row adds the effect of spatial pattern to this model, in which habitat patches are connected to various degrees to a large mainland source population. This landscape model would be appropriate when

**Table 2.1** Conceptual overview of different models for representing landscapes, spatially structured populations, and gene flow. Note that all landscape models can include linear landscape features such as roads, rivers, etc., that might pose barriers to population connectivity and gene flow.

| Data type used for landscape model | Landscape model | Spatial population model | Gene flow model |
|---|---|---|---|
| Categorical: habitat–non-habitat | Single (isolated) habitat fragment/patch | Single (closed) population; classic population ecological model | Single, panmictic population |
| Categorical: habitat–non-habitat | Small "island" population connected by dispersal to large "mainland" population | Island–mainland | Island–mainland |
| Categorical: habitat–non-habitat | Multiple habitat patches imbedded within a matrix of homogeneous non-habitat | Multiple (sub)populations with varying degrees of connectivity, which are largely determined by size of patches and geographical distance among them; classic metapopulation model | Stepping-stone/isolation-by-distance (IBD) among populations |
| Categorical: habitat–corridor–non-habitat | Multiple habitat patches imbedded within a matrix of homogeneous non-habitat, connected by habitat corridors | Multiple (sub)populations with varying degrees of connectivity, which are largely determined by presence/length of corridors among patches | Stepping-stone/isolation-by-distance (IBD) among populations |
| Continuous/categorical: habitat matrix | Multiple habitat patches imbedded within a heterogeneous matrix varying quality | Multiple (sub)populations with varying degrees of connectivity, which are largely determined by the quality of the heterogeneous matrix among patches | Isolation-by-resistance (IBR) among populations |
| Continuous: landscape gradient | Heterogeneous matrix of varying quality; discrete habitat patches cannot be delineated | Subpopulations cannot be delineated; continuously distributed individuals | Isolation-by-resistance (IBR) among individuals |

the study population exhibits an island–mainland structure, for example. The third row represents a situation where there is no clear mainland population and the species is distributed in a mosaic of subpopulation patches that are linked to various degrees by dispersal as functions of patch size and distance among them. In this situation a stepping-stone isolation-by-distance gene flow model would be appropriate. The fourth row is the same as the third, but with the addition of corridors linking some of the subpopulations. The fifth row represents a situation where gene flow is affected by multiple habitat features, including perhaps a combination of categorical and continuous landscape variables, and connectivity among populations is determined by the cumulative cost of movement through this complex matrix. A population-level isolation-by-resistance model would be appropriate in this situation. The sixth row is like the fifth, except that subpopulations cannot be delineated among continuously distributed individuals, and an individual-based isolation-by-resistance framework is appropriate.

## 2.6 FRAMEWORKS FOR DELINEATING LANDSCAPES AND POPULATIONS FOR LANDSCAPE GENETICS

It should be clear from the previous sections that assessing the influences of landscape structure on adaptive evolution or gene flow involves complex processes involving many components. Some of the issues are conceptual and involve establishing clear objectives and perspectives on the hypothesized factors that influence these processes and the scales at which they operate. Other issues are technical and have significant implications for the computation and interpretation of measures of spatial structure of the landscape and its association with population genetic processes. A thorough understanding of these issues is prerequisite to the effective application of the protocol that follows. In this section, we present a five-step protocol for quantifying landscape structure for landscape genetic analysis (Fig. 2.11). The protocol presented is not a cookbook with precise step-by-step instructions. Rather, it is a general process outlining broad steps and important considerations. The details of the analysis, that is, the decisions made at each step in the process, must be tailored to the specific context of the application.

### 2.6.1    Step 1: establish analysis objectives

Ultimately, the analysis must be guided by well-formulated objectives. Thus, the first step is to establish the analysis objectives. In landscape genetics research we typically are interested in one or more of the following questions:
• What are the factors that drive gene flow in a population?
• What are the environmental factors that deterime patterns of genetic diversity across the population?
• What are the environmental factors that are related to adaptive variation in a population?
  The choice of which of these questions are of interest will have fundamental influence on the analytical

approaches that are appropriate (see Wagner & Fortin 2013; Chapter 5). Chapter 5 presents a detailed framework for selecting and implementing statistical modeling approaches depending on the objectives of a landscape genetic analysis. Equally important to proper selection of statistical methods, however, is proper definition and analysis of landscape structure.

### 2.6.2    Step 2: define the landscape

Once the analysis objectives have been established, the next step is to define the landscape in a manner that is relevant to the target species. This involves establishing the spatial extent of the analysis, establishing a model of the landscape structure (including choosing the thematic content and resolution), and establishing a relevant spatial grain and data model for the analysis.

**Define the extent of the landscape**

The first step in defining the landscape is to define the extent of the landscape and delineate its boundary. This is often a difficult task because the relevant ecological boundaries often do not correspond to the superimposed administrative and/or analysis boundaries. To the extent possible, the extent of the landscape should be meaningful ecologically given the scale at which the target population operates. For example, the local range of the focal species or of the local population or metapopulation may be suitable as the basis for delineating the landscape. In many cases, however, there will be other practical considerations that must be taken into account. For example, the landscape extent may have to correspond to particular administrative or ownership parcels given logistics or access. At a minimum, the scope and limitations of the analysis given these scaling considerations should be made explicit.

**Establish a model of the landscape structure**

The second step is to establish a digital model of the landscape structure. It is preferable that *a priori* ecological knowledge has guided the selection of the research subject and objectives for the analysis. For example, in landscape genetics, usually *a priori* hypotheses should focus on factors that are likely to drive gene flow or adaptive variation in the target species. It is critically important to establish a model of landscape structure that reflects these hypotheses and enables evaluation of

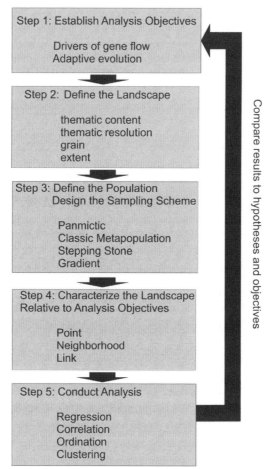

**Fig. 2.11** Schematic of a five-step protocol for quantifying landscape structure for landscape genetic analysis. The first step is to establish the objectives of the analysis, such as identifying the drivers of gene flow or the landscape variables that drive selection and develop *a priori* hypotheses about how landscape features and patterns of these features may affect these processes. The second step is to define the landscape appropriately given the analysis objectives. This involves deciding on the thematic content, thematic resolution, grain and extent at which to represent the landscape variables identified in step 1 as potential drivers of the population process of interest. The third step is to define the structure of the study population and design and implement an appropriate sampling scheme to collect genetic data from that population. Different sampling designs are likely to be appropriate when the population is structured as a classic metapopulation, a stepping-stone spatially-structured metapopulation or a continuous gradient population. The fourth step is to characterize the structure of the landscape relative to the analysis objectives. The appropriate way to characterize landscape structure will depend on whether the objectives suggest point-, neighborhood-, or link-based analysis. For analysis of gene flow link-based analysis will generally be the most appropriate, in which the researcher quantifies the connectivity between each pair of sampled individuals or populations (such as the cumulative cost distance or current flow between them). For analyses of adaptive evolution, analysis will likely need to both consider link-based gene flow and point- or neighborhood-level effects of landscape composition on selection. The fifth step is to conduct the analysis associating genetic patterns to landscape patterns in a manner appropriate to address the proposed hypotheses. This is likely to involve regression, correlation, ordination, or clustering analyses. The final step is to compare the results of the analysis with the hypotheses and analysis objectives.

a range of alternative hypotheses. Depending on the system, focal species, and study objectives a range of models of landscape structure could be appropriate, including linear networks, patch mosaics, or landscape gradients. In addition to correctly deciding whether a network, mosaic or gradient model of landscape structure is appropriate, it is important to carefully decide on the thematic content and resolution at which these features will be represented. This is a broad topic, the full treatment of which is beyond the scope of this chapter, and readers are referred to Chapter 8 for a more in-depth discussion of the process of defining landscape models relative to landscape genetics research. However, a few brief examples here will help illustrate some generally important concepts. If a patch mosaic model is chosen, it is important to decide on the features to be categorized (thematic content; e.g., vegetation cover, elevation classes, or land management designations) and the levels over which they will be categorized (thematic resolution; e.g., how many different classes of vegetation cover). Similarly, if a gradient model is selected to represent the landscape, one needs to decide what continuous environmental variables will be included in the analysis (thematic content; e.g., elevation gradient, precipitation gradient, or canopy cover gradient), and the functional form of relationships hypothesized between these gradients and the response variable (thematic resolution; e.g., linear, unimodal, or logarithmic).

**Establish a relevant grain of analysis**

The last step in defining the landscape is to define a relevant grain (or minimum mapping unit) and digital data model. In some cases, these decisions will be guided by technical considerations owing to the source of the data and the data processing software available. In most cases, a raster data model will be desirable. Additionally, the grain of the data should represent a balance between the desire for accurate calculations of landscape composition and configuration, computational efficiency, and the desire to scale patterns appropriately for the chosen landscape extent. On the one hand, the grain should be kept as fine as possible to ensure that small and narrow, yet meaningful, features of the landscape are preserved in the data model. On the other hand, the grain should be increased in relation to the extent so that unnecessary detail is not confounded with the important coarse-scale patterns over large spatial extents.

### 2.6.3   Step 3: define the population and design the sampling scheme

Once the landscape has been defined relative to objectives, it is important to give equal consideration to the distribution and structure of the study population and to decide how genetic data should be sampled from that population to obtain reliable inferences about relationships between population genetic processes and landscape structure (Chapter 4). There are several important considerations in this step. First, the pattern of population distribution and extent will have large impacts on the genetic processes at play and their interaction with landscape structure. It is important to have reliable knowledge about the extent and distribution of the population. It is particularly important to evaluate whether the study area encompasses an entire population or whether the population extends beyond the boundaries. It is ideal to select a study area of sufficient extent that it encloses a closed population, which allows for more rigorous assessment of population genetic conditions and avoids the edge and boundary effects. However, in most real world landscape genetics studies subject populations extend often far beyond the limits of the study area that can be sampled. It is important to think carefully about the implications of this for the sampling design, analysis approach, and interpretation of results. It is also important to consider whether the population consists of continuously distributed individuals, a territorial mosaic, or local subpopulations linked by dispersal in a metapopulation. These different "population structures" will have a direct influence on the population genetic processes and their interactions with landscape structure, and it is important to think carefully *a priori* about what these are and how they may affect the processes in question, and thus what sampling and analytical approaches are appropriate. Finally, once the extent and structure of the population have been considered, it is important to design an appropriate sampling scheme to collect spatially-references genetic data. Spatially-referenced genetic data are the response variable for landscape genetic analysis, and meaningful results are fundamentally dependent on the sampling scheme used for collecting these genetic data. The complex decision process for designing a sampling scheme is beyond the scope of this chapter, and we refer readers to Chapter 4 for details. However, given the importance of this topic we make a few recommendations. First, the appropriate sampling design will depend on both the landscape model selected (and the thematic content, thematic resolution, grain, and extent) and the population structure present (continuous or metapopulation). The sampling scheme will also depend on whether the objectives are to identify the factors that control neutral gene flow, identify the environmental variables driving adaptive evolution, or both. For example, if the study is intended to identify the factors driving gene flow in a continuous population, and one hypothesizes that gradients in elevation drive landscape resistance to movement, it might be wise to develop a sampling strategy that ensures that a large number of individuals are sampled across the study area, with spatial stratification to ensure equal representation of locations across the elevation gradient. Conversely, if the goal is to identify the environmental variables that drive adaptive evolution in a metapopulation, a suitable sampling strategy might be to sample some number (perhaps 20–30) individuals in each of a large number of subpopulations that vary across the environmental gradients hypothesized to provide selection pressure.

### 2.6.4 Step 4: characterize the landscape relative to analysis objectives

Once the landscape has been defined, the population structure described, and the locations of genetic samples selected, it is necessary to analyze the structure of the landscape relative to the locations of the genetic samples in a manner that is appropriate given the objectives of the project. This analysis will provide the independent (predictor) variables for the analysis, and thus is a critically important part of developing the data set for any landscape genetics project. Depending on the study species, ecological system, and research objective, the pertinent measurements are likely to be node-based, neighborhood-based, or link-based (Wagner & Fortin 2013; Chapter 5). When the objectives of the research are to identify the factors that drive neutral gene flow then generally link-based measures of landscape structure will be appropriate (Wagner & Fortin 2013; Chapter 5). Link-based analyses are based on the landscape attributes that lie between sampling points, rather than the characteristics of the points themselves. This is the appropriate framework for analyses of gene flow because gene flow is governed by the connectivity of the landscape *between* points rather than the attributes of the landscape *at* points. There are a large number of methods commonly used to calculate relevant variables

for link-based analyses, most of which are based on the concepts of cost distance (Adriansen et al. 2003) or circuit theory (McRae 2006). When the objective is related to gene flow neighborhood-based measures of landscape structure may be relevant. For example, instead of basing the analysis on pairwise cost distances among points, an analysis could predict genetic differentiation based on the connectivity of the local neighborhood using measures such as centrality (Estrada & Bodin 2008), patch connectivity indices developed in metapopulation ecology (Moilanen & Hanski 2006), and gravity models (Murphy et al. 2010). In such situations the landscape analysis would involve calculating the relevant neighborhood connectivity metric given the structure and composition of the landscape surrounding each of the sampled locations. When the analysis objective involves identifying relationships between adaptive variation and environmental gradients, the analytical framework will usually be based on point and neighborhood analysis of landscape conditions. This is because when an environmental feature drives directional selection of certain alleles we would expect the frequency of those alleles in any local population to be related to the environmental conditions at or around that site. Relevant landscape analyses to develop variables for an analysis of this kind of question might include sampling of the environmental variable at the sampled location, or perhaps the mean and variability of that environmental variable within differing neighborhood extents surrounding that location. In addition, in some situations the spatial pattern of environmental features may influence the strength and direction of selection. In such cases, calculating landscape pattern metrics (McGarigal et al. 2002) for variables represented as patch mosaics and surface metrics (McGarigal & Cushman 2005; McGarigal et al. 2009) within differing neighborhood extents surrounding the sampled sites may also be important, depending on the hypotheses being tested.

### 2.6.5 Step 5: conduct analysis

The final step in the analysis is to conduct statistical modeling to associate the genetic structure among the sampled locations to the environmental conditions in the landscape at, around, or between those locations. The methods available to analyze these pattern–process relationships are beyond the scope of this chapter, and we refer the readers to Chapter 5 for presentation of a synthetic framework to guide selection and implementation of statistical analysis in landscape genetics. The point we want to raise here is that the success of any analytical methods applied to landscape genetic data will fundamentally depend on correct decisions made in each of the four steps described above. Meaningful objectives and clear *a priori* hypotheses must guide careful delineation of landscape extent, thematic content, thematic resolution and grain of analysis, which must be followed by appropriate sampling design to collect genetic data (response variables) and spatial analysis to characterize the landscape structure at, around, or between those sampling locations (predictor variables) in a manner that provides a robust data set for analysis.

## 2.7   CURRENT CHALLENGES AND FUTURE OPPORTUNITIES

In addition to the inherent challenge of correctly matching data collection, landscape definition, and analyses, there are several issues that pose particular current challenges, as well as future opportunities, in landscape genetics research. One of the most challenging tasks in landscape genetic analysis is to resolve the potential difficulties posed by time lags and past population histories. Many methods currently used in landscape genetics are based on associating genetic differences among populations or individuals with the structure of the landscape at, around, or between those locations. There are several apparent problems related to temporal scale in this effort, such as time lags for the genetic structure of the population to respond to landscape changes (e.g., ghosts of landscape past; Landguth et al. 2010b). Also, different population histories, especially patterns of range expansion or contraction from previous refugia, can create genetic patterns in the current landscape that may mimic the effects of isolation-by-distance and resistance, or directional selection along environmental gradients.

There are a number of analytical challenges that have yet to be resolved which complicate analysis of pattern–process relationships in landscape genetics. For example, this chapter focused on the sensitivity of landscape genetic analysis to thematic content, resolution, grain, and extent of the landscape, but it remains a challenge to decide how to best represent different landscape variables. We argued that *a priori* knowledge should guide these decisions, but often there is insufficient knowledge to make informed choices. A number of efforts have been made to develop predictions of optimal or functional grain, thematic content and thematic resolution (e.g., based on simulations, see Chapter 6) and to develop modeling frameworks that enable the evaluation of a large number of competing hypotheses (e.g., Shirk et al. 2010; Chapter 5).

One of the greatest challenges and opportunities for landscape genetics is the development of theories and methods to study adaptive evolution in complex spatial environments. In adaptive landscape genetics it is not just the landscape extent between locations that matters, but also the local environment (e.g., Wagner & Fortin 2013). Selection is expected to be driven by local or neighborhood environmental conditions, while the effects of this selection will only be expressed as measurable genetic differentiation when gene flow is sufficiently restricted (such as in isolated subpopulations or along gradients of isolation-by-distance or isolation-by-barriers (IBB); e.g., Neimiller et al. 2008; Nosil et al. 2008; Yang et al. 2013). Understanding how to combine the effects of selection and gene flow in complex landscapes remains an area of nascent study with much additional work to be done. Thus far, simulation modeling (e.g., Chapter 6) has provided the strongest framework for exploring spatially dependent selection in its interaction with gene flow in affecting adaptive evolution of populations (e.g., Landguth & Balkenhol 2012; Landguth et al. 2012).

The review and protocol presented in this chapter suggest that correctly delineating and analyzing landscape heterogeneity for a given landscape genetic question is not a trivial matter. In many past applications researchers have not carefully considered thematic content, thematic resolution, grain, and extent of their landscape model prior to analysis. Too often researchers take the spatial data that are available without considering the implications the scale and content of those data have for analysis. Our most important message in this chapter is for researchers to carefully tailor genetic sampling, landscape definition, and analysis to the question, system, and organism under study. We hope that the concepts and protocol proposed in this chapter will facilitate this.

## REFERENCES

Adriansen, F., Chardon, J.P., De Blust, G., Swinnen, E., Gulinck, H., & Matthysen, E. (2003) The application of "least-cost" modeling as a functional landscape model. *Landscape and Urban Planning* **64**, 223–47.

Alverson, W., Waller, D.M., & Solheim, S.L. (1988) Forests too deer: edge effects in northern Wisconsin. *Conservation Biology* **2**, 348–58.

Anderson, C.D., Epperson, B.K., Fortin, M.J., Holderegger, R., James, P., Rosenberg, M.S., & Spear, S. (2010) Considering spatial and temporal scale in landscape-genetic studies of gene flow. *Molecular Ecology* **19**, 3565–75.

Baker, W.L. & Dillon, G.K. (2000) Plant and vegetation responses to edges in the southern Rocky Mountains. In Knight, R.L., Smith, R.W., Buskirk, S.W., Romme, W.H., & Baker, W.L. (ed.), *Forest Fragmentation in the Southern Rocky Mountains*. University Press of Colorado, Boulder.

Bélisle, M., Desrochers, A., & Fortin, M.J. (2001) Influence of forest cover on the movements of forest birds: a homing experiment. *Ecology* **82**, 1893–904.

Bender, D.J., Contreras, T.A., & Fahrig, L. (1998) Habitat loss and population decline: a meta-analysis of the patch size effect. *Ecology* **79**, 517–33.

Bender, D.J., Tischendorf, L., & Fahrig, L. (2003) Evaluation of patch isolation metrics for predicting animal movement in binary landscapes. *Landscape Ecology* **18**, 17–39.

Bergin, T.M., Best, L.B., Freemark, K.E., & Koehler, K.J. (2000) Effects of landscape structure on nest predation in roadsides of a midwestern agroecosystem: a multiscale analysis. *Landscape Ecology* **15**, 131–43.

Bohonak, A.J. (1999) Dispersal, gene flow and population structure. *Quarterly Review of Biology* **44**, 21–45.

Borcard, D. & Legendre, P. (2012) Is the Mantel correlogram powerful enough to be useful in ecological analysis? *A simulation study. Ecology* **93**, 1473–81.

Bormann, F.H., Likens, G.E., Siccama, T.C., Pierce, R.S., & Eaton J.S. (1974) The export of nutrients and recovery of stable conditions following deforestation at Hubbard Brook. *Ecological Monographs* **44**, 255–77.

Brittingham, M.C. & Temple, S.A. (1983) Have cowbirds caused forest songbirds to decline? *BioScience* **33**, 31–5.

Carlson, A. & Hartman, G. (2001) Tropical forest fragmentation and nest predation – an experimental study in an Eastern Arc montane forest, Tanzania. *Biodiversity and Conservation* **10**, 1077–85.

Charlesworth, D. & Charlesworth, B. (1987) Inbreeding depression and its evolutionary consequences. *Annual Review of Ecology and Systematics* **18**, 237–68.

Chasko, G.G. & Gates, J.E. (1982) Avian habitat suitability along a transmission line corridor in an oak-hickory forest region. *Wildlife Monographs* **82**, 1–41.

Chen, J., Franklin, J.F., & Spies, T.A. (1992) Vegetation response to edge environments in old-growth Douglas-fir forests. *Ecological Applications* **2**, 387–96.

Cushman, S.A. & Landguth, E.L. (2010) Scale dependent inference in landscape genetics. *Landscape Ecology* **25**, 967–79.

Cushman, S.A. & Landguth, E.L. (2012) Multi-taxa population connectivity in the Northern Rocky Mountains. *Ecological Modeling* **231**, 101–12.

Cushman, S.A., McKelvey, K.S., Hayden, J., & Schwartz, M.K. (2006) Gene flow in complex landscapes: testing multiple hypotheses with causal modeling. *The American Naturalist* **168**, 486–99.

Dale, V.H., O'Neill, R.V., Southworth, F., & Pedlowski, M. (1994) Modeling effects of land management in the Brazilian Amazonia settlement of Rondonia. *Conservation Biology* **8**, 196–206.

Debinski, D.M. & Holt, R.D. (2000) A survey and overview of habitat fragmentation experiments. *Conservation Biology* **14**, 342–55.

de Casenave, J.L., Pelotto, J.P., & Protomastro, J. (1995) Edge-interior differences in vegetation structure and composition in a chaco semiarid forest, Argentina. *Forest Ecology and Management* **72**, 61–9.

Donovan, T.M. & Flather, C.H. (2002) Relationships among North American songbird trends, habitat fragmentation and landscape occupancy. *Ecological Applications* **12**, 364–74.

Endler, J.A. (1973) Gene flow and population differentiation studies of clines suggest that differentiation along environmental gradients may be independent of gene flow. *Science* **179**, 243–50.

Epperson, B.K. (1993) Recent advances in correlation studies of spatial patterns of genetic variation. In: Hecht, M.K., MacIntyre, R.J., & Clegg, M.T. (eds.), *Evolutionary Biology*, Springer.

Estrada, E. & Bodin, Ö. (2008) Using network centrality measures to manage landscape connectivity. *Ecological Applications* **18**, 1810–25.

Fahrig, L. (2002) Effect of habitat fragmentation on the extinction threshold: a synthesis. *Ecological Applications* **12**, 346–53.

Fahrig, L. (2003) Effects of habitat fragmentation on biodiversity. *Annual Review of Ecology, Evolution and Systematics*, pp. 487–515.

Findlay, C.S. & Houlahan, J. (1997) Anthropogenic correlates of species richness in southeastern Ontario wetlands. *Conservation Biology* **11**, 1000–9.

Flather, C.H. & Bevers, M. (2002) Patchy reaction-diffusion and population abundance: the relative importance of habitat amount and arrangement. *The American Naturalist* **159**, 40–56.

Franklin, J.F. & Forman, R.T.T. (1987) Creating landscape configuration by forest cutting: ecological consequences and principles. *Landscape Ecology* **1**, 5–18.

Ganzhorn, J.U. & Eisenbeiß, B. (2001) The concept of nested species assemblages and its utility for understanding effects of habitat fragmentation. *Basic and Applied Ecology* **2**, 87–99.

Gates, J.E. & Mosher, J.A. (1981) A functional approach to estimating habitat edge width for birds. *The American Midland Naturalist* **105**, 189–192.

Geiger, R. (1965) *The Climate near the Ground*. Harvard University Press, Cambridge, MA.

Gibbs, J.P. & Stanton, E.J. (2001) Habitat fragmentation and arthropod community change: carrion beetles, phoretic mites and flies. *Ecological Applications* **11**, 79–85.

Govindaraju, D.R. (1988) Relationship between dispersal ability and levels of gene flow in plants *Oikos* **52**, 31–5.

Grace, J. (1977) *Plant Response to Wind*. Academic Press, London.

Grace, J. (1983) *Plant–Atmosphere Relationships: Outline of Studies in Ecology*. Chapman & Hall, London.

Gulve, P.S. (1994) Distribution and extinction patterns within a northern metapopulation of the pool frog, *Rana lessonae*. *Ecology* **75**, 1357–67.

Haines-Young, R., Green, D.R., & Cousins, S.H. (eds.) (2003) *Landscape Ecology and Geographical Information Systems*. CRC Press, London.

Hanski, I. & Gilpin, M. (1991) Metapopulation dynamics: brief history and conceptual domain. *Biological Journal of the Linnean Society* **42**, 3–16.

Harrison, S. (1991) Local extinction in a metapopulation context: an empirical evaluation. *Biological Journal of the Linnean Society* **42**, 73–88.

Harrison, S. (1994) Metapopulations and conservation. In: Edwards, P.J., May, R.M., & Webb, N.R. (eds.), *Large-Scale Ecology and Conservation Biology*. Blackwell, Oxford.

Harrison S. & Taylor, A.D. (1997) Empirical evidence for metapopulation dynamics. In: Hanski, I.A.& Gilpin, M.E. (eds.), *Metapopulation Biology: Ecology, Genetics and Evolution*. Academic Press, San Diego, CA.

Hartl, D.L. & Clark, A.G. (1997) *Principles of Population Genetics*. Sinauer associates Inc., Sunderland.

Hill, M.F. & Caswell, H. (1999) Habitat fragmentation and extinction thresholds on fractal landscapes. *Ecology Letters* **2**, 121–7.

Hornbeck, J.W. (1973) Storm flow from hardwood-forested and cleared watersheds in New Hampshire. *Water Resources Research* **9**, 346–54.

Johnson, A.R., Wiens, J.A., Milne, B.T., & Crist, T.O. (1992) Animal movements and population dynamics in heterogeneous landscapes. *Landscape Ecology* **7**, 63–75.

Kapos, V. (1989) Effects of isolation on the water status of forest patches in the Brazilian Amazon. *Journal of Tropical Ecology* **5**, 173–85.

Kareiva, P. (1987) Habitat fragmentation and the stability of predator–prey interactions. *Nature* **326**, 388–90.

Kareiva, P. & Wennergren, U. (1995) Connecting landscape patterns to ecosystem and population processes. *Nature* **373**, 299–302.

King, A.W. & With, K.A. (2002) Dispersal success on spatially structured landscapes: when do spatial pattern and dispersal behavior really matter? *Ecological Modeling* **147**, 23–39.

Koen, E.L., Garroway, C.J., Wilson, P.J., & Bowman, J. (2010) The effect of map boundary on estimates of landscape resistance to animal movement. *PLoS One* **5**, e11785.

Kolozsvary, M.B. & Swihart, R.K. (1999) Habitat fragmentation and the distribution of amphibians: patch and landscape correlates in farmland. *Canadian Journal of Zoology* **77**, 1288–99.

Kremsater, L. & Bunnell, F.L. (1999) Edge effects: theory, evidence and implications to management of western North American forests. In: Rochelle, J.A., Lehmann, L.A., & Wiesniewski, J. (eds.), *Forest Fragmentation: Wildlife and Management Implications*. Brill, Leiden, The Netherlands, pp. 117–53.

Kroodsma, R.L. (1982) Edge effect on breeding forest birds along a power-line corridor. *Journal of Applied Ecology* **19**, 361–70.

Kurki, S., Nikula, A., Helle, P., & Linden, H. (2000) Landscape fragmentation and forest composition effects on grouse breeding success in boreal forests. *Ecology* **81**, 1985–97.

Lande, R. (1987) Extinction thresholds in demographic models of territorial populations. *The American Naturalist* **130**, 624–35.

Lande, R. (1988) Genetics and demography in biological conservation. *Science* **241**, 1455–60.

Landguth, E.L. & Balkenhol, N. (2012) Relative sensitivity of neutral versus adaptive genetic data for assessing population differentiation. *Conservation Genetics* **13**, 1421–6.

Landguth, E.L., Cushman, S.A., Murphy, M.A., & Luikart, G. (2010a). Relationships between migration rates and landscape resistance assessed using individual-based simulations. *Molecular Ecology Resources* **10**, 854–62.

Landguth, E.L., Cushman, S.A., Schwartz, M.K., McKelvey, K.S., Murphy, M., & Luikart, G. (2010b) Quantifying the lag time to detect barriers in landscape genetics. *Molecular Ecology* **19**, 4179–91.

Landguth, E.L., Cushman, S.A., & Johnson, N.A. (2012) Simulating natural selection in landscape genetics. *Molecular Ecology Resources* **12**, 363–8.

Laurance, W.F., Pérez-Salicrup, D., Delamônica, P., Fearnside, P.M., D'Angelo, S., Jerozolinski, A., Pohl, L., & Lovejoy, T.E. (2001) Rain forest fragmentation and the structure of Amazonian liana communities. *Ecology* **82**, 105–16.

Leopold, A. (1933) *Game Management*. Charles Scribners, New York.

Levin, S.A. (1974) Dispersion and population interactions. *The American Naturalist* **108**, 207–28.

Levin, S.A. (1992) The problem of pattern and scale in ecology: the Robert H. MacArthur award lecture. *Ecology* **73**, 1943–67.

Likens, G.E., Bormann, F.H., Johnson, N.M., Fisher, D.W., & Pierce, R.S. (1970) Effects of forest cutting and herbicide treatment on nutrient budgets in the Hubbard Brook watershed-ecosystem. *Ecological Monographs* **40**, 23–47.

Lovejoy, T.E., Bierregaard, R.O. Jr. Rylands, A.B., Malcolm, J.R., Quintela, C.E., Harper, L.H., Brown, K.S. Jr. Powell, A.H., Powell, G.V.N., Schubart, H.O.R., & Hays, M.B. (1986) Edge and other effects of isolation on Amazon forest fragments. In Soule, M.E. (ed.), *Conservation Biology: The Science of Scarcity and Diversity*. Sinauer Associates Inc., Massachusetts.

Mahan, C.G. & Yahner, R.H. (2000) Effects of forest fragmentation on behaviour patterns in the eastern chipmunk (*Tamias striatus*). *Canadian Journal of Zoology* **77**, 1991–7.

Manel, S., Schwartz, M.K., Luikart, G., & Taberlet, P. (2003) Landscape genetics: combining landscape ecology and population genetics. *Trends in Ecology and Evolution* **18**, 189–97.

Mangel, M. & Clark, C.W. (1986) Towards a unified foraging theory. *Ecology* **67**, 1127–38.

Marsh, D.M. & Trenham, P.C. (2001) Metapopulation dynamics and amphibian conservation. *Conservation Biology* **15**, 40–9.

Marsh, D.M., Thakur, K.A., Bulka, K.C., & Clark, L.B. (2004) Dispersal and colonization through open fields by a terrestrial, woodland salamander. *Ecology* **85**, 3396–405.

McCoy, E.D. & Mushinsky, H.R. (1999) Habitat fragmentation and the abundances of vertebrates in the Florida scrub. *Ecology* **80**, 2526–38.

McGarigal, K. & Cushman, S.A. (2002) Comparative evaluation of experimental approaches to the study of habitat fragmentation effects. *Ecological Applications* **12**, 335–45.

McGarigal, K. & Cushman, S.A. (2005) *The Gradient Concept of Landscape Structure. Issues and Perspectives in Landscape Ecology*. Cambridge University Press, Cambridge.

McGarigal, K., Cushman, S.A., Neel, M.C., & Ene, E. (2002) *FRAGSTATS: Spatial Pattern Analysis Program for Categorical Maps* [Online]. Available at: http://www.umass.edu/landeco/research/fragstats/fragstats.html.

McGarigal, K., Tagil, S., & Cushman, S.A. (2009) Surface metrics: an alternative to patch metrics for the quantification of landscape structure. *Landscape Ecology* **24**, 433–50.

McKinney, M.L. & Lockwood, J.L. (1999) Biotic homogenization: a few winners replacing many losers in the next mass extinction. *Trends in Ecology and Evolution* **14**, 450–3.

McRae, B.H. (2006) Isolation by resistance. *Evolution* **60**, 1551–61.

McRae, B.H., Dickson, B.G., Keitt, T.H., & Shah, V.B. (2008) Using circuit theory to model connectivity in ecology, evolution and conservation. *Ecology* **89**, 2712–24.

Mills, L.S. (1995) Edge effects and isolation: red-backed voles on forest remnants. *Conservation Biology* **9**, 395–403.

Moen, A.N. (1974) Turbulence and visualization of wind flow. *Ecology* **55**, 1420–4.

Moilanen, A. & Hanski, I. (2006) Connectivity and metapopulation dynamics in highly fragmented landscapes. In Crooks, K.R. & Sanjayan, M. (Eds.), *Connectivity Conservation*. Cambridge University Press, Cambridge, UK.

Monteith, J.L. (1975) *Vegetation and the Atmosphere*. Academic Press, London.

Mooney, H.A.& Hobbs, R.J. (eds.) (2000) *Invasive Species in a Changing World*. Island Press, Washington.

Murphy, M.A., Evans, J.S., & Storfer, A. (2010) Quantifying *Bufo boreas* connectivity in Yellowstone National Park with landscape genetics. *Ecology* **91**, 252–61.

Neimiller, M.L., Fitzpatrick, B.M., & Miller, B.T. (2008) Recent divergence with gene flow in Tennessee cave salamanders (Plethodontidae; Gyrinophylus) inferred from gene genealogies. *Molecular Ecology* **17**, 2258–75.

Nosil, P. (2008) Speciation with gene flow could be common. *Molecular Ecology* **17**, 2103–6.

Pearson, S.M., Turner, M.G., Gardner, R.H., and O'Neill, R.V. (1996) An organism-based perspective of habitat fragmentation. In: Szaro, R.C. and Johnston, D.W. (eds.), *Biodiversity in Managed Landscapes: Theory and Practice*, Oxford University Press, New York.

Pickett, S.T.A. & Cadenasso, M.L. (1995) Landscape ecology: spatial heterogeneity in ecological systems. *Science* **269**, 331–4.

Ranney, J.W., Bruner, M.C., & Levenson, J.B. (1981) The importance of edge in the structure and dynamics of forest islands. In Burgess, R.L.& Sharpe, D.M. (eds.), *Forest Island Dynamics in Man-Dominated Landscapes*. Springer-Verlag, New York.

Robbins, C.S., Dawson, D.K., & Dowell, B.A. (1989) Habitat area requirements of breeding forest birds of the middle Atlantic states. *Wildlife Monographs* **103**, 1–34.

Rothermel, B.B. & Semlitsch, R.D. (2002) An experimental investigation of landscape resistance of forest versus old-field habitats to emigrating juvenile amphibian. *Conservation Biology* **16**, 1324–32.

Rousset, F. (1997) Genetic differentiation and estimation of gene flow from F-statistics under isolation-by-distance. *Genetics* **145**, 1219–28.

Rukke, B.A. (2000) Effects of habitat fragmentation: increased isolation and reduced habitat size reduces the incidence of dead wood fungi beetles in a fragmented forest landscape. *Ecography* **23**, 492–502.

Sánchez-Zapata, J. & Calvo, J.F. (1999) Raptor distribution in relation to landscape composition in semi-arid Mediterranean habitats. *Journal of Applied Ecology* **36**, 254–62.

Saunders, D., Hobbs, R.J., & Margules, C.R. (1991) Biological consequences of ecosystem fragmentation: a review. *Conservation Biology* **5**, 18–32.

Schmiegelow, F.K. & Mönkkönen, M. (2002) Habitat loss and fragmentation in dynamic landscapes: avian perspectives from the boreal forest. *Ecological Applications* **12**, 375–89.

Schoener, T.W. & Spiller, D.A. (1992) Stabilimenta characteristics of the spider *Argiope argentata* on small islands – support of the predator-defense hypothesis. *Behavioral Ecology and Sociobiology* **31**, 309–18.

Segelbacher, G., Cushman, S.A., Epperson, B.K., Fortin, M.J., Francois, O., Hardy, D.J., Holderegger, R., Taberlet, P., Waits, L.P., & Manel, S. (2010) Applications of landscape genetics in conservation biology: concepts and challenges. *Conservation Genetics* **11**, 375–85.

Shirk, A.J., Wallin, D.O., Cushman, S.A., Rice, C.G., & Warheit, K.I. (2010) Inferring landscape effects on gene flow: a new model selection framework. *Molecular Ecology* **19**, 3603–19.

Short Bull, R.A., Cushman, S.A., Mace, R., Chilton, T., Kendall, K.C., Landguth, E.L., Schwartz, M.K., McKelvey, K., Allendorf, F.W., & Luikart, G. (2011) Why replication is important in landscape genetics: American black bear in the Rocky Mountains. *Molecular Ecology* **20**, 1092–107.

Sklar, F.H. & Costanza, R. (1991) The development of dynamic spatial models for landscape ecology: a review and prognosis. In: Turner, M.G. & Gardner, R.H. (eds.), *Quantitative Methods in Landscape Ecology: The Analysis and Interpretation of Landscape Heterogeneity*. Springer-Verlag.

Spear, S.F., Balkenhol, N., Fortin, M.J., McRae, B.H., & Scribner, K.I.M. (2010) Use of resistance surfaces for landscape genetic studies: considerations for parameterization and analysis. *Molecular Ecology* **19**, 3576–91.

Steffan-Dewenter, I., Münzenberg, U., Bürger, C., Thies, C., & Tscharntke, T. (2002) Scale-dependent effects of landscape context on three pollinator guilds. Ecology **83**, 1421–32.

Strelke, W.K. & Dickson, J.G. (1980) Effect of forest clearcut edge on breeding birds in Texas. *Journal of Wildlife Management* **44**, 559–67.

Tansley, A.G. (1935) The use and abuse of vegetational concepts and terms. *Ecology* **16**, 284–307.

Templeton, A.R. (1981) Mechanisms of speciation – a population genetic approach. *Annual Review of Ecology and Systematics* **12**, 23–48.

Thomas, C.D., Cameron, A., Green, R.E., Bakkenes, M., Beaumont, L.J., Collingham, Y.C., Erasmus, B.F.N., de Siqueira, M.F., Grainger, A., Hannah, L., Hughes, L., Huntley, B., van Jaarsveld, A.S., Midgley, G.F., Miles, L., Ortega-Huerta, M.A., Peterson, A.T., Phillips, O.L., & Williams, S.E. (2004) Extinction risk from climate change. *Nature* **427**, 145–8.

Tischendorf, L., Bender, D.L., & Fahrig, L. (2003) Evaluation of patch isolation metrics in mosaic landscapes for specialist vs. generalist dispersers. *Landscape Ecology* **18**, 41–50.

Tranquillini, W. (1979) *Physiological Ecology of the Alpine Timberline*. Springer-Verlag, Berlin.

Turner, M.G. (ed.) (1987) *Landscape Heterogeneity and Disturbance*. Springer-Verlag.

Turner, M.G. (1989) Landscape ecology: the effect of pattern on process. *Annual Review of Ecology and Systematics* **20**, 171–97.

Turner, M.G. (2005) Landscape ecology: what is the state of the science?. *Annual Review of Ecology, Evolution, and Systematics* 319–44.

Turner, M.G., Gardner, R.H., & O'Neill, R.V. (2001) *Landscape Ecology in Theory and Practice: Pattern and Process*. Springer.

Vallan, D. (2000) Influence of forest fragmentation on amphibian diversity in the nature reserve of Ambohitantely, highland Madagascar. *Biological Conservation* **96**, 31–43.

Vance, M.D., Fahrig, L., & Flather, C.H. (2003) Effect of reproductive rate on minimum habitat requirements of forest-breeding birds. *Ecology* **84**, 2643–53.

Virgos, E. (2001) Role of isolation and habitat quality in shaping species abundance: a test with bagers (*Meles meles* L.) in a gradient of forest fragmentation. *Journal of Biogeography* **28**, 381–9.

Wagner, H.H. & Fortin, M.J. (2013) A conceptual framework for the spatial analysis of landscape genetic data. *Conservation Genetics* **14**, 253–61.

Wales, B.A. (1972) Vegetation analysis of north and south edges in a mature oak-hickory forest. *Ecological Monographs* **42**, 451–71.

Wasserman, T.N., Cushman, S.A., Schwartz, M.K., & Wallin, D.O. (2010) Spatial scaling and multi-model inference in landscape genetics: *Martes americana* in northern Idaho. *Landscape Ecology* **25**, 1601–12.

Watt, A.S. (1947) Pattern and process in the plant community. *Journal of Ecology* **35**, 1–22.

Wegner, J.F. & Merriam, G. (1979) Movements by birds and small mammals between a wood and adjoining farmland habitats. *Journal of Applied Ecology* **16**, 349–57.

Wiens, J.A. (1989) Spatial scaling in ecology. *Functional Ecology* **3**, 385–97.

Wiens, J.A. (1992) What is landscape ecology, really? *Landscape Ecology* **7**, 149–50.

Wiens, J.A., Crawford, C.S., & Gosz, J.R. (1985) Boundary dynamics: a conceptual framework for studying landscape ecosystems. *Oikos* **45**, 421–7.

Wiens, J.A., Stenseth, N.C., van Horne, B., & Ims, R.A. (1993) Ecological mechanisms and landscape ecology. *Oikos* **66**, 369–80.

With, K.A. & Crist, T.O. (1995) Critical thresholds in species' responses to landscape structure. *Ecology* **76**, 2446–59.

With, K.A. & King, A.W. (1999) Extinction thresholds for species in fractal landscapes. *Conservation Biology* **13**, 314–26.

With, K.A., Gardner, R.H., & Turner, M.G. (1997) Landscape connectivity and population distributions in heterogeneous environments. *Oikos* **78**, 151–69.

Wright, S. (1943) Isolation-by-distance. *Genetics* **28**, 114–38.

Yahner, R.H. (1986) Microhabitat use by small mammals in even-aged forest stands. *The American Midland Naturalist* **115**, 174–80.

Yang, J., Cushman, S.A., Yang, J., Yang, M., & Bao, T. (2013) Effects of climatic gradients on genetic differentiation of Caragana in the Ordos Plateau, China. *Landscape Ecology* **28**, 1729–41.

Zen, M.C. (1995) Growth and morphological development of Douglas-fir and western hemlock on forest edges in the Alberni Valley. MSc thesis, University of Washington, Seattle, WA.

# Chapter 3

# BASICS OF POPULATION GENETICS: QUANTIFYING NEUTRAL AND ADAPTIVE GENETIC VARIATION FOR LANDSCAPE GENETIC STUDIES

*Lisette P. Waits[1] and Andrew Storfer[2]*

[1] *Fish and Wildlife Sciences, University of Idaho, USA*
[2] *School of Biological Sciences, Washington State University, USA*

## 3.1 INTRODUCTION

The field of landscape genetics integrates concepts and methods from landscape ecology and population genetics. Population genetics has a rich history focused around the study of the genetic composition of populations and changes in genetic variation that are driven by the processes of mutation, selection, gene flow and drift. This chapter is designed to provide an overview of molecular methods and population genetic theory for readers who have no or little background in population genetics. We introduce a conceptual framework describing how landscapes influence genetic variation and highlight the major research questions in landscape genetics. We review molecular genetic terms and molecular methods for evaluating genetic variation and present a summary of population genetic theories that are particularly relevant for landscape genetics. This chapter provides an overview of analytical methods for measuring genetic diversity of populations and individuals, and the methods used to assess genetic structure and gene flow in natural populations. We end with a discussion of future challenges in population and landscape genetics, with particular attention to interpretation and analysis of models and analyses for harnessing the power of next-generation sequencing data.

*Landscape Genetics: Concepts, Methods, Applications*, First Edition. Edited by Niko Balkenhol, Samuel A. Cushman, Andrew T. Storfer, and Lisette P. Waits.

## 3.2 OVERVIEW OF LANDSCAPE INFLUENCES ON GENETIC VARIATION

Deoxyribonucleic acid, or DNA, is the genetic material that codes for all of the diversity of life. The information in DNA is stored in four chemical bases: adenine (A), guanine (G), cytosine (C), and thymine (T). DNA bases pair up with each other, A with T and C with G, to form units called base pairs. Each base is attached to a sugar molecule and a phosphate molecule to make up the double helix backbone of DNA. The order, or sequence, of bases provides the information for building and maintaining individual organisms. DNA contains *genes* that code for proteins, which are created through the processes of transcription and translation. During transcription, DNA is translated to messenger RNA (mRNA) and during translation the sequence of mRNA is read. Each sequence of three bases codes for an amino acid and mRNA is converted to a protein one amino acid at a time. Overall, the DNA sequence is very similar among individuals within a species. For example, human DNA consists of about 3 billion base pairs and less than 1% is different among individuals. However, these small numbers of differences, known as genetic variation, are very important.

At its most basic level, genetic variation can be defined as DNA sequence differences at the same physical location in the genome. This variation can be observed at locations in the genome that are under selection (***adaptive loci*** or ***non-neutral loci***) or locations that are not under selection (***neutral loci***). Genetic variation of a population or species is influenced by four main processes – mutation, selection, gene flow, and genetic drift (Hedrick 2011; Allendorf et al. 2012). ***Mutation*** is the process that creates new genetic variants or ***alleles*** due to errors in DNA replication. Selection impacts genetic variation by increasing the frequencies of alleles that are favorable in a particular environment and decreasing the frequencies of alleles that are less favorable in that environment. If a ***locus*** or a mutation is neutral, selection does not directly influence the frequency of alleles. Gene flow occurs as a result of migration (dispersal and subsequent reproduction) and this process moves alleles between populations and tends to make them more similar genetically. Genetic drift is the change in allele frequencies due to random sampling effects as alleles are passed on from one generation to the next (Wright 1931). Genetic drift is thus influenced by the number of breeding individuals and the variance in reproductive success among breeders.

Genetic drift is directly linked to ***effective population size*** ($N_e$). Sewall Wright (Wright 1931, 1938) introduced the concept of $N_e$ to mathematically and conceptually represent the effects of genetic drift, and $N_e$ is the population genetic analog to census size used in ecology. Wright described $N_e$ as the number of breeding individuals in an idealized population that contribute genetically to the next generation. A simplified but useful approach for understanding $N_e$ is to think about it as the number of individuals that are able to successfully pass on their genes to the next generation. Populations with larger $N_e$ will be able to maintain higher amounts of genetic variation because more new alleles are created by mutation due to a larger number of breeding events and fewer alleles are lost to genetic drift since there is a larger sample of breeders each generation. $N_e$ also influences the relative impacts of genetic drift and selection in natural populations (Fisher 1930; Wright 1931). For example, as $N_e$ decreases, the effect of selection decreases relative to the effect of genetic drift and important adaptive genetic variation can be lost (Hedrick 2011).

Population geneticists often evaluate two main components of genetic variation: genetic diversity and genetic structure. Genetic diversity is the amount of genetic variation found within an individual, a population, or a spatial area. On the other hand, genetic structure is the distribution of that variation among individuals, populations, or areas. Landscapes influence the amount and distribution of genetic variation in many ways. First, the landscape influences where an organism can live and how many individuals live in a particular location that affects the distribution and amount of genetic variation (Fig. 3.1). The landscape also influences the amount of movement and

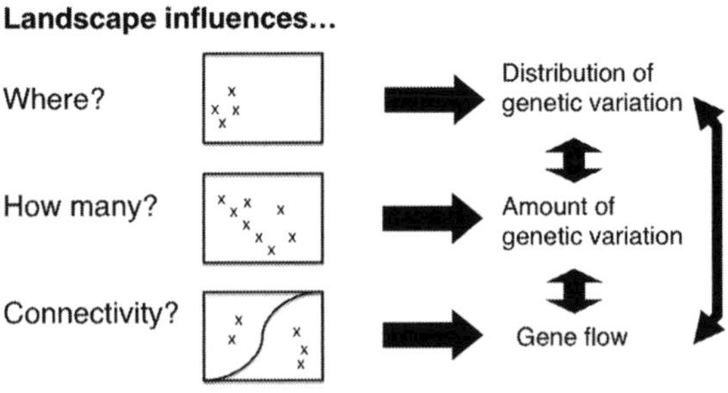

**Fig. 3.1** Influence of the landscape on individual organisms and genetic variation.

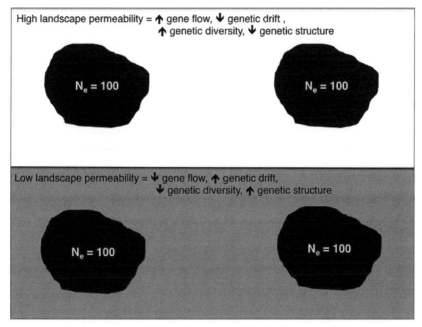

**Fig. 3.2** The effect of landscape permeability on genetic diversity and structure with equal effective population sizes ($N_e$). The white landscape matrix in the top panel has high permeability while the gray landscape matrix in the bottom panel has low permeability. Thus, gene flow is higher and genetic drift is lower in the top panel, leading to higher levels of genetic diversity and lower levels of genetic structure in the white landscape compared to the gray landscape.

subsequent gene flow that occurs among individuals in different locations, which directly affects $N_e$ and the amount and distribution of genetic variation (Fig. 3.1).

For example, when two populations of equal $N_e$ are imbedded in habitat matrices of differing levels of permeability, the two populations separated by the more permeable matrix will retain more genetic variation due to increased gene flow and reduced genetic drift (Fig. 3.2). This difference in matrix permeability also creates different levels of genetic structure. With the less permeable matrix (Fig. 3.2), a greater amount of genetic structure is created due to the increased restriction of gene flow, which increases genetic drift and causes allele frequencies to diverge at a faster rate. In an alternative scenario, where matrix permeability is the same but the two populations have different $N_e$, the population with the smaller $N_e$ will lose genetic variation more quickly and genetic structure will increase more rapidly due to increased genetic drift (Fig. 3.3). In this scenario, differences in $N_e$ could be due to differences in the habitat quality of each patch, and the

patch with the higher habitat quality could support a higher $N_e$.

For loci that are under selection, the amount of genetic variation and the frequency of particular alleles are directly affected by the environmental conditions at a site. For example, the melanocortin-1 receptor (*Mc1r*) gene has a large effect on coat color in beach mice (Hoekstra et al. 2006). One allele for this gene codes for dark color mice and is in very high frequency in the forest habitats, where dark mice match the forest floor. Another allele codes for light color and is found in high frequency in mice that match the sandy beach habitats where they reside (Linnen et al. 2009).

In the following sections, we will (1) describe molecular markers and methods used to generate genetic data for landscape genetic studies, (2) review population genetic theory and terminology, (3) overview the metrics and methods used to measure genetic diversity, genetic structure, and gene flow, and (4) discuss future directions and challenges for genetic analysis.

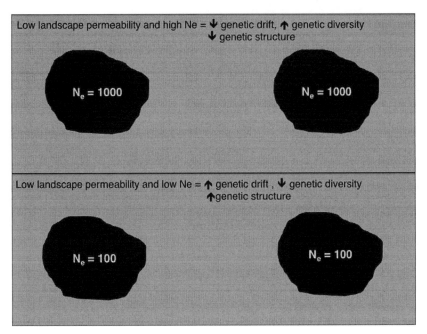

**Fig. 3.3** The effect of effective population size ($N_e$) on genetic diversity and structure. The landscape has low permeability in the top and bottom panels but the top panel has populations with much higher $N_e$ than the bottom panel. Thus, genetic drift is lower in the top panel, leading to higher levels of genetic diversity and less genetic structure compared to the example in the bottom panel.

### 3.3  OVERVIEW OF DNA TYPES AND MOLECULAR METHODS

#### 3.3.1  Types of DNA

There are two main types of DNA in animal cells: *mitochondrial DNA* (mtDNA) and *nuclear DNA (nDNA)*. Plant cells contain mtDNA, nDNA, and *chloroplast DNA* (cpDNA). MtDNA is a circular DNA molecule of the mitochondrion, an organelle that is the energy powerhouse of cells. MtDNA has a uniparental mode of inheritance in animals and plants as it is generally passed only from mother to offspring (Gillham 1974; Birky 1978), but paternal transmission has been documented in some animals (Zouros 2000; Zhao et al. 2004; Wolff et al. 2013) and conifer species (Neale et al. 1989). Nuclear DNA is the DNA of chromosomes found in the nucleus of a cell and has a biparental mode of inheritance since genetic material is inherited from both the mother and the father. The cpDNA is contained in the chloroplast molecule responsible for photosynthesis in plants and its mode of inheritance differs by species

and can be maternal, paternal, or biparental (Gillham 1974; Neale et al. 1989). On average, the $N_e$ of nDNA loci is four times higher than the $N_e$ of cpDNA and mtDNA because organellar DNA is most often uni-parentally inherited and *haploid*. The type of DNA, its mutation rate, and mode of inheritance influence the research questions that can be addressed and the temporal inference that can be obtained from genetic data. In general, nDNA loci are used more frequently in landscape genetic studies (90%) than mtDNA or cpDNA loci (Storfer et al. 2010) because they provide better resolution for detecting recent changes to the landscape.

#### 3.3.2  Adaptive versus neutral loci

At noted above, landscape genetic researchers can collect data at two main types of loci: adaptive loci, which are under selection, and neutral loci, which are not affected by selection. Adaptive loci directly or indirectly affect the phenotype and fitness of an organism, and landscape or environmental factors at a site

**Table 3.1** Examples of landscape genetic research questions that can be addressed with neutral and adaptive loci.

| Locus type | Research question |
| --- | --- |
| Neutral | What landscape features are barriers to movement and gene flow? |
| Neutral | What landscape features facilitate movement and gene flow? |
| Neutral | What spatial areas or patches are sources and sinks? |
| Neutral | How does the landscape influence the amount of genetic variation and has it changed over time? |
| Adaptive | Which loci are under selection? |
| Adaptive | Are there changes in allele frequencies at certain loci across an environmental gradient? |
| Adaptive | Which loci are associated with phenotypic differences among populations? |
| Adaptive | How do environmental effects on adaptive genetic variation influence individual fitness and persistence of populations and species? |

will influence allele frequencies through selection (see Chapter 9). In contrast, allele frequencies at neutral loci are primarily affected by gene flow and drift. This inherent difference between neutral and adaptive loci determines the types of research questions that can be addressed with neutral and adaptive loci (Table 3.1). When using neutral loci, researchers can address research questions related to how the landscape influences gene flow, the effective population size, or neutral genetic diversity. Thus, neutral genetic data are particularly well suited for research questions on impacts of past and present landscapes on genetic connectivity (e.g., corridor and reserve design) and for predicting

how future landscape change would affect this connectivity and resulting genetic variation (see Chapters 10 to 12 for empirical examples). Using adaptive loci, we can address how environmental factors, such as temperature and elevation, affect the geographic distribution of genetic variation (Chapter 9). Currently, neutral loci are used more commonly than adaptive loci in landscape genetics (Holderegger and Wagner 2008; Sork and Waits 2010), but the number of studies using adaptive loci is increasing.

### 3.3.3 Molecular methods

After choosing to collect data for a particular type of DNA and locus, a variety of methods are used to obtain genetic data. There are two main types of nDNA molecular methods – codominant and dominant approaches. In codominant methods like microsatellite analysis or single-nucleotide polymorphism (SNP) analysis, it is possible to visualize both alleles at a particular locus in the form of a *genotype* (AA versus AB, for example). While dominant loci methods, such as amplified fragment length polymorphism (AFLP), create banding patterns of tens to hundreds of bands for each individual that resemble a barcode, both alleles from a particular locus cannot be identified. Dominant loci data are generally recorded in a binary presence/absence format, where 1 indicates presence and 0 indicates absence (0, 1, 0, 0, 0, 1, for example). Mitochondrial DNA and cpDNA are generally analyzed using DNA sequencing or restriction fragment length polymorphism (RFLP), approaches that generate *haplotype* data rather than genotype data since the loci are haploid. Fully describing these methods is beyond the scope of this chapter, but the most commonly used approaches, including protein allozyme analysis, nDNA microsatellite analysis, AFLP analysis, SNP analysis, and cpDNA/mtDNA sequencing, are summarized in Box 3.1 and Table 3.2. For more detailed descriptions of each molecular method, see Lowe et al. (2004) or Allendorf et al. (2012).

### Box 3.1   Overview of molecular methods used in landscape genetic analyses

**Allozyme analysis** – Allozyme analysis detects allelic variants of protein enzymes encoded by genes.
**Polymerase chain reaction (PCR)** – PCR is the chemical process used to make millions of copies of a particular target DNA region or locus using short single-stranded pieces of DNA known as primers (Mullis and Faloona 1987). PCR has revolutionized molecular genetics and is used in all methods described below.

**Sanger DNA sequencing** – Sanger DNA sequencing is the chemical process used to read the sequence of nucleotides (DNA base pairs – A, G, C, T) at a particular DNA region (Sanger et al. 1977).
**Next-generation sequencing (NGS)** – This group of DNA sequencing approaches parallelize the sequencing process and can produce a greater volume of sequence data in a shorter period of time (Shendure and Ji 2008) than the original Sanger sequencing method.
**Microsatellite analysis** – Microsatellite loci (also known as short tandem repeats (STRs) or simple sequence repeats (SSRs)) or "μsats" are repeating sequences of 2–6 base pairs (CACACA or GTCGTCGTC, for example). Nuclear DNA microsatellite analysis has been the most commonly used data collection approach in landscape genetics (Storfer et al. 2010).
**Amplified fragment length polymorphism (AFLP) analysis** – AFLP analysis is a PCR-based tool for assaying genetic variation that does not require knowledge of the DNA sequence of the target species. The amplified DNA fragments are separated by size on a polyacrylamide gel and visualized using fluorescence or autoradiography (Vos et al. 1995).
**Single nucleotide polymorphism (SNP) analysis** – SNP analysis is a newer molecular method that surveys single base pair genetic polymorphisms in many locations throughout the genome. This method is rarely used in landscape genetics, but the application of this method is predicted to expand rapidly in coming years because it surveys a greater proportion of the genome and can be automated for high throughput analyses (Morin et al. 2004; Garvin et al. 2010).

The most important methodological advance in genetics was the development of the polymerase chain reaction or PCR (Mullis & Faloona 1987). This technique makes it possible to survey genetic variation by making many copies of a particular DNA locus. A recent review of landscape genetics quantified the frequency of application of different molecular methods and found that 70% of animal and 31% of plant studies used nDNA microsatellite loci, approximately 10% of animal and plant studies used mtDNA or cpDNA sequence data, 21% of plant and 5% of animal studies used AFLP analysis, and <2% of studies used SNPs (Storfer et al. 2010).

### 3.3.4   Unit of analysis

The two main units of analysis for genetic data in landscape genetics are individuals and populations (i.e., groups of individuals). In general, the population is chosen as the unit of analysis for species that are patchily distributed and the individual is chosen as the unit of analysis when the species is more continuously distributed across a landscape and there are no clear population boundaries. However, recent simulation work has shown that it can also be effective to use individuals as the unit of analysis for patchily distributed organisms (Prunier et al. 2013). Choosing the unit of analysis is an important study design decision for landscape genetics research because it influences the sampling strategy (especially the sampling level, see Chapter 4) as well as the metrics and analyses methods that can be applied. In the following sections, we will describe population genetic metrics that can be applied at either the population or individual level.

**Table 3.2** Characteristics of the most commonly used molecular methods in landscape genetics.

| Molecular method | Inheritance | Variability | Temporal scale of Inference |
| --- | --- | --- | --- |
| Nuclear DNA microsatellite | Biparental | High | Recent |
| Amplified fragment length polymorphism (AFLP) | Dominant* | High | Recent |
| Single nucleotide polymorphism (SNP) | Biparental | Moderate | Intermediate |
| Mitochondrial DNA sequence | Uniparental | Low–moderate | Historic |
| Chloroplast DNA sequence | Uniparental | Low | Historic |

*Dominant methods provide only the presence or absence of a DNA fragment.

## 3.4 IMPORTANT POPULATION GENETIC MODELS

### 3.4.1 Hardy–Weinberg equilibrium

The basic mathematical formulation for population genetics was developed by William Weinberg (1908), a German physician, and Godfrey Hardy (1908), an English mathematician, in 1908. Although Weinberg published his findings approximately six months before Hardy, their major contribution is now known as the "Hardy–Weinberg" rule. This rule states that for a diploid organism with two alleles at one locus, if $p$ represents the frequency of one allele and $q$ represents the frequency of the other allele, then genotype frequencies will be $p^2 + 2pq + q^2 = 1$, whereby $p^2$ represents the $pp$ genotype or the $p$ **homozygote** frequency, $2pq$ represents the **heterozygote** frequency, and $q^2$ represents the $q$ **homozygote** frequency. These allele frequencies are expected under Hardy–Weinberg equilibrium (HWE) in one generation of random mating in a population, that: (1) is infinite in size; (2) has no selection; (3) has no mutation; (4) has no migration; and (5) has random mating. Although these conditions are unrealistic expectations for natural populations, Hardy–Weinberg equilibrium is a useful starting assumption. That is, if predictions of allele frequencies under HWE are violated with empirical data, one can assume that at least one of the five evolutionary processes above is operating. Much of population genetics theory then deals with allele frequency expectations under models such as selection, mutation, migration, or combinations of two to three of these processes. For detailed overviews of population genetics theory, texts such as Hartl and Clark (2006) and Hedrick (2011) are recommended. We discuss some of the models most relevant to landscape genetic studies below.

### 3.4.2 Linkage equilibrium

Under Mendel's law of independent assortment, it is assumed that the cross-generational transmission of alleles at any particular locus is independent of alleles at other loci and that the fitnesses of possible pairs of alleles are decoupled. This idea is called linkage equilibrium or gametic equilibrium. It follows, then, that **linkage disequilibrium** exists when there is a statistical (non-random) association between alleles at different loci. These

alleles can be physically linked on chromosomes, whereby they are in near proximity to one another and those inherited together. Alternatively, other evolutionary processes such as genetic drift in small populations can create linkage disequilibrium among pairs of alleles. Moreover, selection can cause genetic "hitchhiking", whereby alleles of little-to-no effect on fitness are inherited together with alleles favored under selection.

Because landscape genetics studies often rely on a dozen or so loci, the assumption of linkage equilibrium among alleles and loci is important for estimations of population structure. That is, loci are assumed to be representative of genomic levels of variation and significant linkage disequilibrium among loci means they cannot be treated independently because they tend to be inherited together. As such, for studies with relatively few loci (e.g., fewer than 20), linkage disequilibrium among loci essentially reduces overall power to assess population structure. This is less of an issue for population genomic studies based on next-generation sequencing technologies (see Chapter 9) where hundreds to thousands of loci are generated. Loci under significant linkage disequilibrium can thus be discarded in these cases, whereby overall sample size is proportionally less affected. Most basic population genetics software packages, such FSTAT (Goudet 1995) and GENEPOP (Rousset 2008), will estimate linkage disequilibrium among loci.

### 3.4.3 Effective population size and genetic drift

Populations are not infinite in size and $N_e$ is almost always smaller than the census population size or the number of individuals estimated from field or other surveys. Unequal sex ratio, assortative mating, and overlapping generations are all examples of demographic factors that make effective population sizes smaller than census population sizes. A review by Frankham (1995) suggests that effective population sizes are generally 10% (or lower) of the census population size across different wildlife species. There are two commonly estimated measures of $N_e$. Variance effective size ($N_{eV}$) is the size of an ideal population experiencing drift at the same rate as the actual population. **Inbreeding** effective size ($N_{eI}$) is the size of an ideal population losing heterozygosity, due to increased relatedness, at the same rate as the actual population. For stable, large populations $N_{eI}$ and $N_{eV}$ are similar, but when populations are growing or shrinking they can differ greatly (Crandall et al. 1999). Detailed descriptions of methods for estimating $N_e$ are beyond the scope of this chapter but good reviews of methods can be found in Leberg (2005), Wang (2005), and Hare et al. (2011).

Genetic drift results in random changes in allele frequencies across generations resulting from sampling of gametes, and thus has larger effects on smaller populations than larger populations. Consider, for example, a population of 10 haploid individuals, with an allele frequency of 0.1 for allele A at a locus versus a population of 100 individuals with the same allele frequency. In the smaller population, allele A is present in 1 individual, while in the larger population, it is present in 10 individuals. Therefore, in the smaller population allele A can be lost easily if the single individual with that allele does not breed for some reason. It is harder to imagine that a stochastic event will cause all 10 individuals in the larger population to fail to breed, however. In fact, the probability that an allele is lost in a population is 1 minus its starting allele frequency, so rare alleles in small populations tend to be lost.

In diploid organisms, heterozygosity (the proportion of individuals with 2 different alleles at a locus) is lost at a rate of $1/(2N_e)$ at each generation due to genetic drift. As such, the reduction of genetic diversity due to drift has become a management concern for small populations or populations of endangered species because maintenance of genetic diversity is recognized as important for the evolutionary potential of populations to respond to future environmental change. Extreme and rapid declines of (effective) population size result in population **bottlenecks** and thereby often rapid reduction in genetic diversity resulting from drift. Conservation biologists have proposed maintaining $N_e$ above 50 to preserve short-term viability and avoid inbreeding depression (e.g., Franklin 1980) and at least 500 to avoid loss of evolutionary potential (Franklin 1980; Lande 1995). However, this general guideline has been controversial (Jamieson & Allendorf 2012; Frankham et al. 2013) and recent studies have suggested higher thresholds are needed to avoid inbreeding depression and maintain minimum viable populations (Lynch & Lande 1998; Frankham et al. 2013).

### 3.4.4 Mutation

Mutation is the ultimate source of genetic variation in populations and species. Mutations occur when mistakes are made during DNA replication and can take two main forms: point mutations or insertions or deletions of segments of DNA. Point mutations are changes of a single nucleotide, such as A to G or G to C. As much of the genome is non-coding, point mutations are often considered "silent" as they do not result in changes in protein structure. Even when point mutations do occur in segments of DNA that code for amino acids (i.e., exons), they often do not result in changes of amino acid sequences (and are called **synonymous substitutions**) due to the conservation of the genetic code (i.e., 64 possible codons code for only 20 amino acids). Point mutations that do change amino acid sequences are called **non-synonymous substitutions**. Insertions or deletions change the length of a strand of DNA and are hence called frameshift mutations when they occur in exons because they shift the reading frame that is transcribed, often resulting in major changes in amino acid sequences or even premature termination of protein formation.

Before we even knew what DNA was, mutations were already considered in population genetic models. In the 1930s, Fisher (1930) and Wright (1931) modeled the fate of neutral mutations, or those that had no fitness effect on the individual with a particular allele. The frequency of an allele in a generation ranges from 0 to $2N_e$ for diploid organisms. Mutation introduces variation at a rate of $2N_e\mu$, where $\mu$ = the mutation rate. As above, genetic drift reduces genetic variation at a rate of $1/(2N_e)$ per generation. Since the average number of new mutations entering the population each generation is $2N_e\mu$ and the chance of allelic extinction (loss of an allele) is $1/(2N_e)$, then the average substitution rate is $2N_e\mu \times 1/(2N_e) = \mu$.

Later, the idea of the "neutral theory" of molecular evolution (Kimura 1968) embraced the neutrality of mutations and therefore considered genetic variation in populations as a balance between the rate at which new alleles were introduced into new populations via mutation and the rate at which they were lost due to genetic drift. This view regards selection as a minor evolutionary force and heated debates about the relative influence of mutation, selection, and genetic drift ensued. A revision to the neutral theory called "nearly neutral theory" (Ohta 1973) was later developed. Nearly neutral theory suggests that, while most mutations can be considered selectively neutral, mutations that are slightly deleterious or advantageous are also allowed in the resulting models. In these models, alleles are chosen at random from the previous generation and extinction or fixation of alleles happens much more quickly in small populations than larger populations due to genetic drift. This also means that even slightly advantageous alleles can be lost in small populations due to chance (i.e., genetic drift).

In population and landscape genetics studies, the model of mutation can be important because it introduces new alleles and thus results in allele frequency changes in populations, thereby affecting how genetically diverged or similar the study populations are. Three main models of mutation are commonly used in landscape genetics. The first is the infinite alleles model (IAM). The infinite alleles model assumes essentially that an infinite number of alleles are possible at a particular locus and, as such, any new mutation results in a new allele. At equilibrium, the number of alleles in a population remains constant because mutation introduces new alleles at the same rate at which they are lost by genetic drift. A more recently developed model that applies to microsatellites is the stepwise mutation model (SMM) (Ohta & Kimura 1973). Under the SMM, it is assumed that mutations occur stepwise to the next possible allelic state. Thus, for microsatellites, for example, an allele with 8 repeats can only mutate to 7 repeats or 9 repeats. As microsatellite studies revealed that mutations can occur at greater than one mutational step, the two-phase model (TPM) was developed (Di Rienzo et al. 1994). The two-phase model addresses this by considering mutations of 1 repeat (one-phase) occurring with probability $p$ and mutations of $\geq 1$ repeat (two-phase) with probability $1 - p$, with a geometric distribution of lengths greater than 1 repeat.

### 3.4.5  Migration (gene flow)

Although the term "migration" has a few different definitions, in population genetics it is synonymous with gene flow. Gene flow is defined as the movement of genes (via successful mating) among populations. High rates of gene flow tend to homogenize populations, thereby resulting in similar allele frequencies. Conversely, low rates of gene flow tend to result in population isolation and genetic differentiation. In turn, if populations have a small $N_e$, genetic drift will act to differentiate these populations genetically via random changes in allele frequencies. However, it only takes small amounts of migration (about 1 effective migrant per generation in theory) to overcome drift, which is a relatively weak evolutionary force (Wright 1969).

In reality, maintenance of population connectivity from a management standpoint may require artificial movements of greater than one actual individual per generation, depending on the effective size of the population(s) being managed (Hedrick 1995). For example, when the Florida panther reached a very small population size and showed evidence of inbreeding depression, 20 Texas cougars were introduced as a one-time measure to increase genetic diversity in the panther population (Hedrick 1995). The introduction was successful, but required detailed analyses and, in this case, involved concern about introgression of a separate subspecies. These concerns were addressed with a single, rather than repeated introductions. Thus, management strategies for maintaining genetic diversity in small or declining populations, while based on Wright's (1969) original theory, need to be modified with current data that are often likely to be species-specific.

Several models of migration have been developed for use in population genetics and application in landscape genetics studies. The original "mainland-island" model of migration was developed by Wright in 1931 (Fig. 3.4a). In essence, the island model assumes that there is a mainland of infinite size exchanging migrants with an island of finite size at a rate of $m$. Given the larger size of the mainland, the allele frequencies on the island will come to resemble those of the mainland because the mainland has a disproportionate effect on the island. Alternatively, the island model can be thought of exchange of an infinite number of genes (as under the IAM) among an infinite number of islands, with the chance of any particular allele moving at rate $m$ and that allele not moving as $1 - m$. Essentially the mainland island model can be thought of as an ideal situation or null model because real populations generally do not conform to its expectations. Later, a more restricted migration model called the "stepping stone" model (Fig. 3.4b) was developed (Kimura & Weiss 1964). The stepping stone model only allows migration between nearest neighbor populations. Consider a two-dimensional lattice of populations. For migration to occur from a population to another population two steps away in the lattice, individuals must move to an intermediate population one step away first before making the second step. In contrast to the mainland island model, this is at the other end of the continuum, with highly restricted gene flow among populations.

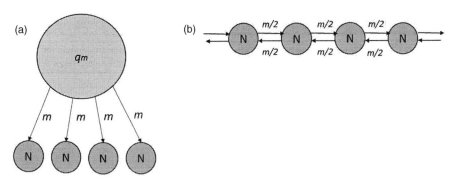

**Fig. 3.4** Representations of different models of population structure. (a) Continent-island (or island) models. Shows an island of essentially infinite population size (*q*) that sends out migrants unidirectional at an equal rate (*m*) to subpopulations with finite population size (*N*). (b) Stepping-stone model. Populations of finite size (*N*) share migrants only with their nearest neighbor population at a rate of *m*/2 since gene flow is bidirectional and symmetrical.

### 3.4.6 Isolation-by-distance and landscape

The above "classical" methods of conceiving gene flow formed the basis for later tests of isolation-by-distance (IBD; Wright 1969). The concept of isolation-by-distance was first envisioned by Sewall Wright (1943, 1951) and assumes that genetic distance among populations is positively correlated with geographic distance among those populations. Implicit in this theoretical framework is that populations are finite in size and subdivided. The main approach for testing for this correlation has utilized Mantel tests (Mantel 1967) of matrix correlations between straight-line geographic distance and genetic distance. In general, most population genetics studies find significant positive correlations of genetic distance and geographic distance (Slatkin 1995; Whitlock & MacCauley 1999); this spatial scale generally correlates with the dispersal capabilities of the study organism. However, strict isolation-by-distance assumes movement in straight lines ("as the crow flies") and fails to account for which environmental characteristics might influence movement of individuals among populations. Landscape genetics as a discipline directly addressed this gap, recognizing that different components of the landscape, such as mountains, rivers and forests, can affect dispersal and consequent of different species in different ways (Manel et al. 2003; Storfer et al. 2007). For example, a mountain might act as a barrier to a species of vole, but not to a hawk. As such, the resulting general approach of "isolation-by-landscape" (IBL) considers the influence of intervening landscape

characteristics on dispersal patterns and consequent gene flow of individuals or **demes** (see Chapter 8). Nearly all landscape genetics studies have shown that inclusion of landscape variables explains significantly more genetic variation among demes than (what is now most often considered) the null model of isolation-by-distance (Storfer et al. 2010).

A commonly used application of the isolation-by-landscape approach considers isolation-by-resistance (IBR) as a way of considering the effects of landscape features on the distribution of genetic variation in an explicitly spatial manner (Spear et al. 2010). The landscape under study is converted to a resistance surface, and resistance values are calculated between pairs of individuals or populations. Generally, resistance values are assigned to each landscape variable or environmental feature contained in a spatial layer of a raster GIS environment (Spear et al. 2010). Higher assigned values translate to greater hypothesized resistance to gene flow by that particular variable, whereas lower values suggest lower resistance. Thus, each raster cell in a GIS layer is assigned a cumulative value based on hypothesized habitat use and movement through each of the landscape variables contained in that cell. A resistance surface is comprised of all cells that comprise the intervening habitat matrix between samples collected across a study area.

Two major modeling frameworks have been used to convert resistance values into measures of population connectivity – least-cost paths and circuit-based analyses. Least-cost paths are based on the underlying assumption that costs, in terms of cumulative resistance values between populations, are minimized such that individuals tend to move along an optimal path (Adriaensen et al. 2003). In contrast, circuit theory-based models simultaneously integrate all possible pathways that connect populations, while still assuming that genetic distance is positively correlated with resistance (McRae 2006). Additional details and pros and cons of these two modeling frameworks will be discussed in more detail in Chapter 8.

A more recent approach is "isolation-by-environment" (IBE; Wang et al. 2013), which predicts a correlation between genetic distance and environmental dissimilarity. It is expected that the more dissimilar habitats are among populations, where the stronger divergent selection will be consequently reducing fitness of dispersers (Lee & Mitchell-Olds 2011). This approach shows

promise, as results of IBE analysis explained more spatial variation in gene flow than IBL analyses for 17 species of Caribbean *Anolis* lizards (Wang 2013).

## 3.5 MEASURING GENETIC DIVERSITY

### 3.5.1 Population level

There are a variety of ways that genetic diversity can be quantified (Table 3.3). When the population is the unit of analysis, there are four main summary metrics for codominant nDNA loci: (1) number of alleles, (2) allelic richness, (3) heterozyosity, and (4) proportion of polymorphic loci. The average number of alleles per locus

**Table 3.3** Commonly used genetic diversity metrics for (a) populations and (b) individuals.

**(a) Population level**

| Metric | Symbol | Equation | Method type |
| --- | --- | --- | --- |
| Polymorphism | $P$ | Number of polymorphic loci/number of loci | nDNA SNP or μsat, AFLP[*] |
| Observed heterozygosity | $H_o$ | Number of heterozygotes/number of individuals | nDNA SNP or μsat |
| Expected heterozygosity | $H_e$ | $2n(1 - \Sigma p_i^2)/(2n - 1)$, $n$ = sample size, $p_i$ freq $i$th allele | nDNA SNP or μsat, AFLP[*] |
| Allelic richness | $A_r$ | Rarefaction at smallest sample size | nDNA SNP or μsat |
| Haplotypic diversity | $h$ | $n(1 - \Sigma p_i^2)/(n - 1)$, $n$ = sample size, $p_i$ freq $i$th allele | mtDNA or cpDNA sequence |
| Nucleotide diversity | $\pi$ | Average number of nucleotide differences per site | mtDNA, cpDNA or nDNA sequence, AFLP[*] |

**(b) Individual level**

| Metric | Symbol | Equation | Method type |
| --- | --- | --- | --- |
| Heterozygosity | H | Number of heterozygosity loci/number of loci | nDNA SNP or μsat |
| Standardized heterozygosity | SH | Proportion of heterozygous loci/mean H of loci | nDNA SNP or μsat |
| Internal relatedness | IR | $(2H - \Sigma f_i)/(2N - \Sigma f_i)$, H = homozygous loci, $N$ = number of loci, $f_i$ = freq $i$th allele | nDNA SNP or μsat |
| Homozygosity per locus | HL | $\Sigma E_h/(\Sigma E_h + \Sigma E_j)$, $E_h$ is expected H of homozygous loci and $E_j$ is expected H of heterozygous loci | nDNA SNP or μsat |

AFLP = amplified fragment length polymorphism, cpDNA = chloroplast DNA, mtDNA = mitochondrial DNA, nDNA = nuclear DNA, SNP = single nucleotide polymorphism, μsat = microsatellite.
[*] An analogue to this metric can be estimated for dominant loci like AFLP when making specific assumptions.

(A) can be used to evaluate genetic diversity; however, this metric is very sensitive to differences in sample size. This problem can be avoided by using allelic richness ($A_r$), which can correct for differences in sample sizes using subsampling or rarefaction techniques (El Mousadik & Petit 1996; Leberg 2002; Kalinowski 2004). There are two main metrics for heterozygosity: observed and expected heterozygosity. Observed heterozygosity ($H_o$) is the proportion of individuals that are heterozygous at a particular locus. Expected heterozygosity ($H_e$) is calculated based on the observed frequency of alleles, assuming Hardy–Weinberg equilibrium, and can be corrected for sample size differences among populations (Nei 1978). Both $H_o$ and $H_e$ are estimated for each locus for a population sample and then averaged across loci. The polymorphism metric (P) estimates the proportion of loci out of the total analyzed that contain more than one allele. The polymorphism metric is used less frequently because loci are generally selected for landscape genetic studies because they are known to be polymorphic and thus informative for measuring diversity. The $A_r$ metric is the most sensitive for detecting recent losses of genetic diversity (Allendorf et al. 2012). When sampling is more continuous and the location of populations is less clear, an alternative approach is to calculate these metrics in a spatially-explicit manner based on grouping individuals into genetic neighborhoods that match the population structure (Shirk & Cushman 2011).

For cpDNA and mtDNA sequence data, the main diversity metrics are the number of haplotypes, haplotypic diversity, and nucleotide diversity. A haplotype is the haploid version of a genotype and for cpDNA and mtDNA sequence data refers to a unique DNA sequence. Haplotypic diversity (h), also known as gene diversity, is the haploid version of expected heterozygosity. Nucleotide diversity ($\pi$) is the average number of differences in the nucleotides at each position along a DNA sequence. For dominant loci, like AFLPs, the standard diversity metric is the average number of bands per population, but analogues of $H_e$ and $\pi$ can be estimated under specific assumptions (Bonin et al. 2007).

### 3.5.2  Individual level

There are many metrics used to estimate individual heterozygosity using codominant loci: heterozygosity (H), standardized heterozygosity (SH), internal relatedness (IR), and homozygosity by loci (HL). Heterozygosity is just the number of heterozygous loci divided by the total number of loci analyzed for an individual sample. However, this simple metric can be biased when loci differ in number or frequency of alleles or when all individuals are not genotyped for the same set of loci. Thus, other metrics have been derived to address these potential biases. For standardized heterozygosity, the score for each locus is weighted by the average heterozygosity at that locus (Coltman et al. 1999). Amos et al. (2001) proposed a different measure, IR, based on allele sharing, where the frequency of each allele counts towards the final score, thereby allowing the sharing of rare alleles to be weighted more than the sharing of common alleles. The IR and SH metrics are highly correlated and both have been shown to perform well in studies of individual genetic diversity and fitness (Amos et al. 2001). However, IR underestimates heterozygosity of individuals carrying rare alleles. Homozyosity per locus (HL) has been developed as an alternative metric that avoids this weakness and reduces the sample sizes required to achieve a given statistical power (Aparacio et al. 2006). All of these metrics can be calculated in the R function GENHET (Coulon 2005).

## 3.6  EVALUATING GENETIC STRUCTURE AND DETECTING BARRIERS

One of the key components of a landscape genetic analysis is evaluating genetic structure (i.e., differentiation) among populations and individuals. When populations are *panmictic*, or completely randomly mating, there are no discernable differences in allele frequencies and estimates of gene flow become infinite. However, as noted above, geographic distance plus landscape and environmental variables generally limit gene flow and over multiple generations this will produce specific patterns of population subdivision and genetic structure. These patterns can be evaluated using population-based or individual-based measures of genetic structure or distance, Bayesian clustering analysis, and barrier detection methods.

### 3.6.1  Population-based measures

The oldest and most widely used population-level metric of genetic structure is Wright's $F_{ST}$, one of three F-statistics derived by Sewall Wright. F-statistics were developed to describe the partitioning of genetic variation within a species (Wright 1931, 1951). Three different F-statistics were derived to describe the partitioning of genetic variability among the total population ($T$) or all populations being considered, the subpopulation ($S$) or individuals within a subpopulation ($I$). $F_{ST}$ is a measure of subpopulation-level genetic differentiation relative to the total population that ranges from 0 (***panmixia***) to 1 (complete genetic isolation) and measures allele frequency divergence among subpopulations:

$$F_{ST} = \frac{H_T - H_S}{H_T}$$

where $H_T$ is the estimated heterozygosity if all subpopulations were randomly mating and $H_S$ is the average heterozygosity among subpopulations. Based on Wright's island model, $F_{ST}$ is estimated to be inversely correlated with the number of migrants per generation using the formula

$$F_{ST} = \frac{1}{1 + 4N_e m}$$

where $m$ is the number of migrants per generation. Due to the unrealistic assumptions of the continent-island model, such as the infinite alleles model and infinite continent population size, however, the utility of this direct calculation of $m$ from $F_{ST}$ has been widely cautioned (Whitlock & MacCauley 1999). The other two F-statistics, $F_{IS}$ and $F_{IT}$, are thought of as inbreeding estimates within and among populations, respectively, and are also standard population genetic estimates. For more detailed discussion on Wright's F-statistics and their derivations, see Hedrick (2011).

Since Wright, several $F_{ST}$ analogs have been developed to estimate genetic differentiation among populations (Table 3.4). Nei (1972) developed $G_{ST}$, an extension of $F_{ST}$ that accounts for multiple alleles at a given locus instead of two, as in Wright's original model. Weir and Cockerham's (1984) developed an analysis of variance (ANOVA) approach to estimate an $F_{ST}$ analog theta ($\theta$). Assuming a step-wise mutational model, Slatkin (1995) developed $R_{ST}$ specifically for loci like microsatellites with high numbers of alleles and heterozygosity. However, $R_{ST}$ does not perform well unless the loci strictly conform to the step-wise mutational model (Balloux et al. 2000). Since very few loci conform to the assumptions of this model, this metric is no longer recommended (Meirmans & Hedrick 2011). More recently, additional $F_{ST}$

analogs like $G'_{ST}$ and $G''_{ST}$ and alternative metrics (Jost's $D$) have been developed to improve the performance for data sets with high levels of heterozygosity and/or a small number of populations, or those likely to be out of equilibrium (Hedrick 2005; Meirmans 2006; Jost 2008; Meirmans & Hedrick 2011).

Many additional population-level measures of genetic differentiation, or genetic distance, have been developed and used in contemporary population genetic studies including: Cavalli-Sforza chord distance ($D_c$) (Cavalli-Sforza & Edwards 1967), Nei's genetic distance (Nei's $D$) (Nei 1972), and Bowcock et al.'s (1994) proportion of shared alleles ($D_{ps}$). These statistics rely on estimates of heterozygosity or differences in allelic similarity among populations. A more recently derived statistic, conditional genetic distance (or cGD) is based on graph theory and creating networks of populations with edges (connections) and nodes (populations) (Dyer et al. 2010). For a detailed discussion of graph theory and network modeling, see Chapter 10. The choice of a genetic distance statistic is an important decision in landscape genetic studies and results can differ depending on the metric chosen (Lindsay et al. 2008; Pavlacky et al. 2009; Spear & Storfer 2010; Goldberg & Waits 2010), making it important for researchers to understand the assumptions of each metric and evaluate whether results are consistent across multiple metrics. See Table 3.4 for a summary of these statistics and their assumptions. For landscape genetics studies, the most commonly used statistic has been $F_{ST}$ and its analogs followed by $D$ and $D_c$ (Storfer et al. 2010). Commonly available free software can be used to estimate different statistics associated with gene flow among populations, including: FSTAT (Goudet 1995), GENEPOP (Rousset 2008), and GENALEX (Peakall & Smouse 2006). A large number of $R$ software packages also exist for population genetics calculations, including: gstudio (Dyer 2014), pegas (Paradis et al. 2015), and genetics (Warnes 2015).

### 3.6.2 Individual-based genetic distance metrics

The population-based genetic distance metrics discussed above provide pairwise estimates of genetic connectivity between populations that consist of groups of individuals. However, for species that are more continuously distributed on the landscape, such as many plants and highly mobile animals (e.g., grizzly

**Table 3.4** Estimators of gene flow and genetic differentiation among (a) populations and (b) individuals.

**(a) Population-level metrics**

| Statistic | Description and/or assumptions | Citations |
| --- | --- | --- |
| $F_{ST}$ | Island model of migration | Wright (1951) |
| $D_c$ | Chord distance – geometric estimate of genetic distances | Cavalli-Sforza and Edwards (1967) |
| $G_{ST}$ | Island model, accounts for multiple alleles | Nei (1972) |
| $D$ (Nei's) | Related to the number of changes per locus, assumes mutation rate is constant among loci. Genetic differences arise among populations due to mutation and genetic drift | Nei and Chesser (1983) |
| $D$ (Jost's) | Measures the fraction of allelic variation among populations as an absolute rather than a relative measure (such as $F_{ST}$). | Jost (2008) |
| $\Theta$ (theta) | Accounts for variation in sample size among populations; accounts for error in incomplete population sampling | Weir and Cockerham (1984) |
| $D_{ps}$ | Proportion of shared alleles – measures similarity in allelic identity among populations | Bowcock et al. (1994) |
| $R_{ST}$ | Developed especially for microsatellites; accounts for mutation rates at microsatellite loci | Slatkin (1995) |
| $G'_{ST}$ | Corrects for the bias associated with hypervariable loci, such as microsatellites, and normalizes the standard $F_{ST}$ estimate by dividing by maximum $F_{ST}$ | Hedrick (2005) |
| $G''_{ST}$ | Corrects for the bias in $G'_{ST}$ associated with sampling a small number of populations; should be used for estimating Hedrick's (2005) standardized measure whenever the number of sampled populations is small | Meirmans and Hedrick (2011) |

**(b) Individual-level metrics**

| Statistic | Description and/or assumptions | Citations |
| --- | --- | --- |
| $a_r$ | Rousset's $a$ assumes that individuals are drawn from a two-dimensional population at migration-drift equilibrium, analogous to the $F_{ST}/(1 - F_{ST})$ ratio using pairs of individuals | Rousset (2000) |
| $D_{ps}$ | Proportion of shared alleles – measures the degree of similarity in allelic identity among individuals | Bowcock et al. (1994) |
| $r$ | Relatedness is a measure of identity by descent or shared ancestry between individuals; $r$ is the genetic similarity between two individuals relative to that between random individuals from some reference population. Multiple estimators have been derived | Queller and Goodnight (1989), Smouse and Peakall (1999), Lynch and Ritland (1999) |
| $B_C$ | Bray Curtis dissimilarity metric was originally derived to measure dissimilarity between species composition and abundance at two sites, but can also be used to measure dissimilarity of alleles between two individuals | Legendre and Legendre (1998) |

bears or cougars), it is often more appropriate to use the individual as the unit of analysis when estimating genetic connectivity. A review of landscape genetic studies (Storfer et al. 2010) showed that the most commonly used individual-based genetic distance metrics are pairwise estimates of relatedness or kinship (Queller & Goodnight 1989; Loiselle et al. 1995; Lynch & Ritland 1999; Smouse & Peakall 1999; Wang 2002). Other commonly used metrics include: the proportion of shared alleles (Bowcock et al. 1994), Rousset's a (2000), and Bray Curtis dissimilarity metric (Legendre & Legendre 1998) (Table 3.4). These individual-based genetic distances can be calculated between all (or a subset of) pairs of individuals and then used as a response variable in analyses that evaluate the relationships between genetic distance and geographic or effective landscape distance. Methods to generate individual-based genetic distances between any two individual genotypes are available in software packages such as: SPAGeDI (Hardy & Vekemans 2002), GENALEX (Peakall & Smouse 2006), and ALLELES IN SPACE (Miller 2005).

### 3.6.3   Bayesian clustering methods

Another approach to detecting genetic structure in population and landscape genetics is Bayesian clustering analysis. In these methods, the individual is the unit of analysis and the number of distinct genetic clusters is inferred from a set of individual multilocus genotypes by estimating genetic ancestry from unobserved source populations. The inferred clusters can then be used for assigning group membership to genotypes and individuals. This methodological approach was introduced by Pritchard et al. (2000) and used to evaluate genetic structure among Taita thrush (*Turdus helleri*) in Africa. This new individual-based approach to evaluating genetic structure and defining populations is implemented in the software, STRUCTURE, and has become one of the most commonly used methods in population and landscape genetic studies. Additional Bayesian clustering methods have been developed that explicitly include spatial locations of individuals in the model such as GENELAND (Guillot et al. 2005), BAPS5 (Corander et al. 2008), and TESS (Chen et al. 2007). These approaches can be particularly effective for defining populations and detecting landscape barriers when genetic differentiation among subpopulations is moderate to high (Latch et al. 2006; Chen et al. 2007;

Safner et al. 2011; Blair et al. 2012), but can provide misleading results if samples are not collected evenly across space (Frantz et al. 2009; Schwartz & McKelvey 2009). For a more detailed description of Bayesian clustering and other clustering methods used to evaluate genetic structure see Chapter 7.

### 3.6.4   Barrier detection methods

In a 2010 review, approximately 35% of all landscape genetics papers had a focus on detecting barriers or genetic breaks among groups of individuals. There is particular interest in identifying barriers from a basic research standpoint to better understand evolution of different populations via genetic structuring (Manel et al. 2003; Storfer et al. 2007, 2010). In addition, identification of barriers is important from a conservation and management perspective because barriers cause genetic isolation among populations and potential loss of evolutionary potential via loss of diversity due to genetic drift. Two main methods used to detect barriers in population and landscape genetics studies are Bayesian clustering methods and edge detection methods (Safner et al. 2011). As discussed above, Bayesian clustering methods determine the most likely groups of genetically related individuals based on their multilocus genotypes. When explicit spatial (i.e., georeferenced) locations of individual genotypes are included in clustering analyses, barriers between resulting clusters of individuals are included. For example, GENELAND and TESS both use tessellation, which creates polygons around genotype locations that partition the study area without leaving gaps. Barriers are then drawn along edges of these polygons where genotypes on one side of the polygon belong to one cluster and genotypes on the other side belong to the other (Guillot et al. 2005).

Edge detection methods are established methods in spatial statistics that can be used to detect barriers among groups of genotypes. For example, Wombling (Womble 1951) and Monmonier's maximum difference algorithm (Monmonier 1973) essentially assign likely barriers to geographic areas where there are maximum differences in allele frequencies or genetic distances over short geographic distances. In addition to population or group level analyses, applications of individual-based estimators of gene flow (Table 3.4) have been developed to help detect barriers among continuously distributed genotypes. For example,

Manel et al. (2007) developed an individual-based method to detecting genetic breaks or discontinuities within populations. This method uses **assignment tests** that are applied in a moving window over individuals across a study area to generate a probability surface of finding that genotype across the landscape. Genetic discontinuities are then found by finding the areas of highest slope of the probability surfaces across all individuals (Manel et al. 2007). A related approach uses individual genotypes and their relationship to nearby genotypes to create a heat map of possible membership to one population or another (Murphy et al. 2008). These heat maps can show areas of genetic boundaries where individuals located on either side have a high probability of belonging to one population or another.

A recent comparison of edge detection methods and Bayesian clustering algorithms on both simulated data and actual empirical data showed better performance of Bayesian clustering methods (Safner et al. 2011). Specifically, with small dispersal distances in the simulated data, the software TESS performed best and was able to most accurately detect the number of barriers. At higher dispersal distances, GENELAND had almost 100% accuracy in detecting barriers (Safner et al. 2011). In addition, Bayesian methods were superior to edge detection methods with particular regard to detecting permeable barriers.

## 3.7 ESTIMATING GENE FLOW USING INDIRECT AND DIRECT METHODS

Before the development of molecular techniques, researchers estimated gene flow indirectly with mark–recapture studies of animals and pollen and seed dispersal studies in plants. These estimates, while measuring direct dispersal rates and exchange of migrants among populations, fail to account for whether these individuals have successfully mated in those new populations. Estimates of gene flow require successful mating in addition to successful dispersal because only then can the dispersing individual's genes influence allele frequencies in the new population. Today, most studies use molecular estimates of gene flow due to the time it takes to conduct a successful mark–recapture study or the difficulty associated with tracking pollen from plant to plant, combined with the rapidly decreasing costs associated with molecular studies. Nonetheless, the combination of molecular data with mark–recapture dispersal estimates make inferences about dispersal most meaningful.

The two main approaches to estimating gene flow among populations using molecular methods can be classified as indirect or direct methods. Indirect methods estimate the number of migrants per generation between populations using different model-based approaches. Many of the indirect methods have been discussed above. When estimating genetic distance, it is assumed that gene flow is inversely related to genetic distance. An explicit calculation for estimating gene flow from Wright's $F_{ST}$ statistic is discussed above. In contrast, direct methods can directly detect migrants or offspring of migrants, and identify the parents of offspring to document direct movement and gene flow. The following sections provide an overview of additional indirect estimates using **coalescent** methods, as well as direct methods using assignment tests and parentage analysis to estimate dispersal and gene flow.

### 3.7.1 Indirect measures of gene flow – coalescent approaches

Gene flow and migration rates among populations can be estimated with coalescent methods. Modern formulations of coalescent theory extend from Wright–Fisher models of identity by descent (Wright 1931; Fisher 1930). These models assume the conditions for Hardy–Weinberg equilibrium, except that population size is finite ($N$). The models assume that, for each neutral locus, alleles are chosen at random among breeding adults to form the gene pool of the next generation. Thus, in a population of size $N$, there are $2N$ gene copies and the probability an allele chosen at random in the next generation coming from a particular parent is $1/(2N_e)$. Coalescent theory considers allele frequencies in contemporary populations and back-calculates the time since alleles coalesce to a common ancestor from which they descended. Alleles coming from the same ancestor are thus *identical by descent*, based on the Wright–Fisher model (Kingman 1982; Hudson 1990). In generation 1, the probability of two alleles bring identical by descent is $1/(2N_e)$ and the probability of two alleles *not* being identical by descent is $1 - 1/(2N_e)$. Therefore, after $t$ generations, the probability that any two alleles coalesce is

$$P_c(t) = \left(1 - \frac{1}{2N_e}\right)^{t-1} \left(\frac{1}{2N_e}\right)$$

This model assumes a diploid population with a constant effective population size and no selection. However, more recent mathematical formulations have allowed for selection, recombination, fluctuations in population size, etc. (Kingman 1982; Hudson 1990; Nielsen & Wakeley 2001; Rannala & Yang 2003).

When two populations become separated, allelic differences between them may arise through the process of "lineage sorting", whereby different alleles are chosen at random (assuming selective neutrality) to form the gene pool in each successive generation in different populations (Hudson 1990). In these cases, coalescent theory can thus be used to estimate time since divergence of these populations and gene flow between them.

In modern molecular studies, coalescent theory can be applied in a number of different ways, including estimation of effective population sizes within, as well as gene flow among, populations. Commonly used software packages to accomplish these tasks include MIGRATE-N (Beerli & Palczewski 2010) and IMa2 (Hey & Neilsen 2007; Hey 2010), both of which use MCMC randomization methods to sample gene geneaologies and estimate model parameters with Bayesian inference to estimate posterior probabilities. Note that due to computational intensity, there are limits to using the coalescent approach (Wakeley 2005), which is evident in the fact that IMa2 can only handle up to 10 total populations.

### 3.7.2   Direct measures – assignment tests

Assignment tests are analytical methods that use genotypic information to ascertain population membership of individuals or groups of individuals. There are two main groups of assignment tests. The first group of methods assumes that population units are known and defined across a landscape. In the original ***likelihood-based*** assignment test (Paetkau et al. 1995), an individual is assigned to its putative population of origin, based on its multilocus genotype and the expected probabilities of that genotype in each of the potential source populations under the assumptions of Hardy–Weinberg and linkage equilibrium. Later derivations of assignment methods also used Bayesian models (Rannala & Mountain 1997) and included a modification that can statistically exclude potential source populations for a given individual rather than attempting to assign the individual to the most likely source

population (Cornuet et al. 1999). These methods can be used for direct estimates of dispersal between populations and geographic areas since they can identify individuals that have dispersed away from their natal site. For example, Dixon et al. (2006) used assignment tests to evaluate the effectiveness of a regional corridor in connecting two Florida black bear (*Ursus americanus floridanus*) populations and found that the corridor did provide functional connectivity but that the movement was primarily unidirectional. All of the above methods can be implemented in the software GENECLASS2 (Piry et al. 2004).

The second group of assignment test approaches is the aspatial and spatial Bayesian clustering methods that were discussed above. These methods do not require *a priori* definitions of populations, but population memberships are estimated for each individual in the form of ancestry coefficients for the inferred population units. Thus, these methods can also be used to detect dispersers and offspring with mixed ancestry from two or more genetic groups documenting evidence of gene flow (Manel et al. 2005). Multiple studies have shown that assignment test approaches can be very successful in detecting individuals that disperse between populations (Paetkau et al. 1995; Rannala & Mountain 1997; Manel et al. 2002, 2005; Hauser et al. 2006) and that accuracy improves with increases in (i) the number of loci, (ii) the number of alleles per locus, (iii) the levels of genetic differentiation, and (iv) sample sizes from contributing populations (Cornuet et al. 1999; Paetkau et al. 2004; Berry et al. 2004; Latch & Rhodes 2005; Hauser et al. 2006; Chapter 7).

### 3.7.3   Parentage analysis

Another direct approach to studying dispersal and gene flow across a landscape is parentage analysis. In this approach, multilocus genotypes are used to determine the parents of progeny from a pool of candidate parents. Two main approaches are used in parentage analysis: parentage exclusion and statistically based parentage assignment (Marshall et al. 1998; Jones & Arden 2003; Jones et al. 2010). For example, if the offspring does not share alleles with the putative parent at one or more loci then this parent can be excluded (Table 3.5). However, exclusion methods are rarely used in natural populations since it often not possible to sample all potential parents, loci may not be powerful enough to exclude all but one parent, and mutation and

**Table 3.5** Example of paternity analysis using exclusion. Genotypes are provided for a known mother, offspring, and two possible fathers. Alleles contributed by the mother are highlighted in bold and inferred alleles of the father are in italic. Father 1 can be excluded because of mismatches at loci 1 and 3. Father 2 cannot be excluded as a possible father.

|  | Locus 1 | Locus 2 | Locus 3 | Locus 4 |
|---|---|---|---|---|
| Mother | 122/128 | 130/134 | 222/240 | 188/190 |
| Offspring | **128**/*130* | **134**/*140* | *218*/**222** | **190**/*194* |
| Father 1 | 124/126 | 134/140 | 220/232 | 188/194 |
| Father 2 | 126/130 | 132/140 | 216/218 | 186/194 |

genotyping errors can cause genotype mismatches between parent and offspring that can exclude true parents. More commonly, a statistically based parentage assignment using maximum likelihood or Bayesian methods is used to determine the most likely mother or father (for a review see Jones & Ardren 2010). In general, 10–15 microsatellite loci and 40–60 SNPs are needed to determine the father when the mother is known and twice this number of loci would be needed if neither parent is known (Allendorf et al. 2012).

While this technique has great potential in landscape genetics, it has been used primarily for plants (see Chapter 10) because it is much easier to sample potential mothers and fathers compared to animals, which can be more difficult to locate and capture. In plants, there are two dispersal processes – seed dispersal and pollen dispersal. Pollen is the paternal contribution to gene flow and dispersal distances with pollen are generally longer than seed dispersal distances, but seed dispersal is critical for demographic connectivity in plants (Sork & Smouse 2006). To quantify pollen sources and pollen movement across the landscape, paternity analysis and two-generation pollen pool structure analysis, known as Two-Gener, are used (Smouse et al. 2001; Smouse & Sork 2004). Paternity analysis can directly reveal the number of fathers and the spatial location of fathers in the landscape. For example, Kamm et al. (2010) genotyped 1183 offspring from 49 mother trees in the rare, insect-pollinated forest tree, *Sorbus domestica*, to determine paternity using maximum likelihood-based parentage assignment (Marshall et al. 1998). In this system, all 167 possible parental trees were known and georeferenced and paternity analysis revealed that 108 trees contributed to the offspring. They also demonstrated that none of the landscape features (settlements, open land, deep valleys, and closed forest) were an impermeable barrier to gene flow from pollen, which moved up to 16 km.

## 3.8 CONCLUSION AND FUTURE DIRECTIONS

The field of population genetics has a long, rich history that is over a century old. In this chapter, we have only been able to give a brief overview of the field and focus on a small subset of the theory and genetic diversity/population structure methods that are most widely used in landscape genetics. Several full textbooks have been written, including *Genetics of Populations* (Hedrick 2011) and *Principles of Population Genetics* (Hartl & Clark 2006), that provide much more comprehensive overviews of the subject.

As we look to the future, the major challenges in population and landscape genetics involve developing predictive models for changes in landscape genetic structure in response to global change, as well as the challenges in harnessing the full potential of data gathered in next-generation, high-throughput DNA sequencing platforms. With our ability to generate thousands of markers (e.g., SNPs) using high-throughput sequencing, we are now able to assess the distribution of neutral and adaptive genetic variation (Joost et al. 2007; Schoville et al. 2012). Models have shown that we can virtually saturate the statistical power for our ability to detect population genetic structure with approximately 1000 alleles from neutral loci (e.g., Schoville et al. 2012). However, complete genomic sequencing has provided new avenues for understanding the genetic basis of functionally adaptive trait variation. For model organisms such as humans and mice, we now know that structural genomic features, such as copy number variation, chromosomal inversions, and transposable elements, are also directly involved in functional adaptive trait variation (Fedoroff 2012; Ellegren 2014). A challenge for population genetics may involve development of new theory to incorporate how these features evolve because they behave quite differently from models of point mutations or slippage mutations that are commonly used in models to estimate genetic distance and population structure. Rapid advances in our ability to obtain genomic data have also caused a paradigm shift in the way we view "genes". Once thought to be directly related to phenotype, genes operate in complex genomic landscapes, rather than in isolation. Genes are also expressed differently in different ecological landscapes. The major challenge for landscape genetics in the future is to integrate data from the complex genomic landscape, as well as the ecological landscapes in which individuals and populations exist (see Chapter 9 for more details).

## REFERENCES

Adriaensen, F., Chardon, J.P., De Blust, G., Swinnen, E., Villalba, S., Gulinck, H., & Matthysen, E. (2003) The application of "least-cost" modeling as a functional landscape model. *Landscape and Urban Planning* **64**, 233–47.

Allendorf, F.W., Luikart, G.H., & Aitken, S.N. (2012) *Conservation and the Genetics of Populations*. Wiley-Blackwell, Oxford.

Amos, W., Worthington Wilmer, J., Fullard, K., Burg, T.M., Croxall, J.P., Bloch, D., & Coulson, T. (2001) The influence of parental relatedness on reproductive success. *Proceedings of the Royal Society London B* **268**, 2021–7.

Aparacio, J.M., Ortego, J., & Cordero, P.J. (2006) What should we weigh to estimate heterozygosity, alleles or loci? *Molecular Ecology* **15**, 4459–6.

Balloux, F., Brünner, H., Lugon-Moulin, N., Hausser, J., & Goudet, J. (2000) Microsatellites can be misleading: an empirical and simulation study. *Evolution* **54**, 1414–22.

Beerli, P. & Palczewski, M. (2010) Unified framework to evaluate panmixia and migration direction among multiple sampling locations. *Genetics* **185**, 313–26.

Berry, O., Tocher, M.D., & Sarre, S.D. (2004) Can assignment tests measure dispersal? *Molecular Ecology* **13**, 551–61.

Birky, C.W. (1978) Transmission genetics of mitochondria and chloroplasts. *Annual Review Genetics* **12**, 471–512.

Blair, C., Weigel, D.E., Balazik, M., Keeley, A.T.H., Walker, F.M., Landguth, E.L., Cushman, S.A., Murphy, M., Waits, L.P., & Balkenhol, N. (2012) A simulation-based evaluation of methods for inferring linear barriers to gene flow. *Molecular Ecology Resources* **12**, 822–33.

Bonin, A., Ehrich, D., & Manel, S. (2007) Statistical analysis of amplified fragment length polymorphism data: a toolbox for molecular ecologists and evolutionists. *Molecular Ecology* **16**, 3737–58.

Bowcock, A.M., Ruiz-Linares, A., Tomfohrde, J., Minch, E., Kidd, J.R., & Cavalli-Sforza, L.L. (1994) High resolution of human evolutionary trees with polymorphic microsatellites. *Nature* **368**, 455–7.

Cavalli-Sforza, L.L. & Edwards, A.W.F. (1967) Phylogenetic analysis – models and estimation procedures. *The American Journal of Human Genetics* **19**, 233–257.

Chen, C., Durand, E. Forbes, F., & François, O. (2007) Bayesian clustering algorithms ascertaining spatial population structure: a new computer program and a comparison study. *Molecular Ecology Notes* **7**, 747–56.

Coltman, D.W., Pilkington, J.G., Smith, J.A., & Pemberton, J.M. (1999) Parasite-mediated selection against inbred Soay sheep in a free-living, island population. *Evolution* **53**, 1259–67.

Corander, J., Marttinen, P., Sirén, J., & Tang, J. (2008) Enhanced Bayesian modeling in BAPS software for learning genetic structures of populations. *BMC Bioinformatics* **9**, 539.

Cornuet, J.M., Piry, S., Luikart, G., Estoup, A., & Solignac, M. (1999) New methods employing multilocus genotypes to select or exclude populations as origins of individuals. *Genetics* **153**, 1989–2000.

Coulon, A. (2010) GENHET: an easy-to-use R function to estimate individual heterozygosity. *Molecular Ecology Resources* **10**, 167–9.

Crandall, K.A., Posada, D., & Vasco, D. (1999) Effective population sizes: missing measures and missing concepts. *Animal Conservation* **2**, 317–19.

Di Rienzo, A., Peterson, A.C., Garza, J.C., Valdes, A.M., Slatkin, M., & Freimen, N.B. (1994) Mutational processes of simple-sequence repeat loci in human populations. *Proceedings of the National Academies of Sciences USA* **91**, 3166–70.

Dixon, J.D., Oli, M.K., Wooten, M.C., Eason, T.H., McCown, J.W., & Paetkau, D. (2006) Effectiveness of a regional corridor in connecting two florida black bear populations. *Conservation Biology* **20**, 155–62.

Dyer, R.J. (2014) gstudio: analyses and functions related to the spatial analysis of genetic marker data. R package, version 1.3.

Dyer, R.J., Nason, J.D., & Garrick, R.C. (2010) Landscape modeling of gene flow: improved power using conditional genetic distance derived from the topology of population networks. *Molecular Ecology* **19**, 3746–59.

El Mousadik, A. & Petit, R.J. (1996) High level of genetic differentiation for allelic richness among populations of the argantree (*Argania spinosa* (L.) Skeels) endemic to Morocco. *Theoretical and Applied Genetics* **92**, 832–9.

Ellegren, H. (2014) Genome sequencing and population genomics in non-model organisms *Trends in Ecology and Evolution* **29**, 51–63.

Fedoroff, N. (2012) Transposable elements, epigenetics, and genome evolution. *Science* **338**, 758–67.

Fisher, R.A. (1930) *The Genetical Theory of Natural Selection*. Oxford University Press, New York.

Frankham, R. (1995) Effective population size/adult population size ratios in wildlife: a review. *Genetics Research* **66**, pp. 95–107.

Frankham, R., Brook, B.W., Bradshaw, C.J.A., Traill, L.W., & Spielman, D. (2013) 50/500 rule and minimum viable populations: response to Jamieson and Allendorf. *Trends in Ecology and Evolution* **28**, 187–8.

Franklin, I.R. (1980), Evolutionary change in small populations. In: Soulé, M.E. & Wilcox, B.A. (eds.), *Conservation Biology: An Evolutionary Ecological Perspective*. Sinauer Associates, Sunderland, MA.

Frantz, A.C., Cellina, S., Krier, A., Schley, L., & Burke, T. (2009) Using spatial Bayesian methods to determine the genetic structure of a continuously distributed population: clusters or isolation-by-distance? *Journal of Applied Ecology* **46**, 493–505.

Garvin, M.R., Saitoh, K., & Gharrett, A.J. (2010) Application of single nucleotide polymorphisms to non-model species: a technical review. *Molecular Ecology Resources* **10**, 915–34.

Gillham, N.W. (1974) Genetic analysis of the chloroplast and mitochondrial genomes. *Annual Review Genetics* **8**, 347–91.

Goldberg, C.S. & Waits, L.P. (2010) Comparative landscape genetics of two pond-breeding amphibian species in a highly modified agricultural landscape. *Molecular Ecology* **19**, 3650–63.

Goudet, J. (1995) FSTAT (Version 1.2): A computer program to calculate F-statistics. *Journal of Heredity* **86**, 485–6.

Guillot, G., Mortier, F., & Estoup, A. (2005) Geneland: a program for landscape genetics. *Molecular Ecology Notes* **5**, 712–15.

Hardy, G.H. (1908) Mendelian proportions in a mixed population. *Science* **28**, 49–50.

Hardy, O.J. & Vekemans, X. (2002) Spagedi: a versatile computer program to analyze spatial genetic structure at individual or population levels. *Molecular Ecology Notes* **2**, 618–20.

Hare, M.P., Nunney, L., Schwartz, M.K., Ruzzante, D.E., Burford, M., Waples, R.S., Ruegg, K., & Palstra, F. (2011) Understanding and estimating effective population size for practical application in marine species management. *Conservation Biology* **25**, 438–49.

Hauser, L., Seamons, T.R., Dauer, M., Naish, K.A., & Quinn, T.P. (2006) An empirical verification of population assignment methods by marking and parentage data: hatchery and wild steelhead (*Oncorhynchus mykiss*) in Forks Creek, Washington, USA. *Molecular Ecology* **15**, 3157–73.

Hartl, D.L. & Clark, A. G. (2006) *Principles of Population Genetics*. Sinauer Associates, Sunderland, MA.

Hedrick, P.W. (1995) Gene flow and genetic restoration: the Florida panther as a case study. *Conservation Biology* **9**, 996–1007.

Hedrick, P.W. (2005) A standardized genetic differentiation measure. *Evolution* **59**, 1633–8.

Hedrick, P.W. (2011) *Genetics of Populations*, 4th edition. Jones & Bartlett Publishers, Boston, MA.

Hey J. (2010) Isolation with migration models for more than two populations. *Molecular Biology and Evolution* **27**, 905–20.

Hey, J. & Nielsen, R. (2007) Integration within the Felsenstein equation for improved Markov chain Monte Carlo methods in population genetics. *Proceedings of the National Academy of Sciences USA* **104**, 2785–90.

Hoekstra, H.E., Hirschmann, R.J., Bundney, R.A., Insel, P.A., & Crossland, J.P. (2006) A single amino acid mutation contributes to adaptive beach mouse color pattern. *Science* **313**, 101–4.

Holderegger, R. & Wagner, H.H. (2008) Landscape genetics. *BioScience* **58**, 199–207.

Hudson, R.R. (1990) Gene genealogies and the coalescent process. In: Futuyma, D. & Antonovics, J. (eds.), *Oxford Surveys in Evolutionary Biology*, Vol. **7**. Oxford University Press, New York.

Jamieson, I.G. & Allendorf, F.W. (2012) How does the 50/500 rule apply to MVPs? *Trends in Ecology and Evolution* **27**, 578–84.

Jones, A.G. & Ardren, W.R. (2003) Methods of parentage analysis in natural populations. *Molecular Ecology* **12**, 2511–23.

Jones, A.G., Small, C.M., Paczolt, K.A., & Ratterman, N.L. (2010) A practical guide to methods of parentage analysis. *Molecular Ecology Resources* **10**, 6–30.

Joost, S., Bonin, A., Bruford, M.W., Després, L., Conord, C., Erhardt, G., & Taberlet, P. (2007) A spatial analysis method (SAM) to detect candidate loci for selection: towards a landscape genomics approach to adaptation. *Molecular Ecology* **16**, 3955–69.

Jost, L. (2008) GST and its relatives do not measure differentiation. *Molecular Ecology* **17**, 4015–26.

Kalinowski, S.T. (2004) Counting alleles with rarefaction: private alleles and hierarchical sampling designs. *Conservation Genetics* **5**, 539–43.

Kamm, U., Gugerli, F., Rotach, P., Edwards, P.J. & Holderegger, R. (2010) Open areas in a landscape enhance pollen-mediated gene flow of a tree species: evidence from northern Switzerland. *Landscape Ecology* **25**, 903–11.

Kimura, M. (1968) Evolutionary rate at the molecular level. *Nature* **217**, 624–6.

Kimura, M. & Weiss, G.H. (1964) The stepping stone model of population structure and the decrease of genetic correlation with distance. *Genetics* **49**, 561–76.

Kingman, J.F.C. (1982) The coalescent. *Stochastic Processes and Their Applications* **13**, 235–48.

Lande, R. (1995) Mutation and conservation. *Conservation Biology* **9**, 782–91.

Latch, E.K. & Rhodes, O.E. Jr. (2005) The effects of gene flow and population isolation on the genetic structure of reintroduced wild turkey populations: Are genetic signatures of source populations retained? *Conservation Genetics* **6**, 981–97.

Latch, E.K., Dharmarajan, G., Glaubitz, J.C., & Rhodes, O.E. Jr. (2006) Relative performance of Bayesian clustering software for inferring population substructure and individual assignment at low levels of population differentiation. *Conservation Genetics* **7**, 295–302.

Leberg, P.L. (2002) Estimating allelic richness: effects of sample size and bottlenecks. *Molecular Ecology* **1**, 2445–9.

Leberg, P.L. (2005) Genetic approaches for estimating the effective size of populations. *Journal of Wildlife Management* **69**, 1385–99.

Lee, C.-R. & Mitchell-Olds, T. (2011) Complex trait divergence contributes to environmental niche differentiation in ecological speciation of *Boechera stricta*. *Molecular Ecology* **22**, 2204–17.

Legendre, P. & Legendre, L. (1998) *Numerical Ecology*. Elsevier, Amsterdam.

Lindsay, D.L., Barr, K.R., Lance, R.F., Tweddale, S.A., Hayden, T.J., & Leberg, P.L. (2008) Habitat fragmentation and genetic diversity of an endangered, migratory songbird, the golden-cheeked warbler (*Dendroica chrysoparia*). *Molecular Ecology* **17**, 2122–33.

Linnen, C.R., Kingsley, E.P., Jensen, J.D., & Hoekstra, H.E. (2009) On the origin and spread of an adaptive allele in deer mice. *Science* **325**, 1095–8.

Loiselle, B.A., Sork, V.L., Nason, J.D., & Graham, C.H. (1995) Spatial genetic structure of a tropical understory shrub, *Psychotria officinalis* (Rubiaceae). *American Journal of Botany* **82**, 1420–5.

Lowe, A.J., Harris, S.A., & Ashton, P.A. (2004) *Ecological Genetics; Design, Analysis and Application*. Wiley-Blackwell, Chichester.

Lynch, M. & Lande, R. (1998) The critical effective size for a genetically secure population. *Animal Conservation* **1**, 701.

Lynch, M. & Ritland, R. (1999) Estimation of pairwise relatedness with molecular markers. *Genetics* vol. **152**, 1753–66.

Manel S., Berthier P., & Luikart G. (2002) Detecting wildlife poaching: identifying the origin of individuals using Bayesian assignment tests and multi-locus genotypes. *Conservation Biology* **16**, 650–9.

Manel, S., Schwartz, M.K., Luikart, G., & Taberlet, P. (2003) Landscape genetics: combining landscape ecology and population genetics. *Tree* **18**, 189–97.

Manel S., Gaggiotti O. & Waples R. (2005) 'Assignment methods: matching biological questions with appropriate techniques', *Trends in Ecology and Evolution* **20**, 136–42.

Manel, S., Berthoud, F., Bellemain, E., Gaudeul, M., Luikart, G., Swenson, J.E., Waits, L.P., & Taberlet, P. (2007) A new individual-based spatial approach for identifying genetic discontinuities in natural populations. *Molecular Ecology* **16**, 2031–43.

Mantel, N. (1967) The detection of disease clustering and a generalized regression approach. *Cancer Research* **27**, 209–20.

Marshall, T.C., Slate, J., Kruuk, L.E.B., & Pemberton, J.M. (1998) Statistical confidence for likelihood-based paternity inference in natural populations. *Molecular Ecology* **7**, 639–55.

McRae, B.H. (2006) Isolation by resistance. *Evolution* **60**, 1551–61.

Meirmans, P.G. (2006) Using the AMOVA framework to estimate a standardized genetic differentiation measure. *Evolution* **60**, 2399–402.

Meirmans, P.G. & Hedrick, P.W. (2011) Assessing population structure: FST and related measures. *Molecular Ecology Resources* **11**, 5–18.

Miller, M.P. (2005) Alleles in space (AIS): computer software for the joint analysis of interindividual spatial and genetic information. *Journal of Heredity* **96**, 722–4.

Monmonier, M.S. (1973) Maximum-difference barriers: an alternative numerical regionalization method. *Geographical Analysis* **5**, 245–61.

Morin, P.A., Luikart, G., & Wayne, R.K. (2004) SNPs in ecology, evolution and conservation. *Trends in Ecology and Evolution* **19**, 208–16.

Mullis, K.B. & Faloona, F.A. (1987) Specific synthesis of DNA in vitro via a polymerase-catalyzed chain reaction. *Methods in Enzymology* **155**, 335–50.

Murphy, M.A., Evans, J.S., Cushman, S.A., & Storfer, A. (2008) Representing genetic variation as continuous surfaces: an approach for identifying spatial dependency in landscape genetic studies. *Ecography* **31**, 685–97.

Neale, D.B., Marshall, K.A., & Sederoff, R.R. (1989) Chloroplast and mitochondrial DNA are paternally inherited in *Sequia semperviren* D. Don Endl. *Proceedings of the National Academy of Sciences USA* **86**, 9347–9.

Nei, M. (1972) Genetic distance between populations. *American Naturalist* **106**, 283–92.

Nei, M. (1978) Estimation of heterozygosity and genetic distance from small number of individuals. *Genetics* **89**, 583–90.

Nei, M. & Chesser, R.K. (1983) Estimation of fixation indices and gene diversities. *Annals of Human Genetics* **47**, 253–9.

Nielsen, R. & Wakeley, J. (2001) Distinguishing migration from isolation: a Markov chain Monte Carlo approach. *Genetics* **158** (June 1, 2001), 885–96.

Ohta, T. (1973) Slightly deleterious mutant substitutions in evolution. *Nature*, **246**, 96–8.

Ohta T. & Kimura, M. (1973) A model of mutation appropriate to estimate the number of electrophoretically detectable alleles in a finite population. *Genetical Research* **33**, 201–4.

Paetkau, D., Calvert, W., Stirling, I., & Strobec C. (1995) Microsatellite analysis of population structure in Canadian polar bears. *Molecular Ecology* **4**, 347–54.

Paetkau, D., Slade, R., Burden, M., & Estoup, A. (2004) Genetic assignment methods in direct, real-time estimation of migration rate: a simulation-based exploration of accuracy and power. *Molecular Ecology* **13**, 55–65.

Paradis, E., Schliep, K., Potts, A., & Winter, D. (2015) Pegas – population and evolutionary genetics analysis system. R package, version 0.6.

Pavlacky, D.C., Goldizen, A.W., Prentis, P.J., Nicholls, J.A., & Lowe, A.J. (2009) A landscape genetics approach for quantifying the relative influence of historic and contemporary habitat heterogeneity on the genetic connectivity of a rainforest bird. *Molecular Ecology* **18**, 2945–60.

Peakall, R. & Smouse, P.E. (2006) Genalex 6: genetic analysis in Excel. Population genetic software for teaching and research. *Molecular Ecology Notes* **6**, 288–95.

Piry, S., Alapetite, A., Cornuet, J.-M., Paetkau, D., Baudouin, L., & Estoup, A. (2004) Geneclass2: a software for genetic assignment and first-generation migrant detection. *Journal of Heredity* **95**, 536–9.

Pritchard, J.K., Stephens, M., & Donnelly, P. (2000) Inference of population structure using multilocus genotype data. *Genetics* **155**, 945–59.

Prunier, J.G., Kaufmann, B., Fenet, S., Picard, D., Pompanon, F., Joly, P., & Lena, J.P. (2013) Optimizing the trade-off between spatial and genetic sampling efforts in patchy populations: towards a better assessment of functional connectivity using an individual-based sampling scheme. *Molecular Ecology* **22**, 5516–30.

Queller, D.C. & Goodnight, K.F. (1989) Estimating relatedness using molecular markers. *Evolution* **43**, 258–75.

Rannala, B. & Mountain, J.L. (1997) Detecting migration using multilocus genotypes. *Proceedings of the National Academy of Sciences USA* **94**, 9197–201.

Rannala, B. & Yang, J. (2003) Bayes estimation of species divergence times and ancestral population sizes using DNA sequences from multiple loci. *Genetics* **164**, 1645–56.

Rousset, F. (2000) Genetic differentiation between individuals. *Journal of Evolutionary Biology* **13**, 58–62.

Rousset, F. (2008) Genepop'007: a complete reimplementation of the Genepop software for Windows and Linux. *Molecular Ecology Resources* **8**, 103–6.

Safner, T., Miller, M.P., McRae, B.H., Fortin, M.J., & Manel, S. (2011) Comparison of Bayesian clustering and edge detection methods for inferring boundaries in landscape genetics. *International Journal of Molecular Sciences* **12**, 865–89.

Sanger, F., Nicklen, S., & Coulson, A.R. (1977) DNA sequencing with chain-terminating inhibitors. *Proceedings of the Royal Academy of Sciences USA* **74**, 5463–7.

Schoville, S.D., Bonin, A., François, O., Lobreaux, S., Melodelima, C., & Manel, S. (2012) Adaptive genetic variation on the landscape: methods and cases. *Annual Review of Ecology, Evolution and Systematics* **43**, 23–43.

Schwartz, M.K. & McKelvey, K.S. (2009) Why sampling scheme matters: the effect of sampling scheme on landscape genetic results. *Conservation Genetics* **10**, 441–52.

Shendure, J. & Ji, N. (2008) Next-generation DNA sequencing. *Nature Biotechnology* **26**, 1135–45.

Shirk, A.J. & Cushman, S.A. (2011) sGD software for estimating spatially explicit indices of genetic diversity. *Molecular Ecology Resources* **11**, 923–34.

Slatkin, M. (1995) A measure of population subdivision based on microsatellite allele frequencies. *Genetics* **139**, 457–62.

Smouse, P.E. & Peakall, R. (1999) Spatial autocorrelation analysis of individual multiallele and multilocus genetic structure. *Heredity* **82**, 561–73.

Smouse, P.E. & Sork, V.L. (2004) Measuring pollen flow in forest trees: an exposition comparison of alternative approaches. *Forest Ecology and Management* **197**, 21–38.

Smouse, P.E., Dyer, R.J., Westfall, R.D., & Sork, V.L. (2001) Two-generation analysis of pollen flow across a landscape I: Male gamete heterogeneity among females. *Evolution* **55**, 260–71.

Sork, V.L. & Smouse, P.E. (2006) Genetic analysis of landscape connectivity in tree populations. *Landscape Ecology* **21**, 821–36.

Sork, V.L. & Waits, L.P. (2010) Contributions of landscape genetics – approaches, insights and future potential. *Molecular Ecology* **19**, 3489–95.

Spear, S.F. & Storfer, A. (2010) Anthropogenic and natural disturbance lead to differing patterns of gene flow in the Rocky Mountain tailed frog *Ascaphus montanus*. *Biological Conservation* **143**, 778–86.

Spear, S. F., Balkenhol, N., Fortin, M.-J., McRae, B. H., & Scribner, K. T. (2010) Use of resistance surfaces for landscape genetic studies: considerations for parameterization and analysis. *Molecular Ecology* **19**, 3576–91.

Storfer, A., Murphy, M.A., Evans, J.S., Goldberg, C.S., Robinson, S., Spear, S.F., Dezzani, R., Delmelle, E., Vierling, L., & Waits, L.P. (2007) Putting the "landscape" in landscape genetics. *Heredity* **98**, 128–42.

Storfer, A., Murphy, M.A., Spear, S.F., Holderegger, R., & Waits, L.P. (2010) Landscape genetics: Where are we now? *Molecular Ecology* **19**, 3496–514.

Vos, P., Hogers, R., Bleeker, M., Reijans, M., van de Lee, T., Hornes, M., Frijters, A., Pot, J. Peleman, J. Kuiper, M., & Zabeau, M. (1995) AFLP: a new technique for DNA fingerprinting. *Nucleic Acid Research* **23**, 4407–14.

Wakeley, J. (2005) The limits of theoretical population genetics. *Genetics* **169**, 1–7.

Wang, J. (2002) An estimator for pairwise relatedness using molecular markers. *Genetics* **160**, 1203–15.

Wang, J. (2005) Estimation of effective population sizes from data on genetic markers. *Philosophical Transactions of the Royal Society B* **360**, 1395–409.

Wang, I.J. (2013) Examining the full effects of landscape heterogeneity on spatial genetic variation: a multiple matrix regression approach for quantifying geographic and ecological isolation. *Evolution* **67**, 3403–11.

Wang, I.J., Glor, R.E., & Losos, J.B. (2013) Quantifying the roles of ecology and geography in spatial genetic divergence. *Ecology Letters* **16**, 175–82.

Warnes, G. (2015) Genetics – population genetics. R package, version 1.3.8.1.

Weinberg, W. (1908) Über den Nachweis der Vererbung beim Menschen. *Jahreshauptversammlung des Vereins für Vaterländische Naturkunde in Württemberg* **64**, 369–82.

Weir, B.S. & Cockerham, C.C. (1984) Estimating F-statistics for the analysis of population structure. *Evolution* **38**, 1358–70.

Whitlock, M.C. & McCauley, D.E. (1999) Indirect measures of gene flow and migration: FST doesn't equal 1/(4*Nm* + 1). *Heredity* **82**, 117–25.

Wolff, J.N., Nafisinia, M., Sutivsky, P., & Ballard, J.W.O. (2013) Paternal transmission of mitochondrial DNA as an integral part of mitochondrial inheritance in metapopulations of *Drosophila simulana*. *Heredity* **110**, 57–62.

Womble, W.H. (1951) Differential systematics. *Science* **114**, 315–22.

Wright, S. (1931) Evolution in Mendelian populations. *Genetics* **16**, 97–159.

Wright, S. (1938) Size of population and breeding structure in relation to evolution. *Science* **87**, 430–1.

Wright, S. (1943) Isolation-by-distance. *Genetics* **28**, 114–38.

Wright, S. (1951) The genetical structure of populations. *Annals of Eugenics*. **15**, 323–54.

Wright, S. (1969) *Evolution and the Genetics of Populations*, Vol. **II**. University of Chicago Press, Chicago.

Zhao, H., Li, R., Wang, Q., Yan, Q., Deng, J.H., Han, D., Bai, Y., Young, W.Y., & Guan, M.X. (2004) Maternally inherited aminoglycoside-induced and nonsyndromic deafness is associated with the novel C1494T mutation in the mitochondrial 12S rRNA gene in a large Chinese family. *American Journal of Genetics* **74**, 139–52.

Zouros, E. (2000) The exceptional mitochondrial DNA system of the mussel family Mytilidae. *Genes and Genetic Systems* **75**, 313–18.

*Chapter 4*

# BASICS OF STUDY DESIGN: SAMPLING LANDSCAPE HETEROGENEITY AND GENETIC VARIATION FOR LANDSCAPE GENETIC STUDIES

*Niko Balkenhol[1] and Marie-Josée Fortin[2]*

*[1]Department of Wildlife Sciences, University of Göttingen, Germany*
*[2]Department of Ecology and Evolutionary Biology, University of Toronto, Canada*

## 4.1 INTRODUCTION

Appropriate study design and effective sampling strategies are crucial aspects for deriving valid and meaningful scientific inferences. Designing optimal landscape genetic studies can be particularly challenging because genetic and landscape data are influenced by processes acting at very different spatial and temporal scales. Statistically relating these different data in meaningful ways is never a simple task, but it becomes even more delicate when studies are not designed to match the scales most relevant for the particular study landscape, the species of interest, underlying key processes, and the exact research question(s). Furthermore, landscape genetic research questions are often addressed within projects that also include many other study goals; hence these projects are not necessarily designed specifically for landscape genetics. Indeed, some landscape genetic studies seem to define exact research questions only after genetic samples have been gathered for some other purpose, and sampling regimes utilized in these projects might consequently not be optimal for addressing specific landscape genetic research questions. Such non-optimal study design not only inhibits our ability to correctly detect all possible landscape effects on neutral and adaptive genetic variation but also means that financial, logistic, and intellectual resources may not be used efficiently. While analyzing entire populations within an experimental framework leads to most accurate and most precise research findings, such ideal

*Landscape Genetics: Concepts, Methods, Applications*, First Edition. Edited by Niko Balkenhol, Samuel A. Cushman, Andrew T. Storfer, and Lisette P. Waits.

conditions can rarely be achieved. Dealing with these challenges requires particularly well-designed studies and a good understanding of the trade-offs involved in balancing various sampling options (e.g., number of sampled populations versus number of sampled individuals per population). In this chapter, we review the basics of study design and suggest guidelines for optimal sampling strategies in landscape genetics. For this, we first provide an overview of study design terminology and discuss general study design considerations that lead to strong scientific inferences. We then highlight the specific design challenges encountered in landscape genetic studies and summarize our current knowledge about sampling effects on landscape genetic conclusions. Based on this summary, we then provide study design recommendations for various landscape genetic research questions and outline future research needs on study design effects in landscape genetics.

## 4.2 STUDY DESIGN TERMINOLOGY USED IN THIS CHAPTER

Generally speaking, a study design is any systematic plan for scientifically investigating one or multiple research questions. More specifically, a study design can be based on various study types (e.g., experimental or observational; see below) and address specific research questions and associated hypotheses. It then requires the determination of the dependent and explanatory variables of interest, the sampling strategy with which data will be collected and the general statistical approach with which the data will be analyzed. All of these aspects of study design are interrelated and should be matched to each other and to the overall goal of the study. The type of study is often modulated by financial and logistical constraints, while hypotheses and relevant variables need to be identified by the researcher for the specific study objectives. Similarly, choosing the best analytical approach can be a complex task and deciding on the exact statistics to be used for final data analysis is often not possible until data have been collected, because the abundance and variability of data affect the type of analyses that can be conducted (see Chapter 5). For example, a study may initially be intended to investigate the effects of four levels of road density on salamander genetic structure, with the study design calling for an ANOVA analysis. However, in the field the researchers may realize that the variable of interest (i.e., road density) acts more as a continuous quantitative variable so that a regression analysis is more

appropriated. This example illustrates that knowledge of the study species and its environmental needs are paramount for sound scientific analyses and that this knowledge is not always available before field data are collected. As the sampling strategy can strongly influence the success of data collection, and largely determines whether research questions can be addressed adequately, the design of the sampling strategy is the most important aspect of study design. We therefore focus on "sampling for landscape genetics" in this chapter. For a more general discussion on how to design a landscape genetic study, we refer readers to Hall and Beissinger (2014).

A sampling strategy includes at least four different components, all of which are highly relevant for landscape genetics.

### 4.2.1 Sampling level

The sampling level determines the analytical unit for which statistics will be calculated and for which conclusions will be drawn. For example, a study could focus on the overall genetic diversity of a species in a study area (e.g., total number of alleles) or on comparing genetic diversity of populations in that area (e.g., population-specific allelic richness). In both cases, genetic data would be gathered for individuals (= sampling unit), but the way to analyze the data would vary. In the first case, data from all individuals would be analyzed together as a single analysis unit. In the latter case, data from individuals would be combined per population before analyses. Correctly identifying the analysis unit and corresponding sampling level *a priori* is important for an efficient sampling strategy because it can help to optimize sampling efforts. For example, sampling thousands of individuals in the study landscape is not necessarily adequate for the second research question if these individuals stem from only two populations. After all, the statistical sample size for the analysis would only be $n = 2$.

### 4.2.2 Sampling intensity

The sampling intensity is simply the number of samples gathered (i.e., sample size) relative to the size of the total population. Sampling a higher proportion of individual samples from a population increases sampling intensity and usually leads to more accurate and more precise statistical results. Importantly, sampling intensity is also related to statistical power, that is, the ability of a statistic to detect a significant effect if that effect actually exists. Note that an exhaustive sampling intensity that includes all individuals of a population no longer represents a sample, but a census. As complete censuses can rarely be conducted and the size of the total study population is seldom known in empirical studies involving field work, sampling intensity can often not be assessed directly. Thus, rather than considering actual sampling intensity, researchers instead focus on sample sizes and use different criteria to ensure that the obtained number of samples is adequate for sufficient power, accuracy, and precision (e.g., at least 30 observations to be close to a normal distribution or at least a symmetrical distribution).

### 4.2.3   Spatial sampling scheme

The spatial distribution of sampling locations can have profound influences on analyses and conclusions of a research project (Anderson et al. 2010). Three major spatial sampling schemes can be distinguished (Fig. 4.1). Samples can be located randomly across the study area or by strata within the study area;

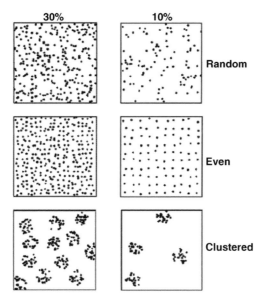

**Fig. 4.1** Example of different sampling intensities and spatial sampling schemes. In the left panel, 300 samples (30%) were collected, while 100 samples (10%) were collected from the same population ($N = 1000$) in the right panel. Samples are distributed randomly in space in the top panel, evenly distributed in the middle row, and spatially clustered in the bottom row.

they can be spaced evenly or uniformly; or they can be distributed in a clumped or clustered way.

When designing a spatial sampling scheme *a priori*, it may prove difficult to actually collect samples from all sampling locations originally anticipated, especially in landscape genetic studies involving elusive and/or mobile species. Nevertheless, it is still vital to try to match the spatial sampling scheme to the study area and research question(s), as we will discuss below. Furthermore, even if the final distribution of samples is not exactly as originally intended, the final spatial distribution of samples should be quantified. For example, the nearest-neighbor-index (NNI) (Dale & Fortin 2014) can be used to determine whether samples are randomly distributed in space, significantly closer than expected under randomness (i.e., clumped), or significantly further apart (i.e., evenly distributed). Considering the spatial distribution of samples is particularly important when analyzing data that show ***spatial auto-correlation*** and for analytically distinguishing this autocorrelation from spatial dependence (see Chapter 5).

### 4.2.4   Temporal sampling scheme

In addition to the spatial distribution of samples, their temporal distribution is also a crucial aspect of sampling strategies. Generally, samples can be gathered (a) at a single point in time (snapshot); (b) repeatedly at the same period over several years, for example, in spring of year 1 and year 2; (c) at multiple but different points in time, for example, spring versus fall; or (d) more or less continuously during a certain time period, for example, from 2015 to 2017. Furthermore, different temporal sampling schemes might be applied in different study areas or populations, or the same area might be sampled with a combination of (a) through (d). Temporal sampling patterns are rarely considered in final data analysis and are usually restricted to tests that assess whether significant differences exist among different sampling periods or whether all data can be pooled.

### 4.3   GENERAL STUDY DESIGN CONSIDERATIONS

The most valid and meaningful scientific inferences can be drawn from study designs based on replicated experiments. An experiment involves the direct manipulation of only one or a few factors that are hypothesized to impact a system or phenomenon of interest, while all other conditions are held constant. Systematically manipulating factors under different experimental conditions makes it possible to assess the relative importance of the factors and to evaluate whether their influences change under varying conditions. Replicating the same experiment multiple times or at multiple locations also makes it possible to quantify the variation of obtained results, so that the certainty of inferences can be estimated. Thus, replicate experiments allow researchers to identify actual cause-and-effect relationships by testing hypotheses under controlled conditions.

In landscape genetics and many other scientific disciplines, various constraints often hamper the application of actual experiments. For example, it is very challenging to repeatedly alter the composition or configuration of a landscape in order to experimentally assess the influence of landscape structure on gene flow in a certain species, at least when the species has a large body size and/or is highly mobile. Furthermore, genetic variation does not immediately respond to changes in the landscape (Anderson et al. 2010), so that landscape genetic experiments are difficult to conduct during the length of typical funding periods.

When actual experiments cannot be conducted, researchers have to rely on observational studies. In observational studies, researchers systematically collect data on certain variables, but do not actively change these variables or directly control the settings in which they are observed. Observational studies should try to mimic the advantages of manipulations and replications by using study designs that are "quasi-experimental". Such quasi-experimental study design aims to identify genetic responses to current (or past) landscapes by distributing samples in a way that captures the spatial heterogeneity of environmental variables hypothesized to directly or indirectly influence genetic variation in the study species. For example, instead of actual manipulations, observational studies should attempt to separate the effects of one potentially important factor (say roads) from the effects of other factors (e.g., elevation or landcover) as much as possible. This could, for example, be achieved by sampling data (1) in different study areas where the factor of interest either is present or not, (2) along an environmental gradient, or (3) in various environmental strata. Similarly, instead of actual replications, observational studies can compare the effects of a specific variable under varying study settings (e.g., different study areas or sampling periods). This makes it

possible to infer the variability of observed effects and allows an assessment of inferential certainty. Thinking about how to mimic controlled settings, actual manipulations, and replications when designing an empirical study can greatly increase its scientific meaningfulness and determine whether conducted research will have actual inferential strength or is purely descriptive.

## 4.4 CONSIDERATIONS FOR LANDSCAPE GENETIC STUDY DESIGN

Achieving optimal (i.e., quasi-experimental) observational study designs in empirical studies is never an easy task. However, several characteristics of typical landscape genetic studies make optimal study design particularly challenging in this field. First, as mentioned above, it is virtually impossible to work with truly replicated landscapes, as each landscape has its specific environmental conditions and series of historical disturbance events that act on the resulting data structure. This is true for most landscape ecological studies as well (Hargrove & Pickering 1992). Second, landscape genetics tries to link processes that usually act at different spatial and temporal scales (e.g., change in land use versus change in genetic variation). Thus, to achieve optimal results in landscape genetics, study design has to consider the three analytical steps described in Chapter 1. Specifically, we have to design studies so that (a) the landscape heterogeneity is sampled adequately, (b) the genetic variation is sampled adequately, and (c) landscape and genetic data can be matched statistically in a meaningful way. For this, the spatial and temporal scales of both the genetic material (historical versus contemporary) and landscape variables need to be considered (Anderson et al. 2010).

### 4.4.1 Considerations for sampling landscape data

Landscape features (mountains, rivers, urbanized areas, farmland, roads, etc.) affect each species differently. Hence the selection of the landscape variables and the sampling design to gather them need to match the species-specific perception of these selected variables and the spatial domain of the species response to these measured landscape variables. Most landscape variables that are known to affect species distribution, movement, and spatial genetic structures are sampled using either remote sensing data or field work. Both data gathering methods are prone to sources of errors that need to be minimized.

Acquisition of landscape data through remotely sensed data has the advantage that data for large study *extents* are available. However, such landscape data are derived from the various remote sensing platforms that have their specific spatial resolution (i.e., *grain*) and variables derived from these platforms may not be the most appropriate ones to which the species respond. Furthermore, with remotely sensed data there are a series of steps through which errors can propagate up to the final data used in landscape studies, ranging from data capture distortions, cloud cover (leading to missing data), to thematic mapping misclassification and combination of categories (Hunsaker et al. 2001; Langford et al. 2006). Thus, both the spatial resolution of the grain and extent of the remotely sensed data may be inappropriate for the system under study (Lechner et al. 2012).

The appropriate functional spatial resolution of the grain (sampling unit size or pixel size of the remotely sensed data) at which data should be sampled will vary according to the species or system studied. Ideally, the functional grain should be smaller than the area used by the targeted species during its daily movements (see Chapter 2). Based on simulations, Graves et al. (2012) showed that quantifying the effects of landscape resistance on genetic structure was not negatively affected by using only a single locations of a mobile species, as long as a functional grain was used that was smaller than the home range of the species.

While spatial mismatch can be mitigated by aggregating pixels at coarser resolution or by using broader categorical classes for the thematic information, what is often lacking is the appropriate time frame to match the time at which the genetic material was sampled. The most common temporal issues with remotely sensed data are the mismatch between the process(es) of interest and available data, non-availability of data for the various key biological processes through the seasons (e.g., seasonal vegetation patterns) and the years, and a lack of series of remotely sensed data that match the actual changes in landcover types in the studied area. Even when remotely sensed data exist more or less for the same extent and pixel resolution, distortion and misclassification issues need to be addressed. Linke and McDermid (2012) developed a methodology to clean and calibrate historical remotely sensed data to quantify first the degree of landscape changes and second how these changes affect species behavior, dispersal, and ultimately their genetic variation spatially. Also, positioning of the study area can affect the estimated values of the landscape characteristics measured (e.g., amount of forest, number of patches, etc.; Plante et al. 2004). Hence, it is useful to have prior knowledge of the dispersal ability of the studied species to determine the size of the study area accordingly (Chapter 2; Keller et al. 2015).

Acquisition of landscape data through field work has the advantage that the sampled data can be obtained at the same location of the genetic material. Also, such plot data may be more relevant to explain species distribution and adaptive genes than data derived via remote sensing. The major disadvantage, however, is that the sampling effort is higher than in the case of remotely sensed data and usually not the entire study area can be covered. Field data can be spatially interpolated using Kriging (Cressie 1993) to obtain estimated values of the landscape variable(s) at unsampled locations. The errors associated with the interpolated values will vary in magnitude according to the sampling design used (e.g., random, systematic, stratify, nested), the sample size, and the theoretical variogram model used (Fortin et al. 1989; Dale & Fortin 2014). Although technology did improve considerably regarding GPS accuracy, there are still some positional errors associated with the sampling locations of the genetic material. Positional errors combined with errors in the spatial and thematic accuracy of remotely sensed data can lead to erroneous matching between the genetic and landscape data. The most commonly used practice to avoid such mismatches is to aggregate thematic categories and pixel resolution.

Of course the specific type of landscape variables measured needs to correspond to landscape proprieties that are known to affect species behavior (variables at and around the sampling locations) and dispersal abilities (among sampling locations). As most species are affected by plot level data (e.g., within home ranges) as well as landscape-level data (e.g., spatial composition and configuration of landscape cover types and landscape features), it is important to sample data *at*, *around*, or *among* sampling locations (Melles et al. 2012) to test the various landscape genetics hypotheses (Wagner & Fortin 2013; Pflüger & Balkenhol 2014). Given that each location in space had its own series of historical events (Hargrove & Pickering 1992), it is important to have replicates per each stratum as well as various spatial configurations of landscape cover types and landscape features. Indeed, a given landcover type may have completely different effects depending on its size and the other landcover type surrounding it (Fahrig 2007). Thus, when habitat patch size is assumed to matter, patches of varying sizes need to be sampled.

### 4.4.2 Considerations for sampling genetic data

Sampling genetic data actually involves two aspects: the gathering of genetic material from individuals of the study species in the field and the sampling of genetic data from the genome of each individual in the laboratory (e.g., type of marker, chosen loci, etc.; see Chapter 3). However, the increasing availability and affordability of genomic approaches (see Chapter 9) will soon make the sampling of the genome the least limiting factor for landscape genetic study design, because these methods can provide highly abundant genetic data for each sampled individual. Thus, we here focus on sampling design considerations for gathering genetic material in the field. The goal of this sampling should be to capture genetic variation accurately and in a way that allows us to detect potential landscape effects on genetic diversity and/or structure.

As pointed out above, deciding on an appropriate sampling *level* should be one of the first considerations when designing a study. Depending on the landscape genetic research question(s), either individual- or population-level inferences may be most adequate for a particular study (see Chapter 3). Thus, the study goals should ideally dictate whether to focus on individuals or populations for sampling. As a second step, the required sampling *intensity* needs to be considered. For population-level analyses, an adequate number of individuals need to be sampled per population, so that genetic variation within and among populations can be quantified accurately. Ideally, power analyses should be used to determine the sample size required for this, but only very few power analyses specifically for (landscape) genetic exist (e.g., Ryman & Palm 2006). Instead, researchers usually rely on rules-of-thumb or previous research findings to determine the sample size needed per population. For example, Hale et al. (2012) concluded that 25 to 30 individuals are generally sufficient to describe genetic variation of a population based on microsatellites, though more individuals per population may be needed when genetic differentiation among populations is very weak (e.g., Kalinowski 2005). Note again that the statistical sample size in population-level analyses is the

number of populations and not the number of individuals sampled per population. Hence, in addition to sampling enough individuals within each population, we also need to include a sufficient number of populations in our sampling so that the intended statistical methods can be applied. Clearly, there will often be a trade-off between the number of populations and the number of individuals per population that can be sampled and genetically analyzed. A similar trade-off exists in individual studies, when we can focus either on the number of sampled individuals or the number of loci used for analyzing each individual. As highlighted in Chapter 6, simulations can be used for guiding study design with respect to these trade-offs, and we summarize our current understanding of them later in this chapter. In addition to sampling level and intensity, the *temporal* sampling design needs to be considered for accurately capturing genetic variation. For example, the spatial-genetic structure of a species could differ strongly among different seasons because of seasonal movement patterns (e.g., Latch & Rhodes 2006) or because pre- and post-dispersal life stages exist at certain points in time and need to be distinguished (e.g., Goldberg & Waits 2010; Rico et al. 2012). Obviously, the spatial sampling design is also very important in landscape genetics, but since it needs to be matched to the heterogeneity of the landscape, it is treated in the next section.

### 4.4.3 Matching landscape and genetic data

The ultimate challenge in landscape genetic study design is to match the sampling of landscape data with that of the genetic data, so that informative statistical analyses can be conducted and meaningful inferences can be drawn. To achieve this, the *spatial* sampling strategy needs to be considered in conjunction with sample size (e.g., Anderson et al. 2010; Schwartz & McKelvey 2009). In spatial statistics, the rule-of-thumb for the number of samples needed to detect a significant spatial pattern is between 30 to 50 samples (Dale & Fortin 2014). Yet the way in which the samples are laid out in space will affect greatly the ability of the subsequent spatial statistical methods to detect the pattern, with higher sample sizes being needed with non-optimal sampling schemes (Fortin et al. 1989; Manel et al. 2012). In landscape genetics, the spatial layout of the samples should be primarily determined according to the hypotheses, and/or processes, of interest, and match the study species and landscape(s). Specifically, the spatial sampling design needs to account both for the biology of the species and the heterogeneity of the landscape, as captured by the available landscape data.

The first thing to realize when considering a spatial sampling design is that there is a fine balance between: (1) enough samples close to each other such that the spatial genetic structure can be detected at high resolution, but not too many samples close to each other, as this would potentially only duplicate information and thus be a waste of sampling effort, and (2) no samples that are too far away as this increases the spatial heterogeneity captured and in turn increases the variance in the signal(s), which can weaken inferential power (Haining 2003; Dale & Fortin 2014). There are several ways to help one deal with these challenges: (1) perform a pilot study (Legendre et al. 2002), (2) use expert knowledge on the system studied, (3) use published literature on the same system, and (4) perform simulation studies to assess the detection power of the synergetic effects of processes. In landscape genetics, the latter approach has been particularly useful (among others: Schwartz & McKelvey 2009; Epperson et al. 2010; Blair et al. 2012; Graves et al. 2012; Landguth et al. 2012; Oyler-McCance et al. 2013).

One effective way to balance the need of having enough samples nearby so that local spatial patterns can be detected (say isolation-by-distance, IBD) and that the samples cover a large enough region to capture the effects of landscape features (say isolation-by-resistance, IBR) is to use a nested sampling design (also referred to as clustered samplings; Fortin et al. 1989; Webster &

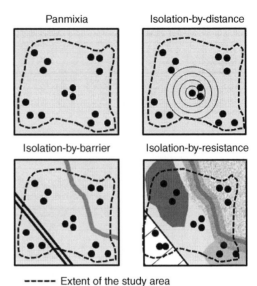

**Fig. 4.2** Example of a nested study design. The black dots represent sampling units for genetic analysis. Using such a nested design, it is possible to detect landscape effects on genetic variation that act at different spatial (and temporal) scales within a single study area.

Oliver 1990; Dungan et al. 2002; Anderson et al. 2010; Manel et al. 2010). In essence, the spatial layout of nested designs is to have groups of samples at short, intermediate, and large distances (Fig. 4.2).

Oher quasi-experimental study designs for landscape genetics include the before-after-control-impact (BACI) design and a stratified sampling (see Box 4.1). It is also

---

**Box 4.1    Examples of strong study designs for landscape genetics**

A well-known example of a quasi-experimental study design is called BACI, which stands for before–after (BA) and control–impact (CI). This design was first described by Green (1979). In essence, a BACI design compares data collected before and after a "treatment" has been applied, and/or data collected from populations that either are, or are not, subjected to the "treatment".

Note that the "treatment" does not have to be a direct manipulation by the researcher, but is more generally any variable of interest. For example, in the first box figure, the "treatment" is the presence or absence of a road, and could be any other variable reflecting landscape fragmentation. Using only a before–after or only a control–impact is less ideal, but is often more applicable in empirical studies than a full BACI, which can be extended in several ways. Specifically, while the different versions of the BACI design provide greater inferential strength than purely descriptive studies, the design usually does not include any replicates in space or time (Hulbert 1984). Thus, from a landscape genetic perspective, an ideal BACI design should be conducted in several replicate landscapes, each including multiple control and impact sites, where data are gathered at multiple points in time.

A study design that makes it possible to have replicates, even within a single landscape, is a stratified random sampling (see the second box figure). In this design, different strata are defined, for example, based on landcover types. In each strata, multiple replicates are ideally sampled.

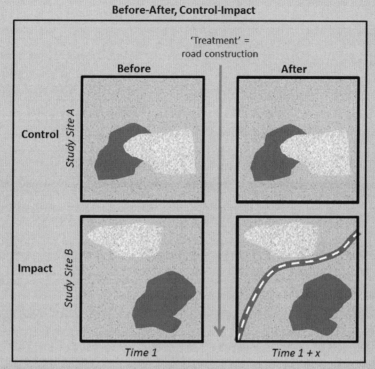

Before–after–control–impact (BACI) study design. In this example, the effect of a road would be assessed by comparing gene flow/genetic differentiation in an impact area before and after road construction with a (similar) control area where no road was built.

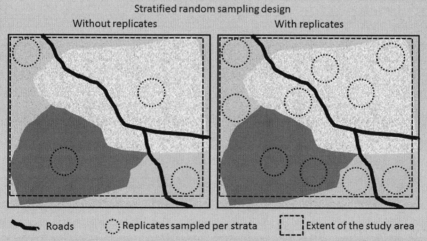

Stratified random sampling designs. In each strata (e.g., based on landcover types), either one (left, no replicates) or several (right, with replicates) areas are sampled. The areas could comprise multiple individuals or populations, and can also be sampled in a nested fashion (see Fig. 4.2).

possible to use the combination of landscape features and landscape cover types as criteria of stratification. Such a stratified sampling design allows us to test the hypothesis of isolation-by-barrier (IBB, e.g., mountain, river, road) at the broad scale, of isolation-by-resistance (e.g., various levels of resistance costs to animal movement; Spear et al. 2010) at the intermediate scale, and of isolation-by-distance at the local scale (Jenkins et al. 2010).

Overall, landscape genetics studies should not strive for perfection in the sampling of one type of data while neglecting the other type. Instead, the trick is to focus sampling on the exact research question(s), the study species, and the predicted landscape–genetic relationships in the study area(s). In the next section, we summarize our current knowledge about how the different factors described above influence landscape genetic results.

## 4.5   CURRENT KNOWLEDGE ABOUT STUDY DESIGN EFFECTS IN LANDSCAPE GENETICS

Studies that have investigated sampling effects in landscape genetics are summarized in Tables 4.1 and 4.2. Table 4.1 lists studies that dealt with sampling effects on the detection of spatial genetic population structure (analytical step 1, see Chapter 1), with a particular

**Table 4.1** Studies investigating sampling effects on the performance of genetic clustering methods.

| Study | Software used* | Main finding(s) |
| --- | --- | --- |
| Evanno et al. (2005) | S | Number of clusters inferred in program STRUCTURE depends on the number of loci scored and the number of individuals sampled from each population |
| Latch et al. (2006) | B, S, P | Performance of methods decreases with decreasing genetic differentiation among clusters |
| Chen et al. (2007) | S, G, GC, T | Spatial methods are more reliable when true clusters are spatially non-overlapping, while aspatial methods are more reliable when spatial domains of clusters overlap |
| Anderson & Dunham (2008) | S | STRUCTURE incorrectly identifies clusters when family groups are included in a data set |
| Frantz et al. (2009) | B, G, S | Under IBD, genetic barriers are detected correctly, but additional (incorrect) clusters are also identified |
| Guillot & Santos (2009) | G | Incorrect number of clusters is inferred under IBD, regardless of whether spatial or aspatial model is used and regardless of spatial sampling design |
| Schwartz & McKelvey (2009) | S | Clusters are incorrectly inferred in a continuous population showing IBD, with numbers of clusters depending on the spatial sampling design. Incorrect inferences are particularly likely with clustered sampling |
| Balkenhol (2009) | B, G, S, T | IBD among individuals may not indicate spatial structure within populations; low sampling intensity may prevent correct inferences |
| Fogelqvist et al. (2010) | S, SR | Methods tend to detect too few clusters with low sample sizes |
| François & Durand (2010) | B, G, S, T | Models without admixture can lead to incorrect inferences when admixed individuals are present in a data set |
| Kalinowski (2011) | S | STRUCTURE tends to detect too few clusters, particularly when sample sizes vary among populations |

* B = BAPS; G = GENELAND; GC = GENECLUST; P = S = STRUCTURE; SR = STRUCTURAMA; T = TESS.

**Table 4.2** Studies investigating sampling effects in landscape genetics.

| Study | Simulated (S) or empirical (E) data | Population- (P) or individual- based (I) analysis | Brief description |
|---|---|---|---|
| Short Bull et al. (2011) | E | I | Correlated individual-based genetic distances with effective distances in 12 different study areas for black bears (*Ursus americanus*) |
| Oyler-McCance et al. (2013) | S | I | Simulated landscape–genetic relationships using spatially-explicit, individual-based simulations and then evaluated whether different sampling schemes, sampling intensities, number of loci, and marker variability affected the detection of landscape–genetic relationships |
| Naujokaitis-Lewis et al. (2013) | E | P | Compared results obtained from a known, complete network of Columbia spotted frog (*Rana luteiventris*) populations, to results obtained after excluding populations from the network |
| Schwartz and McKelvey (2009) | S | I | Simulated continuously distributed populations showing either panmixia or IBD, and then assessed whether six different sampling schemes would correctly assign individuals to a single population |
| Prunier et al. (2013) | S | P, I | Simulated patchy populations and subsampled data using either individuals or populations as sampling units. Then individual- and population-based genetic metrics were used to assess IBD and barrier effects |
| Landguth et al. (2012) | S | I | Simulated using a spatially-explicit model where the number of individuals, number of loci, and number of alleles per locus were found in order to test the power to detect IBD, IBB, and IBR |
| Jaquiéry et al. (2011) | S | P | Simulated populations separated by landscapes of varying permeability to dispersal. Compared three analytical methods for correctly inferring landscape effects on dispersal |
| Luximon et al. (2014) | S | P, I | Simulated continuously distributed populations with varying dispersal parameters |
| Landguth and Schwartz (2014) | S | P, I | Simulated individuals within discrete populations separated by barriers, as well as continuously distributed individuals characterized by IBD. Subsampled the data using four different sampling strategies and used individual- and population-based metrics for analyses |

| Analytical approach | Main conclusions | Study design implications |
|---|---|---|
| (Partial) Mantel tests | Limiting factors that are more variable are more important; finding that a factor does not matter in one area does not mean that it is not important in other areas as it can be context-dependent | Select highly variable landscape variables; conduct analyses in multiple landscapes |

(*continued*)

**Table 4.2**  (*Continued*)

| Analytical approach | Main conclusions | Study design implications |
|---|---|---|
| (Partial) Mantel tests | With large sample sizes (individuals and loci), most sampling schemes will give correct inferences; however, clustered sampling does not work well with continuously distributed populations | Maximize the number of individuals and loci; do not use clustered sampling design in continuously distributed species |
| Network analysis; Mantel tests | Incomplete sampling of a network affects centrality measures of nodes, while measures describing the whole network are more robust. The number of nodes and type of network connectivity used affects the power of the Mantel test to detect the relationship between genetic and landscape distance matrices | Try to sample entire network; conduct sensitivity analyses, especially for centrality measures of nodes |
| Clustering methods | Under IBD, clustering methods incorrectly detect multiple populations when using a clustered sampling design | Avoid using the clustering method when clustered sampling design shows significant IBD. Conduct a preliminary test for spatial autocorrelation and sample at a scale greater than the maximal distance at which significant autocorrelation is found |
| Mantel tests with restricted permutations | Individual-based sampling and analysis are effective for patchy populations | Even in patchy populations, individual-based sampling can be used to detect landscape infuences on gene flow |
| Mantel tests | Overall based on these series of simulations, it was found that 25 loci and 25 alleles from 100 individuals are sufficient to detect genetic structure, yet when polymorphism is low then more individuals are needed | More loci showing more variability are more important than more individuals for increasing power |
| Partial correlation, regression, approximate Bayesian computation | Landscape elements impeding gene flow are easiest to detect. Power and accuracy of methods decrease with increasing landscape complexity and decreasing difference between habitat and matrix permeability | Insignificant landscape effects might simply be low-power methods. To increase power, study designs should focus on landscape features that are likely to be very different from habitat in terms of permeability and impede gene flow |
| (Generalized) linear regression and Mantel test | Compared to sampling groups, individual-based sampling provides more or as much power for detecting IBD, and for inferring underlying dispersal parameters | In continuous populations, sampling individuals is effective for detecting spatial-genetic structure and infer dispersal characteristics |
| Mantel tests | Sampling few individuals from many populations overestimates spatial genetic structure when using population-based metrics. Sampling few populations with many individuals underestimates the strength of IBD with individual-based metrics | Sampling and analysis units need to be matched to the statistical method and the distribution of the species |

focus on clustering methods. These types of methods are particularly popular in landscape genetics (see Chapters 3 and 7) and have therefore received a great deal of attention. Table 4.2 summarizes studies that assessed sampling effects on the detection of landscape–genetic relationships. Note that most of these findings are from simulations rather than empirical studies. This is not surprising, given that simulations are a way to mimic actual experiments (Chapter 6; Epperson et al. 2010). Below, we discuss these studies as well as some others that are relevant, even though they do not fall within the roam of landscape genetics *per se* or did not directly assess sampling effects.

### 4.5.1 Sampling of landscape heterogeneity

The studies show that the sampling design needs to adequately represent the spatial heterogeneity of the landscape, especially with respect to the landscape elements thought to influence spatial genetic structure. Specifically, Cushman et al. (2011, 2012a) showed that landscape elements that are limiting gene flow are easiest to identify. In landscapes where the habitat is abundant and continuous from the perspective of the study species, it will be difficult to detect landscape–genetic relationships (e.g, Reding et al. 2013). This finding is supported by Jaquiéry et al. (2011), who further showed that it is much easier to identify landscape elements that impede gene flow, compared to landscape elements that facilitate it. The same study also demonstrated that the power to detect landscape effects on gene flow decreases with increasing complexity of the landscape and decreasing differences between habitat versus matrix permeability. Similarly, Short Bull et al. (2011) concluded that detecting a significant effect of landscape features is only possible when they show high variability within the study area. Thus, insignificance of a particular landscape variable does not mean that it does not influence movement and gene flow in the stud species, but could simply be caused by a lack of spatial heterogeneity for that variable in the study area. Likewise, DiLeo et al. (2013) concluded that effects of recent landscape changes on genetic connectivity are difficult to detect if the elements coincide with long-standing features, such as mountains, lakes, or rivers. As a remedy, landscape genetic studies in multiple study areas should be performed (e.g., Balkenhol et al. 2013).

Cushman et al. (2012b) concluded that the configuration of habitat is generally more important than habitat amount for explaining the magnitude of genetic differentiation, although the relative importance of habitat amount versus configuration are difficult to disentangle and can be scale-dependent (Millette & Keyghobadi 2015). Furthermore, Keller et al. (2015) pointed out that not only the amount or configuration of landscape elements need to be considered during sampling and analysis but also their shape and spatial location. The implications of this for optimal sampling strategies in landscape genetics have not been formally evaluated.

The extent of the study area and the grain at which variables are measured can of course also influence landscape genetic results, and Galpern and Manseau (2013) developed a method to find the appropriate grain size for resistance surface modeling (Chapter 8). However, Cushman and Landguth (2010) suggested that the thematic resolution of the utilized landscape data is even more important than extent and grain size.

### 4.5.2 Individual- versus population-based sampling

Based on currently available studies, it also seems clear that individual-based sampling and analysis can accurately quantify spatial genetic structures, even in species that show a patchy distribution within the study area. Both Prunier et al. (2013) and Luximon et al. (2014) showed that compared to population-based sampling, individual-based sampling provides more or as much power for detecting isolation-by-distance and for inferring underlying dispersal parameters and landscape influences. Individual-based sampling and analysis also detects barriers to gene flow more quickly than population-based approaches (Landguth et al. 2010; Blair et al. 2012), although the temporal resolution of population-based approaches is greatly increased when using adaptive genetic data that is under locally varying selection pressures (Landguth & Balkenhol 2012). The power of individual-based approaches increases with increasing number of loci used and with increasing variability of these loci (Landguth et al. 2012). In contrast, sampling more individuals does not strongly improve the ability of individual-based metrics to detect the processes underlying gene flow. Indeed, Prunier et al. (2013) concluded that, for patchily distributed species, as few as three to four individuals per patch are sufficient to accurately describe spatial genetic structure using individual-based methods.

### 4.5.3  Spatial sampling design versus sampling intensity

The spatial sampling design appears to have a much stronger impact on estimates of genetic structure than the sheer number of sampled individuals, at least when using individual-based genetic distances and clustering methods. Schwartz and McKelvey (2009) showed that the number of genetic groups detected by clustering methods depends strongly on the spatial sampling regime. Specifically, for continuously distributed species showing IBD, a clustered sampling often leads to the delineation of spurious clusters (see also Frantz et al. 2009). However, Guillot and Santos (2009) actually concluded that an incorrect number of genetic clusters is often inferred under IBD, regardless of the sampling design used.

### 4.5.4  Sampling intensity

While the spatial design of genetic sampling seems to be more important than sample size, sampling intensity has also been shown to be important, especially with respect to underestimating spatial genetic structure. For instance, Balkenhol (2009) concluded that clustering methods were not able to reliably detect simulated genetic clusters under low sampling intensities (i.e., 10% of total population size). Low sampling intensities are particularly problematic for clustering methods when genetic differentiation among populations is low (Evanno et al. 2005; Fogelqvist et al. 2010; Latch et al. 2006). Hence, it is recommended to estimate the fraction of the total population that was sampled and to estimate genetic differentiation among hypothesized populations. If sampling intensity and genetic differentiation are both low, clustering methods may not be the best choice and individual-based genetic distances should be considered instead. While estimating total population size is a challenge in itself and often difficult in empirical studies, the required effort is likely worth it since landscape genetic studies can greatly benefit from incorporating non-genetic information about population sizes, movements, etc.

Sampling design and intensity are also important in population-based studies, which remain useful for various purposes, for example when assessing metapopulation connectivity (see, for example, Pflüger & Balkenhol 2014). When inferences are based on population-based metrics, sampling intensity should be evenly spread among all sampled populations, as varying sampling intensities can influence estimates of genetic diversity and structure (Muirhead et al. 2008). Koen et al. (2013) showed that sampling too few individuals from populations leads to incorrect estimates of genetic structure using a variety of genetic differentiation metrics. Furthermore, when applying network approaches to population-based data (see Chapter 10), results can be influenced by unsampled populations, so that sensitivity analysis is recommended for these approaches (Koen et al. 2013; Naujokaitis-Lewis et al. 2013).

### 4.5.5  Matching sampling and statistical methods

As highlighted by Landguth and Schwartz (2014), it is also important to match the sampling design to the statistical methods applied to the data. It can be challenging to use a population-level sampling design if inferences are to be based on genetic distances among individuals. Similarly, it is difficult to use data from an individual-based genetic sampling for a continuously distributed species if a population-based genetic metric is to be used (e.g., $F_{ST}$-values) and if results are supposed to apply to discrete "populations" (e.g., management units). To make use of the individual- and population-based analysis, we again recommend the nested sampling design shown in Fig. 4.2. This design can make use of the advantages of individual- and population-level analyses and also facilitates the detection of hierarchical genetic structures. Using a clustered sampling design for a continuously distributed species is also not efficient for detecting landscape effects with individual-based genetic distances (Oyler-McCance et al. 2013). For such a species, random, linear (i.e., transect-based), and even sampling regimes are much more adequate, and their power again depends on the number of loci scored as well as the number of sampled individuals.

Overall, the empirical and simulation-based studies draw a complex picture of "optimal" sampling design in landscape genetics. Nevertheless, we can draw several conclusions that have emerged in several studies:
• Individual-based sampling and analysis is very effective for landscape genetic research. While individual-based sampling can be applied in patchy populations, a population-based (i.e., clustered) sampling of

continuously distributed individuals often leads to incorrect inferences.

• When using population-based sampling, it is important to strive for sufficiently high sampling intensities that are similar in all studied populations.

• When the sampling of individuals is already an adequate representation of the spatial distribution of the population and of the environmental heterogeneity in the study area, increasing the number of highly variable loci, rather than the number of sampled individuals, may be more informative for inferring landscape–genetic relationships.

• In highly connected landscapes, accurately quantifying spatial genetic structures and underlying landscape influences will be difficult, even if they exist. Furthermore, factors limiting gene flow can be detected more easily than factors facilitating them, yet this power of detection decreases when landscape complexity increases and when the difference between habitat-matrix permeability decreases. Hence, when spatial genetic structure is weak and/or landscapes are complex with small differences between matrix and habitat permeability, it is particularly important to use a quasi-experimental sampling design that specifically targets landscape–genetic hypotheses, and includes landscape characteristics that are assumed to impede gene flow.

• Knowledge about the limiting factors affecting species is crucial so that the sampling design can capture the spatial heterogeneity of relevant resources.

• Reclassifying the landcover data (i.e., change the thematic resolution) according to the species is more important than having the exact spatial grain of landscape data.

Finally, it is also clear that a single, best sampling strategy for landscape genetics has not been identified and that it is rather unlikely that such a universal study design actually exists. Instead, the sampling of genetic and landscape data have to be carefully matched to the study system, the landscape genetic research questions, and the analytical approaches used for evaluating them.

## 4.6 RECOMMENDATIONS FOR OPTIMAL SAMPLING STRATEGIES IN LANDSCAPE GENETICS

At this point, we cannot develop a simple set of rules for how to best sample landscape and genetic data for certain research questions. However, based on our general study design considerations and our current knowledge of sampling effects in landscape genetics, we can derive several suggestions for the things we need to consider when trying to find an adequate sampling strategy for a landscape genetic study (Fig. 4.3).

First, we caution that a non-targeted sampling strategy might miss landscape–genetic relationships or even lead to erroneous inferences about spatial-genetic structures. Thus, we need to be clear on our research questions and develop exact, testable hypotheses about how landscape heterogeneity might affect genetic variation in our study species, either directly (i.e., via selection; for adaptive genetic data, see Chapter 3) or indirectly (e.g., via dispersal/gene flow, drift, effective population sizes, etc.). These hypotheses should then guide us in selecting sampling and analysis units that also need to match the statistical approaches intended to be used for quantifying spatial genetic patterns and for linking landscape and genetic data. In some cases, these hypotheses also have to be considered for selecting a study area in the first place.

Before going into the field for sampling, we then have to consider the landscape data required for addressing the research questions and hypotheses. If the required data are not already available, we might have to conduct field work to gather it. If available data are to be used, we have to carefully evaluate whether these data fit to our study goals, especially with respect to the thematic representation of limiting factors. Regardless of how the landscape data are obtained, we have to assess the composition and configuration of relevant variables in our study area and estimate how the spatial heterogeneity of these variables might affect the distribution of the species and the likely success of different spatial sampling regimes. The goal of our sampling should be to capture this heterogeneity so that its hypothesized effects on genetic variation can be tested. At this step, we should strive for quasi-experimental study designs that make use of spatially nested replicates, so that landscape–genetic relationships can be inferred with greater certainty.

Once we have chosen one or several designs for sampling genetic data in the field, it is time for a simulation study that assesses the power of chosen analytical methods under different sampling intensities and schemes (Fig. 4.4; adapted from Hoban 2014). In essence, the purpose of the simulation is to determine the minimum sampling intensity (e.g., number of population and/or number of individuals per population) required for detecting hypothesized landscape effects on spatial-genetic structures, given the landscape data

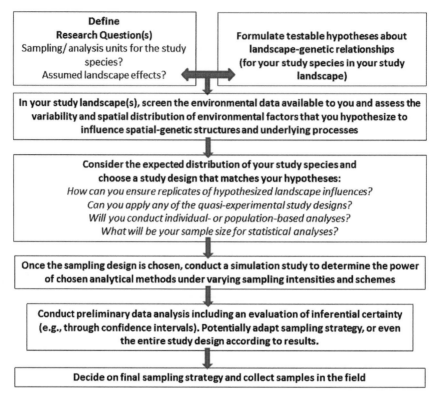

**Fig. 4.3** Schematic of suggested steps for choosing a sampling strategy for landscape genetics.

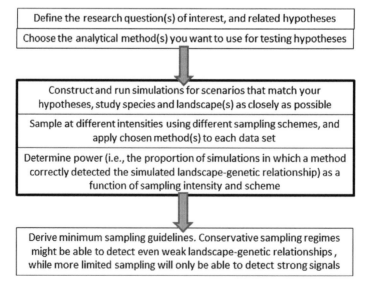

**Fig. 4.4** Schematic of steps in a simulation study for comparing the power of different sampling strategies in landscape genetics.

available for the study area and the traits of the study organism(s).

An in-depth description of how to conduct such a simulation study is not within the scope of this chapter, nor is such a description possible without knowing the details of the intended study. However, general information on landscape genetic simulations can be found in Chapter 6 and examples for evaluating study designs using simulations can be found in Hoban et al. (2013), Prunier et al. (2013), Hoban (2014), and other publications listed in Tables 4.1 and 4.2. Hoban et al. (2013) also provided an online simulation tool (SPOTG) for optimizing the number of individuals and markers to sample for five common types of conservation genetic research topics, including individual assignment and connectivity estimation using $F_{ST}$ values. For the latter task, another software for power analysis is available (POWSIM; Ryman & Palm 2006), but it only simulates two populations of equal size.

After the simulation study has identified a sampling strategy that leads to satisfactory power for detecting hypothesized landscape–genetic relationships, we can finally go into the field and try to sample in the desired way. However, various constraints will often hinder us from accomplishing ideal sampling conditions. Whenever sampling is not possible as intended, we need to re-evaluate whether the actual sampling still reflects our study system and whether it has sufficient power to test our hypotheses. If not, we either have to adjust our sampling strategy, redefine our hypotheses, or give up the study altogether. Such an "adaptive" sampling strategy could also be guided by preliminary data analysis that can help us refine our sampling, for example, when sampling can be conducted in multiple years.

We are fully aware that a perfect study design is often not possible, simply because of the various challenges typically encountered during field work. Nevertheless, we could try to follow the suggestions given above by answering a series of key practical questions, for example:

**1** What are our exact landscape genetic research questions and the precise hypotheses about how the landscape influences genetic variation and underlying processes in the study organism?
**2** Given our hypotheses, how do we define our study area (extent), how do we need to sample landscape heterogeneity (variables, resolution), and what kind of landscape data are already available for the study area?
**3** What is the appropriate sampling level and unit of analysis for our study: individuals or populations?

**4** When is the optimal time (seasons, multiple years) for sampling the organism in the field?
**5** How can we best distribute our sampling units (i.e., individuals or populations) in space?
**6** How can we obtain genetic samples from the organism?
**7** What are our marker options for quantifying genetic variation in the samples?
**8** How do we best allocate our sampling efforts? Should we increase the number of sampled populations, the number of sampled individuals, or the number of loci scored for each individual?
**9** How can we optimize our study design to make it quasi-experimental?

Obviously, our guidelines are meant to be general and scientists will have to make many adjustments and compromises when applying the guidelines to their own studies. For example, the occurrence of a species or the distribution of certain environmental variables may not be known until after the field work has been conducted, so that a well-targeted sampling cannot be chosen *a priori*. When this is the case, a two-stage sampling can be conducted. For example, Prunier et al. (2013) suggested that a landscape-wide individual-based sampling should initially be conducted in order to assess overall genetic patterns and gain a better understanding of landscape influences on genetic variation. Once this information has been gathered, a more targeted sampling could take place that specifically chooses sampling locations to test hypotheses derived from the initial sampling.

In sum, developing strong sampling strategies for landscape genetics is far from trivial and requires us to consider many interrelated aspects and to make many decisions along the way. This is in stark contrast to many current landscape genetic studies that often seem to be based on opportunistic sampling not guided by any of the *a priori* considerations we here deem to be important. We realize that there are many other, good, reasons for sampling genetic data and that even these data can yield interesting and meaningful findings about landscape–genetic relationships. Nevertheless, we hope that our guidelines will encourage researchers to think more critically about the many factors and choices involved in landscape genetic study design.

## 4.7 CONCLUSIONS AND FUTURE DIRECTIONS

We conclude that the most important aspect for optimal sampling design in landscape genetics is to derive testable *a priori* hypotheses about how landscape heterogeneity in the study area(s) is assumed to affect genetic variation and underlying processes in the study species. Clearly, such hypotheses are not always easy to formulate, especially because a true landscape genetic theory has not yet been developed (see Chapter 14). Nevertheless, it will be very beneficial for the field if we start to think more deeply about sampling strategies explicitly focused on landscape genetic research questions. As illustrated throughout this chapter, this means that both genetic variation and landscape heterogeneity need to be sampled in such a way that their statistical comparison allows meaningful conclusions about landscape–genetic relationships. Given the complexity of this task, we believe that a major current research need lies in developing tools for conducting a power analysis in landscape genetics. While such analyses can already be conducted using simulation software (see above and Chapter 6), this is still a rather time-consuming task that often requires some additional programming. Ideally, a simulation platform specifically targeted towards landscape genetic power analysis would give users the ability to evaluate the power of different sampling regimes (e.g., spatial sampling design, sampling intensity) for detecting different landscape genetic hypotheses (e.g., isolation-by-distance/-resistance/-barrier) in the actual study area (i.e., input of available landscape data and species-specific parameters in the simulations), using various statistical options for individual- and population-based analyses. This is without doubt a very challenging task, but certainly one that would be tremendously helpful for many researchers interested in landscape genetics. Indeed, we suggest that a flexible yet easy-to-use tool for landscape genetic power analysis would offer an alternative way to evaluate sampling effects in the future. Currently, studies often evaluate such effects under quite specific settings, but then try to derive recommendations for landscape genetics is general. However, as shown above, many factors and their interactions need to be considered for optimal sampling design in landscape genetics, so that results from a particular sampling study cannot simply be applied to all other instances. Thus, instead of trying to find sampling guidelines that are universally true, we perhaps also need to focus on each study at hand and evaluate different sampling options before conducting field work. This is obviously a very practical approach and may in itself not yield much scientific knowledge. However, it could greatly support our scientific aspirations in landscape genetics, as it could lead to efficient, quasi-experimental sampling designs that will ultimately lead to a better understanding of the emergence and maintenance of genetic variation in heterogeneous environments.

## REFERENCES

Anderson, E. & Dunham, K. (2008) The influence of family groups on inferences made with the program STRUCTURE. *Molecular Ecology Resources* **8**, 1219–29.

Anderson, C.D., Epperson, B.K., Fortin, M.J., Holderegger, R., James, P., Rosenberg, M.S., & Spear, S. (2010) Considering spatial and temporal scale in landscape-genetic studies of gene flow. *Molecular Ecology* **19**, 3565–75.

Balkenhol, N. (2009) Evaluating and improving analytical approaches in landscape genetics through simulations and wildlife case studies. Dissertation, University of Idaho.

Balkenhol, N., Pardini, R., Cornelius, C., Fernandes, F., & Sommer, S. (2013) Landscape-level comparison of genetic diversity and differentiation in a small mammal inhabiting different fragmented landscapes of the Brazilian Atlantic Forest. *Conservation Genetics* **14**, 355–67.

Blair, C., Weigel, D.E., Balazik, M., Keeley, A.T.H., Walker, F.M., Landguth, E.L., Cushman, S.A., Murphy, M., Waits, L.P., & Balkenhol, N. (2012) A simulation-based evaluation of methods for inferring linear barriers to gene flow. *Molecular Ecology Resources* **12**, 822–33.

Chen, C., Durand, E., Forbes, F., & François, O. (2007) Bayesian clustering algorithms ascertaining spatial population structure: a new computer program and a comparison study. *Molecular Ecology Notes* **7**, 747–56.

Cressie, N.A.C. (1993) *Statistics for Spatial Data*. John Wiley & Sons, Inc., New York.

Cushman, S.A. & Landguth, E.L. (2010) Scale dependent inference in landscape genetics. *Landscape Ecology* **25**, 967–79.

Cushman, S.A., Raphael, M.G., Ruggiero, L.F., Shirk, A.S., Wasserman, T.N., & O'Doherty, E.C. (2011) Limiting factors and landscape connectivity: the American marten in the Rocky Mountains. *Landscape Ecology* **26**, 1137–49.

Cushman, S.A., Shirk, A.J., & Landguth, E.L. (2012a) Landscape genetics and limiting factors. *Conservation Genetics* **14**, 263–74.

Cushman, S.A., Shirk, A., & Landguth, E.L. (2012b) Separating the effects of habitat area, fragmentation and matrix resistance on genetic differentiation in complex landscapes. *Landscape Ecology* **27**, 369–80.

Dale, M.R.T. & Fortin, M.-J. (2014) *Spatial Analysis: A Guide for Ecologists*. Cambridge University Press, New York.

DiLeo, M.F., Rouse, J.D., Dávila, J.A., & Lougheed, S.C. (2013) The influence of landscape on gene flow in the eastern Massasauga rattlesnake (*Sistrurus c. catenatus*):

insight from computer simulations. *Molecular Ecology* **22**, 4483–98.

Dungan, J.L., Perry, J.N., Dale, M.R.T., Legendre, P., Citron-Pousty, S., Fortin, M.-J., Jakomulska, A., Miriti, M., & Rosenberg, M.S. (2002) A balanced view of scale in spatial statistical analysis. *Ecography* **25**, 626–40.

Epperson, B.K., McRae, B.H., Scribner, K., Cushman, S.A., Rosenberg, M.S., Fortin, M.-J., James, P.M.A., Murphy, M., Manel, S., Legendre, P., & Dale, M.R.T. (2010) Utility of computer simulations in landscape genetics. *Molecular Ecology* **19**, 3549–64.

Evanno, G., Regnaut, S., & Goudet, J. (2005) Detecting the number of clusters of individuals using the software STRUCTURE: a simulation study. *Molecular Ecology* **14**, 2611–20.

Fahrig, L. (2007) Non-optimal animal movement in human-altered landscapes. *Functional Ecology* **21**, 1003–15.

Fogelqvist, J., Niittyvuopio, A., Ågren, J., Savolainen, O., & Lascoux, M. (2010) Cryptic population genetic structure: the number of inferred clusters depends on sample size. *Molecular Ecology Resources* **10**, 314–23.

Fortin, M.-J., Drapeau, P., & Legendre, P. (1989) Spatial autocorrelation and sampling design. *Vegetation* **83**, 209–22.

François, O. & Durand, E. (2010) Spatially explicit Bayesian clustering models in population genetics. *Molecular Ecology Resources* **10**, 773–84.

Frantz, A.C., Cellina, S., Krier, A., Schley, L., & Burke, T. (2009) Using spatial Bayesian methods to determine the genetic structure of a continuously distributed population: clusters or isolation-by-distance? *Journal of Applied Ecology* **46**, 493–505.

Galpern, P. & Manseau, M. (2013) Modelling the influence of landscape connectivity on animal distribution: a functional grain approach. *Ecography* **36**, 1001–16.

Goldberg, C.S. & Waits, L.P. (2010) Comparative landscape genetics of two pond-breeding amphibian species in a highly modified agricultural landscape. *Molecular Ecology* **19**, 3650–63.

Graves, T.A., Wasserman, T.N., Ribeiro, M., Landguth, E.L., Spear, S.F., Balkenhol, N., Higgins, C.B., Fortin, M.-J., Cushman, S.A., & Waits, L.P. (2012) The influence of landscape characteristics and home-range size on the quantification of landscape–genetics relationships. *Landscape Ecology* **27**, 253–66.

Green, R.H. (1979) *Sampling Design and Statistical Methods for Environmental Biologists*. John Wiley & Sons, Inc., New York.

Guillot, G. & Santos, F. (2009) A computer program to simulate multilocus genotype data with spatially auto-correlated allele frequencies. *Molecular Ecology Resources* **9**, 1112–20.

Haining, R.P. (2003) *Spatial Data Analysis: Theory and Practice*. Cambridge University Press, Cambridge.

Hale, M. L., Burg, T. M., & Steeves, T. E. (2012) Sampling for microsatellite-based population genetic studies: 25 to 30 individuals per population is enough to accurately estimate allele frequencies. *PloS One* **7**, e45170.

Hall, L.A., & Beissinger, S.R. (2014) A practical toolbox for design and analysis of landscape genetics studies. *Landscape Ecology* **29**, 1487–504.

Hargrove, W. & Pickering, J. (1992) Pseudoreplication – a sine-qua-non for regional ecology. *Landscape Ecology* **6**, 251–8.

Hoban, S. (2014) An overview of the utility of population simulation software in molecular ecology. *Molecular Ecology* **23**, 2383–401.

Hoban, S., Gaggiotti, O., Congress Consortium, & Bertorelle, G. (2013) Sample Planning Optimization Tool for conservation and population Genetics (SPOTG): a software for choosing the appropriate number of markers and samples. *Methods in Ecology and Evolution* **4**, 299–303.

Hulbert, S.H. (1984) Pseudoreplication and design if ecological field experiments. *Ecological Monographs* **54**, 187–211.

Hunsaker, C., Goodchild, M., Friedl, M., & Case, T. (eds.) (2001) *Spatial Uncertainty in Ecology. Implications for Remote Sensing and GIS Applications*. Springer-Verlag, New York.

Jaquiéry, J., Broquet, T., Hirzel, A.H., Yearsley, J., & Perrin, N. (2011) Inferring landscape effects on dispersal from genetic distances: how far can we go? *Molecular Ecology* **20**, 692–705.

Jenkins, D.G., Carey, M., Czerniewska, J., Fletcher, J., Hether, T., Jones, A., Knight, S., Knox, J., Long, T., Mannino, M., McGuire, M., Riffle, A., Segelsky, S., Shappell, L., Sterner, A., Strickler, T., & Tursi, R. (2010) A meta-analysis of isolation-by-distance: relic or reference standard for landscape genetics? *Ecography* **33**, 315–20.

Kalinowski, S.T. (2005) Do polymorphic loci require large sample sizes to estimate genetic distance? *Heredity* **94**, 33–6.

Kalinowski, S.T. (2011) The computer program structure does not reliably identify the main genetic clusters within species: simulations and implications for human population structure. *Heredity* **106**, 625–32.

Keller, D., Holderegger, R., van Strien, M. J., & Bolliger, J. (2015) How to make landscape genetics beneficial for conservation management? *Conservation Genetics;* **16**, 503–12.

Koen, E.L., Bowman, J., Garroway, C.J., & Wilson, P.J. (2013) The sensitivity of genetic connectivity measures to unsamples and under-sampled sites. *PLoS One* **8**, e56204.

Landguth, E.L. & Balkenhol, N. (2012) Relative sensitivity of neutral versus adaptive genetic data for assessing population differentiation. *Conservation Genetics* **13**, 1421–6.

Landguth, E.L. & Schwartz, M.K. (2014) Evaluating sample allocation and effort in detecting population differentiation for discrete and continuously distributed individuals. *Conservation Genetics* **15**, 981–92.

Landguth, E.L., Cushman, S.A., Schwartz, M.K., McKelvey, K.S., Murphy, M., & Luikart, G. (2010) Quantifying the lag time to detect barriers in landscape genetics. *Molecular Ecology.* **19**, 4179–91.

Landguth, E.L., Fedy, B., Garey, A., Mumma, M., Emel, S., Oyler-McCance, S., Cushman, S.A., Wagner, H.H., & Fortin,

M.-J. (2012) Effects of sample size, number of markers and allelic richness on the detection of spatial genetic pattern. *Molecular Ecology Resources* **12**, 276–84.

Langford, W.T., Gergel, S.E., Dietterich, T.G., & Cohen, W. (2006) Map misclassification can cause large errors in landscape pattern indices: examples from habitat fragmentation. *Ecosystems* **9**, 474–88.

Latch, E. K. & Rhodes, O. E. (2006) Evidence for bias in estimates of local genetic structure due to sampling scheme. *Animal Conservation* **9**, 308–15.

Latch, E.K., Dharmarajan, G., Glaubitz, J.C., & Rhodes, O.E. Jr. (2006) Relative performance of Bayesian clustering software for inferring population substructure and individual assignment at low levels of population differentiation. *Conservation Genetics* **7**, 295–302.

Lechner, A.M., Langford, W.T., Bekessy, S.A., & Jones, S.D. (2012) Are landscape ecologists addressing uncertainty in their remote sensing data? *Landscape Ecology* **27**, 1249–61.

Legendre, P., Dale, M.R.T., Fortin, M.-J., Gurevitch, J., Hohn, M., & Myers, D. (2002) The consequences of spatial structure for the design and analysis of ecological field surveys. *Ecography* **25**, 601–16.

Linke, J. & McDermid, G.J. (2012) Monitoring landscape change in multi-use west-central Alberta, Canada using the disturbance-inventory framework. *Remote Sensing of Environment* **125**, 112–24.

Luximon, N., Petit, E. J., & Broquet, T. (2014) Performance of individual vs. group sampling for inferring dispersal under isolation-by-distance. *Molecular Ecology Resources* **14**, 745–52.

Manel, S., Joost, S., Epperson, B.K., Holderegger, R., Storfer, A., Rosenberg, M.S., Scribner, K.T., Bonin, A., & Fortin, M.-J. (2010) Perspectives on the use of landscape genetics to detect genetic adaptive variation in the field. *Molecular Ecology* **19**, 3760–72.

Manel, S., Albert, C., & Yoccoz, N.G. (2012) Sampling in landscape genomics. In: Bonin, A. & Pompanon, F. (eds.), *Data Production and Analysis in Population Genomics.* Humana Press, New York.

Melles, S., Fortin, M.-J., Badzinski, D., & Lindsay, K. (2012) Relative importance of nesting habitat and measures of connectivity in predicting the occurrence of a forest songbird in fragmented landscapes. *Avian Conservation and Ecology* **7**, 3. [online] URL: http://www.ace-eco.org/vol7/iss2/art3/

Millette, K. L. & Keyghobadi, N. (2015) The relative influence of habitat amount and configuration on genetic structure across multiple spatial scales. *Ecology and Evolution* **5**, 73–86.

Muirhead, J.R., Gray, D.K., Kelly, D.W., Ellis, S.M., Heath, D.D., & MacIsaac, H.J. (2008) Identifying the source of species invasions: sampling intensity vs. genetic diversity. *Molecular Ecology* **17**, 1020–35.

Naujokaitis-Lewis, I.R., Rico, Y., Lovell, J.L., Fortin, M.-J., & Murphy, M. (2013) Implications of incomplete networks on estimation of landscape genetic connectivity. *Conservation Genetics* **14**, 287–98.

Oyler-McCance, S.J., Fedy, B.C., & Landguth, E.L. (2013) Sample design effects in landscape genetics. *Conservation Genetics* **14**, 275–85.

Pflüger F. & Balkenhol N. (2014) A plea for simultaneously considering matrix quality and local environmental conditions when analyzing landscape impacts on effective dispersal. *Molecular Ecology* **23**, 2146–56.

Plante, M., Lowell, L., Potvin, F., Boots, B., & Fortin, M.-J. (2004) Studying deer habitat on Anticosti Island, Québec: relating animal occurrences and forest map information. *Ecological Modeling* **174**, 387–99.

Prunier, J.G., Kaufmann, B., Fenet, S., Picard, D., Pompanon, F., Joly, P., & Lena, J.P. (2013) Optimizing the trade-off between spatial and genetic sampling efforts in patchy populations: towards a better assessment of functional connectivity using an individual-based sampling scheme. *Molecular Ecology* **22**, 5516–30.

Reding, D.M., Cushman, S.A., Gosselink, T.E., & Clark, W.R. (2013) Linking movement behavior and fine-scale genetic structure to model landscape connectivity for bobcats (*Lynx rufus*). *Landscape Ecology* **28**, 471–86.

Rico, Y., Boehmer, H.J., & Wagner, H.H. (2012) Determinants of actual functional connectivity for calcareous grassland communities linked by rotational sheep grazing. *Landscape Ecology* **27**, 199–209.

Ryman, N. & Palm, S. (2006) POWSIM: a computer program for assessing statistical power when testing for genetic differentiation. *Molecular Ecology Notes* **6**, 600–2.

Schwartz, M.K. & McKelvey, K.S. (2009) Why sampling scheme matters: the effect of sampling scheme on landscape genetic results. *Conservation Genetics* **10**, 441–52.

Short Bull, R.A., Cushman, S.A., Mace, R., Chilton, T., Kendall, K.C., Landguth, E.L., Schwartz, M.K., McKelvey, K., Allendorf, F.W., & Luikart, G. (2011) Why replication is important in landscape genetics: American black bear in the Rocky Mountains. *Molecular Ecology* **20**, 1092–107.

Spear, S.F., Balkenhol, N., Fortin, M.J., McRae, B.H., & Scribner, K.I.M. (2010) Use of resistance surfaces for landscape genetic studies: considerations for parameterization and analysis. *Molecular Ecology* **19**, 3576–91.

Wagner, H. H. & Fortin, M.-J. (2013) A conceptual framework for the spatial analysis of landscape genetic data. *Conservation Genetics* **14**, 253–61.

Webster, R. & Oliver, M.A. (2007) *Geostatistics for Environmental Scientists.* John Wiley & Sons, Inc., New York.

# Chapter 5

# BASICS OF SPATIAL DATA ANALYSIS: LINKING LANDSCAPE AND GENETIC DATA FOR LANDSCAPE GENETIC STUDIES

*Helene H. Wagner and Marie-Josée Fortin*

*Department of Ecology and Evolutionary Biology, University of Toronto, Canada*

## 5.1 INTRODUCTION

The research questions and data analysis methods in landscape genetics stem from a broad range of fields, including population genetics, numerical ecology, metapopulation, landscape ecology, and spatial statistics. Yet landscape genetics has unique research questions (see Chapters 1 to 4 in this book) that can only be addressed through a combination of statistical methods originating from different disciplines. A better integration and unification of these statistical methods may thus be crucial for advancing landscape genetics (Balkenhol et al. 2009a, 2009b).

A core issue is how to explicitly account for the inherent spatial structure of the genetic and landscape data while analyzing their relationship. Indeed, geophysical processes and land-use practices create spatial structure in environmental factors to which the organisms respond, and biotic processes such as spatially restricted mating and dispersal create further spatial structure. Statistically speaking, these processes often create positive ***spatial autocorrelation*** in the genetic data, which means that, on average, nearby observations are more similar than distant ones. Negative spatial autocorrelation, on the other hand, occurs when similar observations are spaced more regularly than random, so that nearby observations are on average more dissimilar than more distant ones. It is helpful to distinguish between types of spatial autocorrelation based on the generating process. ***Induced spatial dependence*** results from the response of organisms to environmental gradients, whereas ***inherent spatial autocorrelation*** arises from spatial ecological or evolutionary processes such as mating, dispersal, and resulting gene flow. Unfortunately, spatial analysis methods cannot discriminate between induced spatial dependence and inherent

*Landscape Genetics: Concepts, Methods, Applications*, First Edition. Edited by Niko Balkenhol, Samuel A. Cushman, Andrew T. Storfer, and Lisette P. Waits.

spatial autocorrelation. This means that spatial auto-correlation in regression residuals may be due to biological processes or a missing landscape predictor. Therefore prior knowledge and hypotheses about the underlying processes responsible for the spatial pattern should be used to design an effective sampling design (see Chapter 4) and to apply the appropriate levels of analysis, which in turn will allow researchers to differentiate between the origins of the spatial patterns.

The presence of spatial autocorrelation in sampled data can be assessed using spatial statistics (see Box 5.1). Like any parametric statistics, spatial statistics are based on assumptions. The main assumption of spatial statistics is that the underlying process that generated the spatial pattern is stationary (i.e., constant mean and variance over the study area) (Fortin and Dale 2005). Non-stationarity can be due to several aspects: (i) large-scale spatial trend in the data such that the mean changes approximately linearly throughout the studied area; (ii) there are spatial changes in mean, variance, or both according to location in the studied area; and (iii) there is directionality (e.g., due to wind direction) in the spatial pattern of the data. When this stationarity assumption is valid, however, spatial statistics (e.g., Moran's $I$, Geary's $c$, semivariance) that measure the average degree of spatial autocorrelation at different spatial lags for the entire study area can be used. These spatial statistics estimate the degree of spatial autocorrelation of a single variable (univariate methods) for the entire study area (i.e., magnitude of spatial autocorrelation, spatial range of the pattern). However, the multivariate nature of genetic data and the many factors that contribute to their spatial structure require multivariate spatial analysis methods (Dray et al. 2012).

**Moran's $I$** is the most widely used statistic to estimate spatial autocorrelation. Moran's $I$ mostly varies between −1 and +1 (deviations may occur for small samples; de Jong et al. 1984), with an expected value of −1/($n$−1) for a sample of $n$ spatially independent observations (Moran 1950). Positive values of Moran's $I$ indicate positive spatial autocorrelation, where nearby observations are more similar on average than distant ones, whereas negative values indicate negative spatial autocorrelation, where nearby samples are more dissimilar than distant ones. A Moran's $I$ correlogram for a single variable **y** is constructed by calculating Moran's $I(d)$ for each spatial lag $d$, with weights $w_{ij(d)} = 1$ if the pair of sites $i$ and $j$ fall into spatial lag $d$ and $w_{ij(d)} = 0$ otherwise. Note that a global Moran's $I$ index refers to the value of a Moran's $I(d)$ correlogram for the first lag, $d = 1$.

A spatial pattern is typically quantified in one of two ways that differ in the interpretation of spatial lags: (1) as a function of the *distance lag* between pairs of observations or (2) through a *neighbor matrix* and weights associated with these neighbors (see Box 5.1). The distance lag approach is compatible with an isolation-by-distance model of gene flow, where rates of gene flow depend largely on the total distance between sampling locations. The neighbor matrix approach is compatible with a stepping stone model, where organisms are expected to disperse to neighboring patches only, and gene flow over larger distances is the result of such stepwise dispersal (migration) events over multiple generations. In each case, we need to be clear about the null and alternative hypothesis we aim to test. The conceptual difference between these two approaches is best illustrated with the calculation and interpretation of a Moran's $I$ correlogram (Epperson 2003; Guillot et al. 2009; see Box 5.1).

When the stationarity assumption of spatial statistics is not valid, spatial autocorrelation can be measured locally at each sampling location with "local spatial statistics" (e.g., local Moran, local Getis; Anselin 1995; Sokal et al. 1998), using only neighboring samples (i.e., first neighbors, $d = 1$). As landscape genetics studies are often designed over large regions, the likelihood that several ecological and environmental factors are acting on the genetic spatial structure is high, which makes the assumption of stationarity unlikely. With such genetic data, local spatial statistics can be used to detect local spatial heterogeneity (Sokal et al. 1998).

Quantification of the spatial structure in the respective data sets (landscape and genetic), on the one hand, provides key information on the potential generating processes and the scales at which they affect the genetic structure of the data (e.g., isolation-by-distance (IBD), Wright 1943; isolation-by-resistance (IBR), McRae 2006; or isolation-by-barrier (IBB), Vignieri 2005). On the other hand, not accounting for the presence of spatial autocorrelation in the data may invalidate inferential statistical tests of the relationship between genetic and landscape data, as these statistical tests assume that the data are independent (see Box 5.1).

Most ecological and genetic data show inherent structure in space (e.g., nearby samples usually have similar values; see Box 5.1), time (e.g., population fluctuations), or phylogeny (e.g., species relatedness) (Fortin & Dale 2005; Peres-Neto 2006). In this chapter we focus solely on spatial structure, acknowledging that the other types of dependency occur. Also, the power of landscape

## Box 5.1  Spatial statistics

### Distance lag approach

A correlogram based on distance lags is constructed by dividing all unique pairs of observations into spatial distance classes (i.e., spatial lags). The null hypothesis is the absence of spatial autocorrelation. The alternative hypothesis is that under isolation-by-distance, we expect increased rates of gene flow among nearby sampling locations, resulting in positive autocorrelation for the first spatial lags. Technically, we can assign each pair of observations at locations $i$ and $j$ a weight $w_{ij(d)} = 1$ if it falls into distance class or lag $d$ and $w_{ij(d)} = 0$ otherwise. For each lag, the autocorrelation is estimated by dividing the (weighted) mean covariance for all pairs within the distance class by the mean covariance among all pairs in the data set. The spatial correlogram is a plot of these autocorrelation estimates against lag distance. In this case, a progressive, one-sided test is appropriate, where we test the first lag and only progress to further lags if all previous lags showed significant positive spatial autocorrelation (Legendre & Legendre 2012). In contrast, testing each lag individually may provide a single lag with positive or negative spatial autocorrelation at a larger distance, which is difficult to interpret in biological terms. Similarly, negative autocorrelation, where nearby observations are genetically more different than distant ones, will rarely be expected. The shape of the correlogram may be used to select and fit an appropriate function for modeling the correlation among regression errors in generalized least squares (GLS) regression or generalized linear mixed models (GLMMs).

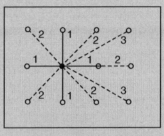

Distance lag approach. Each line indicates a link between the focal site $i$ (filled circle) and a nearby site $j$. Numbers indicate distance classes (lags) and depend on lag definition. Solid lines indicate links in the first lag.

### Neighbor matrix approach

Alternatively, we can start by constructing an $n \times n$ neighbor matrix, where "1" indicates that the row and column observations are neighbors and "0" that they are not neighbors. Neighbors can be defined using a distance threshold or based on a specific graph model (see Chapter 9; Dale & Fortin 2010; Spear et al. 2010). The first spatial lag is defined by first neighbors (those indicated by "1" in the neighbor matrix). The second lag is defined by second neighbors (i.e., locations that could be reached in two generations), etc. If using binary weights, each pair of observations receives the same weight and we proceed as above. However, other weights may be used: we may want to adjust for the number of neighbors $j$ of each observation $i$, so that the weights $w_{ij}$ of all neighbors of $i$ sum to one (such row-standardized weights should be used for spatial regression). For an irregular spatial sampling design, we may want to account for the physical distance between neighbors, so that close neighbors receive more weight than distant ones. In a stepping stone model, we are typically interested in the autocorrelation among first neighbors only (i.e., global Moran's $I$ index calculated for the first spatial lag, $d = 1$), as rates of gene flow among neighbors that are further remote follow from the connectivity among first neighbors. The exact estimate and $p$-value of global Moran's $I$ index will vary with different definitions of first neighbors and weights. Note that when modeling spatial dependence, e.g., using spatial regression methods, the ideal number of neighbors is $4 - 6$, whereas higher numbers of neighbors are inefficient (Florax & Ray 1995; Griffith 1996). The dependence between a sampling location and its second, third, etc., neighbors is indirectly modeled through the dependence on their common neighbours.

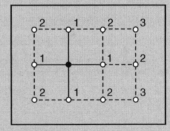

Neighbor matrix approach. Each line indicates a connection between nearest neighbors, thus depending on the definition of neighbors. Numbers indicate the lag, defined by the minimum number of steps between the focal site $i$ (filled circle) and a nearby site $j$. Solid lines indicate connections between site $i$ and its nearest neighbors.

genetics analyses to detect significant spatial relationships between genetic and landscape data is directly linked to the sampling design (Muirhead et al. 2008; Chapter 4), the spatial analysis methods (Fortin & Dale 2005; Epperson 2003), and the genetic markers used (Ryman et al. 2006). Here we focus only on the statistical aspects; discussion about the importance of the sampling

design and genetic markers can be found elsewhere (e.g., Selkoe & Toonen 2006; Chapters 2 and 4).

While we may gain new insights through characterizing spatial structure in the data with spatial statistics, the presence of spatial autocorrelation may invalidate statistical results (Legendre 1993). The example in Box 5.2 illustrates that spatial autocorrelation in either the

---

### Box 5.2  Spatial autocorrelation and regression inference

Our example for illustrating the effects of spatial autocorrelation on regression inference uses a linear simple regression with a single response variable **y**, a single predictor **x**, and an error **u**, simulated on a $20 \times 20$ grid ($n = 400$) under four scenarios. The scenarios differ in the spatial structure of predictor **x** and error **u** and consequently of the response **y** simulated as $\mathbf{y} = 0.5\mathbf{x} + 0.5\mathbf{u}$. In all simulations, the true slope is $b = 0.5$. Under each scenario, we calculated Moran's $I$ of the residuals of a regression of **y** on **x** and estimated the slope ($b$) and its standard error ($SE$). We repeated the simulation and regression analysis 10,000 times under each scenario to estimate the true $SE$ of slope $b$ from the standard deviation of all replicate slope estimates. For each simulated data set we also regressed **u** on **x** to determine type I error rates. Values for Moran's $I$ of residuals, slope $b$, and its estimated $SE$ are means over 10,000 replicates.

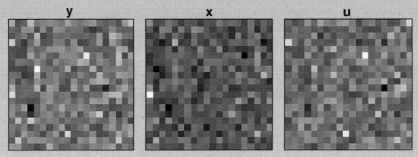

Scenario 1: **x** random, **u** random. The landscape predictor and the response are spatially independent and linear regression results are correct: Moran's $I$ of residuals $= -0.002$, slope estimate ($b \pm SE$) $= 0.500 \pm 0.025$, true $SE$ of slope: 0.025, type I error rate: 0.050.

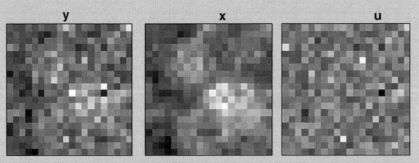

Scenario 2: **x** autocorrelated, **u** random. The landscape predictor is spatially structured, but the response is spatially independent. Linear regression results are correct: Moran's $I$ of residuals $= -0.003$, slope estimate ($b \pm SE$) $= 0.500 \pm 0.025$, true SE of slope: 0.025, type I error rate: 0.048.

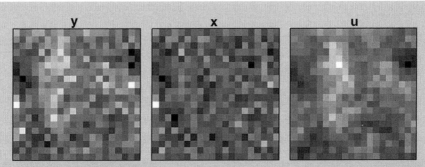

Scenario 3: **x** random, **u** autocorrelated. The landscape predictor is spatially independent, but the response is spatially structured, which may be due to a biotic process such as dispersal or a missing landscape factor that is spatially structured. The linear regression results are correct, although the residuals are spatially autocorrelated: Moran's $I$ of residuals = 0.640, slope estimate $(b \pm SE)$ = 0.501 ± 0.025, true SE of slope: 0.025, type I error rate: 0.054.

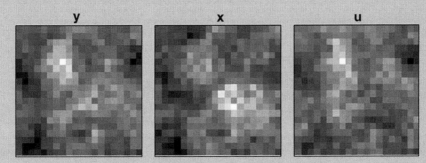

Scenario 4: **x** autocorrelated, **u** autocorrelated. The landscape predictor is spatially structured and the response to it shows further spatial dependence. Linear regression results are incorrect: Moran's $I$ of residuals = 0.629, slope estimate $(b \pm SE)$ = 0.501 ± 0.025, true $SE$ of slope: 0.071, type I error rate: 0.504.

predictor or the error of the response (i.e., residuals) does not necessarily invalidate regression results. If both the landscape predictor and the error are spatially autocorrelated, however, the effect on regression results can be severe, biasing the estimation of the parameters of the regression and their significance. Indeed, type I error rates may be considerably inflated (in the example ranging from 0.05 to 0.5), which means that on average one in two significant slope coefficients may be spurious results. In addition, the standard error of the slope coefficient $b$ may be underestimated (in the example in Box 5.2 by a factor of $0.071/0.025 = 2.84$). This means that confidence intervals for the slope constructed with the estimated standard error would be almost three times too narrow on average.

Thus, both statistical hypothesis testing and parameter estimation in simple linear regression are invalid in this situation (Bini et al. 2009).

This chapter discusses the main issues of relating genetic variation to landscape predictors (Foll & Gaggiotti 2006; Bradburd et al. 2013) in the familiar context of regression analysis or, more generally, in the framework of the linear model. We start with considerations for choosing an appropriate model depending on assumptions about the main underlying evolutionary process, i.e., selection or gene flow (Fig. 5.1). We then present different approaches through which space can be incorporated into the linear model (Fig. 5.2). Finally, we focus on two common goals of spatial analysis in landscape genetics: first, how to test for

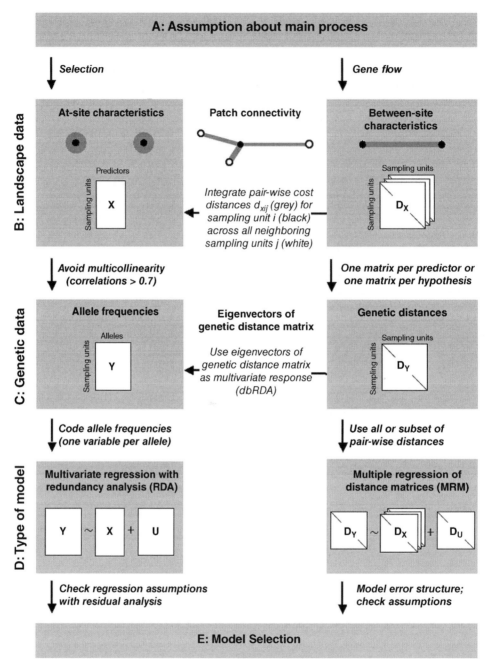

**Fig. 5.1** Flowchart of the statistical model that can be used to relate genetic to landscape data depending on whether one assumes selection or gene flow to be the main underlying evolutionary process. In either case five steps are needed. (A) Determining implicitly or explicitly the main assumptions of the processes. (B) Determining how the landscape data will be analyzed. (C) Determining how the genetic data will be analyzed. (D) Selecting the appropriate regression framework. (E) Selecting the appropriate model.

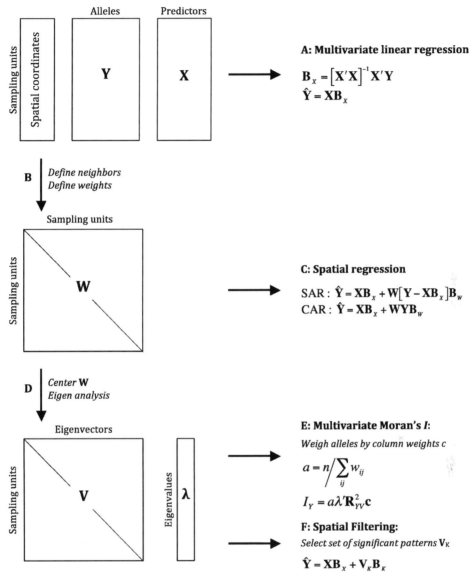

**Fig. 5.2** Flowchart illustrating the steps involved when incorporating space into the multiple linear regression model. (A) Non-spatial regression predicts response **Y** from predictors **X** without reference to spatial coordinate information. (B) Many methods for incorporating space into the regression model start with creating a spatial weight matrix **W**. (C) Spatial regression methods use **W** directly to add a spatial neighborhood term for the response **Y** (conditional autoregressive model, CAR) or the error **U** (simultaneous autoregressive model, SAR). (D) Alternatively, the spatial weight matrix **W** can be centered and subjected to eigenanalysis to extract matrix **V** of spatial eigenvectors. (E) The eigenvalues $\lambda$ corresponding to the spatial eigenvectors **V** can be used to calculate univariate or multivariate Moran's $I$ as a measure of the degree of spatial autocorrelation in the data. (F) Spatial filtering methods add a set $\mathbf{V}_K$ of spatial eigenvectors that are significantly associated with the response **Y** to the regression model to control for spatial autocorrelation.

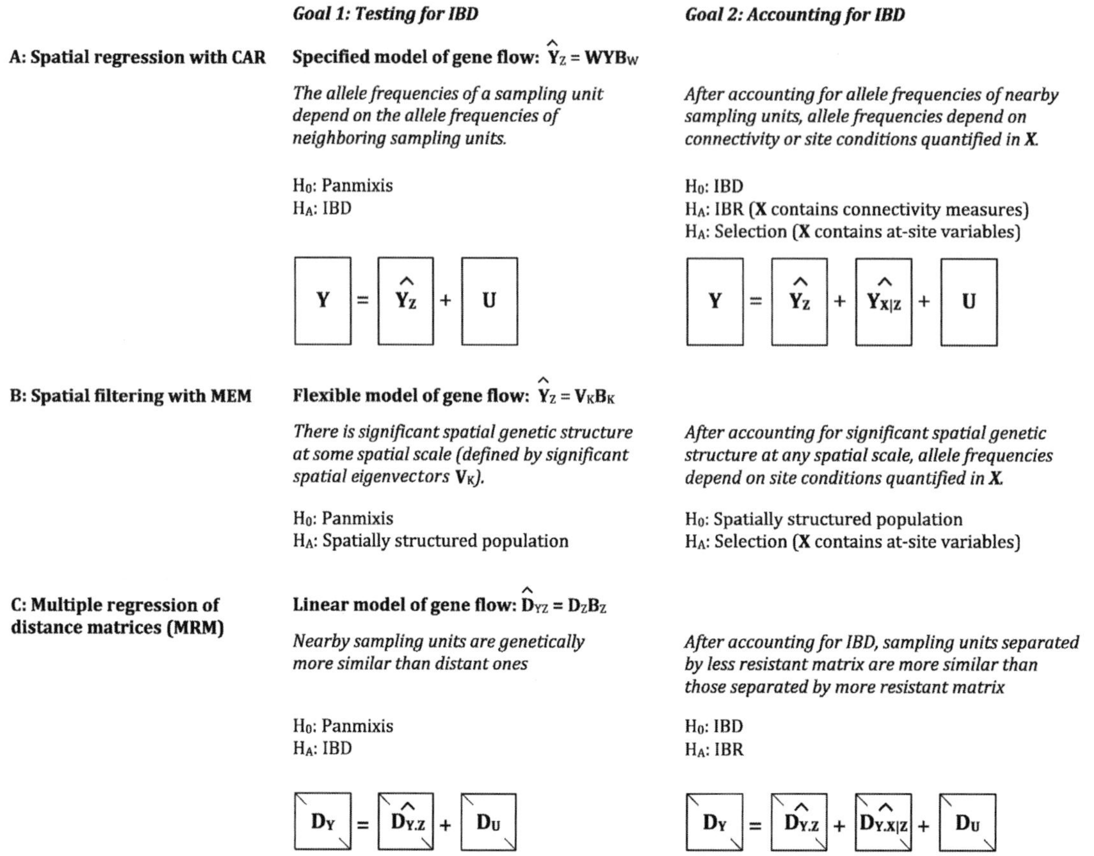

|  | Goal 1: Testing for IBD | Goal 2: Accounting for IBD |
| --- | --- | --- |

**A: Spatial regression with CAR**

**Specified model of gene flow:** $\hat{\mathbf{Y}}_Z = \mathbf{WYB}_W$

*The allele frequencies of a sampling unit depend on the allele frequencies of neighboring sampling units.*

$H_0$: Panmixis
$H_A$: IBD

$$\boxed{\mathbf{Y}} = \boxed{\hat{\mathbf{Y}}_Z} + \boxed{\mathbf{U}}$$

*After accounting for allele frequencies of nearby sampling units, allele frequencies depend on connectivity or site conditions quantified in $\mathbf{X}$.*

$H_0$: IBD
$H_A$: IBR ($\mathbf{X}$ contains connectivity measures)
$H_A$: Selection ($\mathbf{X}$ contains at-site variables)

$$\boxed{\mathbf{Y}} = \boxed{\hat{\mathbf{Y}}_Z} + \boxed{\hat{\mathbf{Y}}_{X|Z}} + \boxed{\mathbf{U}}$$

**B: Spatial filtering with MEM**

**Flexible model of gene flow:** $\hat{\mathbf{Y}}_Z = \mathbf{V}_K\mathbf{B}_K$

*There is significant spatial genetic structure at some spatial scale (defined by significant spatial eigenvectors $\mathbf{V}_K$).*

$H_0$: Panmixis
$H_A$: Spatially structured population

*After accounting for significant spatial genetic structure at any spatial scale, allele frequencies depend on site conditions quantified in $\mathbf{X}$.*

$H_0$: Spatially structured population
$H_A$: Selection ($\mathbf{X}$ contains at-site variables)

**C: Multiple regression of distance matrices (MRM)**

**Linear model of gene flow:** $\hat{\mathbf{D}}_{YZ} = \mathbf{D}_Z\mathbf{B}_Z$

*Nearby sampling units are genetically more similar than distant ones*

$H_0$: Panmixis
$H_A$: IBD

$$\boxed{\mathbf{D}_Y} = \boxed{\hat{\mathbf{D}}_{Y.Z}} + \boxed{\mathbf{D}_U}$$

*After accounting for IBD, sampling units separated by less resistant matrix are more similar than those separated by more resistant matrix*

$H_0$: IBD
$H_A$: IBR

$$\boxed{\mathbf{D}_Y} = \boxed{\hat{\mathbf{D}}_{Y.Z}} + \boxed{\hat{\mathbf{D}}_{Y.X|Z}} + \boxed{\mathbf{D}_U}$$

**Fig. 5.3** Summary of how the null and alternative hypotheses change when the goal is either to test for IBD (left column) or to account for IBD (right column) when testing other landscape predictors, depending on the statistical approach used. (A) Spatial regression using the conditional autoregressive (CAR) model. (B) Spatial filtering using spatial eigenvectors (MEM). (C) Multiple regression of distance matrices (MRM).

the presence of significant IBD and then how to account for IBD by incorporating it into the null model when testing for other landscape effects (Fig. 5.3). We hope that this presentation will aid the current efforts of landscape geneticists to develop new and more integrated methods for linking genetic and landscape data to address the complexity of landscape genetic research questions (Wagner & Fortin 2013).

## 5.2  HOW TO MODEL LANDSCAPE EFFECTS ON GENETIC VARIATION

To describe the relationship between landscape predictors and genetic data, we need an appropriate statistical model. The type of model depends on the data types, which again will depend on how we think about the underlying processes (Fig. 5.1A).

### 5.2.1  Type of landscape data

Organism behavior may depend on (i) the local site conditions at the sampling locations (*at-site characteristics*, i.e., conditions at the grid cell where an individual was sampled or the patch where a discrete local population was sampled), (ii) the local neighborhood (e.g., proximity of a road or resource availability within a distance threshold), or (iii) the intervening landscape matrix between locations with suitable habitat (*between-site characteristics*, e.g., the presence of barriers or the mortality and energetic cost of movement associated with different cover types).

If selection is the main evolutionary process that structures the population of interest, we would expect that genetic variation depends mostly on at-site characteristics, although between-site characteristics may affect the rate of spread of adaptive genetic variation across the landscape. If, however, gene flow is the dominant process, genetic variation would depend to a large degree on between-site characteristics that are likely to affect rates of gene flow, although the at-site characteristics may also affect the probability of individuals leaving a patch, finding a patch, and settling in it (Fig. 5.1B).

In the case of selection, we will represent a set of $p$ landscape predictors observed for $n$ sampling units (representing $n$ individuals in a continuous population or $n$ demes as spatially discrete populations) in a predictor matrix $\mathbf{X}$ with $n$ rows and $p$ columns (*node-level analysis*, Box 5.3; Wagner & Fortin 2013). In the case of gene flow, the values of the landscape predictors refer to the pairwise distances between sampling units and are best represented as a set of distance matrices $\mathbf{D_X}$, each with $n$ rows and $n$ columns (*link-level analysis*, Box 5.3). The terminology is borrowed from graph theory (see Chapter 10), where one would describe a set of habitat patches as *nodes* connected by links along which organisms may move across the intervening matrix (Wagner & Fortin 2013).

For link-level analysis, either each predictor is represented by its own distance matrix (e.g., one matrix for roads, one for forest), leading to a set of $p$ matrices $\mathbf{D_{X1}}$ to $\mathbf{D_{Xp}}$, or each matrix represents a complex hypothesis of landscape resistance that assigns a specific set of resistance values to all landscape features (e.g., high resistance to roads and low resistance to forest). The main difference in terms of statistical analysis is that, if the set of $p$ matrices $\mathbf{D_X}$ represents $p$ landscape predictors, more than one may be required to explain the genetic data (e.g., roads and forest). However, if $\mathbf{D_X}$ represents $p$ competing hypotheses of landscape resistance, any one hypothesis would exclude all the others; hence we would not want to use more than one matrix $\mathbf{D_X}$ as the predictor in the same regression model.

If we are going to use multiple predictors in a multiple regression model, we need to avoid *multicollinearity*. In the strict sense, perfect multicollinearity refers to the case where one predictor variable is a linear combination of some other predictors. An example would be

coding a categorical factor with four levels A to D, corresponding to different cover types, into $q = 4$ dummy variables $\mathbf{X_A}$ to $\mathbf{X_D}$, one for each factor level. A value of $\mathbf{X_D} = 1$ thus indicates that the sampling unit was classified as cover type D and $\mathbf{X_D} = 0$ that it was not classified as type D. However, if we know variables $\mathbf{X_A}$ to $\mathbf{X_C}$, we know the value of $\mathbf{X_D}$ because the four dummy variables must sum to one for each sampling unit; hence $\mathbf{X_D}$ is a linear combination of $\mathbf{X_A}$ to $\mathbf{X_C}$. This problem can be avoided by representing $q$ factor levels by $q - 1$ dummy variables, thus omitting one level.

In the broader sense, the term (multi-)collinearity refers to a high degree of linear correlation among the $p$ predictors. This can be screened by checking a matrix of pairwise correlations among predictors, where, as a rule of thumb, correlations above 0.7 are regarded as problematic (Dormann et al. 2013). If there are two or more highly correlated predictors (e.g., $\mathbf{X_1}$, $\mathbf{X_2}$, and $\mathbf{X_3}$), they should not be used in the same regression model. This can be avoided, either by retaining only one of the intercorrelated predictors (e.g., $\mathbf{X_3}$) or with latent variable methods, where one or more eigenvectors (Box 5.4) are extracted from the set of intercorrelated predictors $\mathbf{X_1}$, $\mathbf{X_2}$, and $\mathbf{X_3}$, thus capturing their joint variation, and the eigenvectors are then used as predictors (Dormann et al. 2013).

*Neighborhood-level analysis* (Box 5.2; Wagner & Fortin 2013) presents alternative ways of analyzing between-site characteristics, which allows them to be combined with at-site characteristics in the same statistical model (Balkenhol et al. 2009a; James et al. 2011). Patch connectivity indices developed in metapopulation ecology can be used to transform pairwise distances of landscape predictors into node-level measures of potential functional connectivity. In this approach, instead of focusing on individual links, connectivity indices are computed that integrate across all links connecting the focal patch with any of its neighbors (Fig. 5.1B). In metapopulation ecology, patch connectivity $S_i$ (representing the unknown number of migrants from all neighboring source patches $j$ into focal patch $i$) is commonly modeled with an incidence function model (Hanski 1994; Moilanen & Nieminen 2002):

$$S_i = \sum_j o_j \, \mathbf{A}_j^b \exp(-\alpha d_{ij})$$

where $o_j$ is a binary indicator whether the species (or allele) being modeled is present in source patch $j$, $A_j$ refers to the source patch area or another patch characteristic,

## Box 5.3 Analytical levels

The main approaches of landscape genetics studies can be classified into four analytical levels. The following illustrations are adapted from Wagner and Fortin (2013).

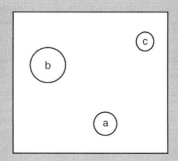

### 1 Node-level analysis

This relates adaptive variation to local landscape factors at sites *a*, *b*, and *c* while accounting for isolation-by-distance (Schoville et al. 2012). The node-level methods include multivariate ordination methods (e.g., RDA; Dray et al. 2012; Manel et al. 2012) and general linear models (Bolker 2008).

### 2 Link-level analysis

This relates neutral variation between sites *a*, *b*, and *c* to between-site landscape factors observed along links *ab*, *ac*, and *bc* to test hypotheses on isolation-by-distance (IBD), isolation-by-resistance (IBR), or isolation-by-barrier (IBB). The most commonly used link-level method is the Mantel test (Mantel 1967; Smouse et al. 1986; Cushman & Landguth 2010), which for multiple predictors extends to multiple regression on distance matrices (MRMs) (Smouse et al. 1986). Partial Mantel tests (Smouse et al. 1986) and causal modeling (Cushman et al. 2006) have been used to account for one process (e.g., IBD) while testing for another process (e.g., IBR). However, several studies have showed the relative lower power of the Mantel test to detect significant relationships and other inferential problems (Dutilleul et al. 2000: Legendre & Fortin 2010: Guillot & Rousset 2013).

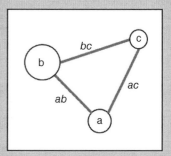

### 3 Neighborhood-level analysis

This relates the relative contribution of all neighboring sampled locations (here *b* and *c*) to the genetic variation observed at a given sampling location *a*. Connectivity measures (Keyghobadi et al. 2005; James et al. 2011) and gravity models (Murphy et al. 2010) can be used in neighborhood-level analyses to assess neighborhood effects on spatial genetic structure.

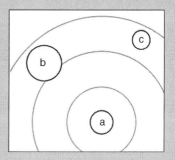

### 4 Boundary-level analysis

This relates genetic groups *a*, *b*, and *c* to landscape barriers. Once spatial groups are identified based on either Bayesian clustering algorithms or edge-detection techniques (see Chapter 7; Guillot et al. 2005; François & Durand 2010; Safner et al. 2011), the next step is to relate these genetic barriers to environmental and landscape barriers using spatial boundary overlap methods (Fortin et al. 1996) or POPS (Prediction of Population genetic Structure Program) (Jay 2011).

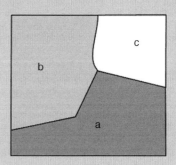

## Box 5.4  Eigenanalysis

Similar to ecological species composition data, genetic allele frequency data often have a large number $m$ of variables (one variable per allele) that were observed at the same $n$ sampling locations (individuals or demes). Analysis of such data is difficult because the signal of common structure in the data set is obscured by noise related to random variation in each variable. Also, the variables may be correlated among themselves so they should not be analyzed independently. Eigenanalysis methods such as *principal component analysis* (PCA) can help to reduce data, thus separating signal from noise, and to replace the original, intercorrelated variables with a set of synthetic variables (eigenvectors) that are orthogonal and uncorrelated among themselves.

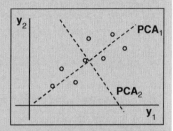

The scatterplot (top) shows a simple example of two correlated variables $y_1$ and $y_2$ (solid lines), such as the frequencies of two alleles, with values observed at eight sampling locations (circles). Eigenanalysis with PCA defines a set of new synthetic variables, known as PCA axes (dashed lines). The first PCA axis, called $PCA_1$, is defined to capture a maximum of the variation in the original variables $y_1$ and $y_2$. The next PCA axis, $PCA_2$, is chosen so that it is orthogonal to PCA1 and captures a maximum of the remaining variation in the data.

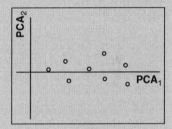

In this example with only two variables $y_1$ and $y_2$, two PCA axes are sufficient to fully capture the variation in the data. A scatterplot (bottom) of the values (or scores) for $PCA_1$ and $PCA_2$ recreates the same point cloud as the scatterplot (top) of the values of $y_1$ and $y_2$ for each sampling location, except for a rotational shift. Note that adding a third variable $y_3$ would result in a third axis $PCA_3$ orthogonal to both $PCA_1$ and $PCA_2$, and so forth, so that $m$ variables result in an $m$-dimensional PCA space.

More generally, a set of $m$ PCA axes $PCA_1$ to $PCA_m$ will capture all variation in a data set with $m$ variables $y_1$ to $y_m$, but contrary to the original variables $y_1$ to $y_m$, the new synthetic variables $PCA_1$ to $PCA_m$ are orthogonal and their pairwise correlation is exactly zero. The first axis, $PCA_1$, will have the highest variance as it contains the largest fraction of the variance in the original data, and each further PCA axis will have a lower variance than the previous ones. In fact, every eigenvector (i.e., PCA axis) has an associated eigenvalue $\lambda$ that is proportional to the variance in the original data that the eigenvector represents. Note that, if the original variables have been centered so that they each have a mean of zero, the last PCA axis, $PCA_m$, will have zero variance and an eigenvalue of $\lambda_m = 0$. Data reduction with PCA is based on the idea that the first few PCA axes contain the multivariate signal, i.e., the variance shared among the variables in the data set, whereas the remaining PCA axes contain largely noise.

PCA is the basic and most common method of eigenanalysis, also referred to as ordination methods. While PCA extracts eigenvectors of a variance–covariance matrix or a matrix of Euclidean distances between observations, principal coordinate analysis (PCoA) and non-metric multidimensional scaling (NMDS) will extract eigenvectors from any measure of resemblance and thus can be used with various measures of genetic distance (Legendre & Legendre 2012). Constrained ordination methods (also known as direct ordination methods) such as redundancy analysis (RDA) extract eigenvectors separately for the fitted values $\hat{Y}$ and for the residuals $U$ of a regression-type model, where the variation in a multivariate response $Y$ is explained by a set of predictors $X$ (Legendre & Legendre 2012). A special case is distance-based redundancy analysis (dbRDA) (Legendre & Legendre 2012), where, in a first step, a genetic distance matrix $D_Y$ is subjected to PCoA to extract a matrix of eigenvectors, which in a second step serves as the response matrix $Y$ in redundancy analysis RDA (Fig. 5.1C).

such as local population size or habitat quality, $d_{ij}$ is the distance between patches $i$ and $j$, and $b$ and $\alpha$ (with $\alpha > 0$) are scaling constants related to emigration and dispersal distance. Other connectivity indices can include some particular attributes of the patches (Moilanen & Nieminen 2002; Saura & Rubio 2010).

Such connectivity indices quantify how connected (as opposed to isolated) each patch is, based on an explicit model of connectivity. Connectivity indices have been used to model univariate genetic data such as genetic diversity (Keyghobadi et al. 2005; Rico et al. 2014) and multivariate allele frequencies (James et al. 2011). Gravity models (Murphy et al. 2010) present an alternative way of combining at-site and between-site variables, and are discussed in Chapter 10.

### 5.2.2 Type of genetic data

Once we have identified the best landscape data and analytical level for our analyses, we need to consider the different types of genetic data (Fig. 5.1C). The original data consist of a set of loci genotyped for all sampled individuals. In most cases, however, analysis will be based either on a table of allele frequencies or a matrix of genetic distances (see Chapter 3) derived from the genotype data.

For multivariate analysis, genotype data need to be coded in a table of allele frequencies for $n$ sampling units (rows) observed at $m$ alleles (columns), thus using one variable per allele (Smouse & Peakall 1999). Technically this will introduce multicollinearity (in the strict sense, see above) in $\mathbf{Y}$, which some analysis methods can handle, while for others, one allele per locus may need to be dropped.

If genotype data are converted to a matrix of pairwise genetic distances (i.e., dissimilarities), analysis may be based on all links or on a subset of meaningful links only. Such a subset may be based on different types of graph models (Chapter 10), which in essence are algorithms to define which sampling units are neighbors (Dale & Fortin 2010). Alternatively, Dyer et al. (2010) proposed testing each link to evaluate whether it shows statistically significant higher similarity (and thus lower genetic distance) than expected, given all other indirect paths connecting the two sampling units (conditional genetic distance). For the remainder of this chapter, however, we will assume that all links are retained.

A genetic distance matrix can be converted into a node-based framework by subjecting it to principal coordinate analysis (PcoA) (equivalent to non-linear multidimensional scaling; Legendre & Legendre 2012), which is similar to principal component analysis (PCA) (see Glossary and Box 5.4) but allows for non-Euclidean distance measures and thus can accommodate any measure of genetic distance. PCoA will return a table of $n$ rows and up to $m$ columns (eigenvectors of the genetic distance matrix), which in essence are a set of perfectly uncorrelated and orthogonal synthetic variables. This table of synthetic variables contains all the variation in the genetic data, which means that it can be used as response matrix $\mathbf{Y}$ in subsequent analysis (see distance-based redundancy analysis, dbRDA, in Box 5.4).

### 5.2.3 Type of statistical model

When modeling landscape effects on genetic variation, we typically assume a *directed relationship* (regression-type model), where the genetic data (response $\mathbf{Y}$) are constrained (i.e., determined) by the landscape predictors $\mathbf{X}$, with potentially high stochasticity, which we will represent by an error $\mathbf{U}$ (i.e., residuals; Fig. 5.1D). It is important to note that we are not talking about establishing a causal mechanism (which can only be done in a properly controlled experiment, which will rarely apply to landscape genetic data), but we have a clear hypothesis that $\mathbf{X}$ affects $\mathbf{Y}$. This is opposed to an *undirected relationship* (correlation-type model), where we would merely assume an association between genetic and landscape data without presuming that one depends on the other.

Ordinary least squares (OLS) regression relies on a set of assumptions. First, the sample must be representative of the population. If this is not the case, statistical tests are not applicable due to unknown sampling bias; hence the sampling design is of key importance (Chapter 4). Second, the predictors $\mathbf{X}$ should be measured without error and there must not be any multicollinearity in the strict sense (see above) among predictors. Third, each of the $n$ residuals in $\mathbf{u}_i = \{u_{i1}, u_{i2}, \ldots, u_{in}\}$ of a single response variable $\mathbf{y}_i$ is itself an outcome of a random variable, and these random variables must be independent of each other and follow the same distribution (often a normal distribution is assumed) with a mean of zero and constant variance (homoscedasticity). A range of ***residual analysis*** tools facilitates checking for violations of regression assumptions for a single response variable **y**. A normal probability plot shows whether the residuals follow a normal distribution; a plot of residuals against the predicted values may reveal problems with non-constant variance (heteroscedasticity) or a non-linear response, and other plots may help identify ***influential points*** that may have a strong influence on parameter estimates and model fit. Transformations such as log(**y**) may be used to stabilize the error variance. For binary response data, such as dominant alleles coded as $1 =$ present and $0 =$ absent, we may expect different error distributions than the normal distribution, which can be accommodated by extensions of the basic OLS model to the generalized linear model (GLM).

If the residuals are not independent of each other, with nearby residuals tending to be more similar than distant ones, this indicates positive spatial autocorrelation, such as generated by isolation-by-distance. A correlogram or a test of Moran's $I$ can be used to check for spatial autocorrelation in the residuals (Box 5.1). If the autocorrelation structure is constant across the study area (i.e., stationary), it may be modeled with spatial regression (see below). Genetic data typically consist of many response variables (e.g., one for each allele of each locus), which calls for ***multivariate regression***. Note that *multivariate regression* means multiple response variables (**Y**), whereas *multiple regression* means multiple predictor variables (**X**). Multivariate regression involves fitting a regression model to each response variable, i.e., each allele. The regression assumptions apply for each allele; however, they are rarely checked individually and there is a lack of multivariate residual analysis tools. To avoid multiplying error rates and to account for correlation among response variables, multivariate significance tests should be used instead of testing each allele individually. Similarly, regression results are not necessarily interpreted individually but across all alleles, which means that the results are a weighted average across alleles. Model fit can be summarized by the canonical $R^2$, the proportion of variance of **Y** (alleles) explained by a linear model of the variables in **X** (landscape predictors), which is a weighted mean of the coefficient of determination $R^2$ of each allele weighted by its variance.

Constrained ordination methods such as redundancy analysis (RDA) facilitate the interpretation of multivariate regression analysis. RDA combines multivariate linear regression ($\mathbf{Y} = \mathbf{XB} + \mathbf{U}$, where **U** is a table of residuals) with principal component analysis (PCA) (see Box 5.3) of the table of fitted values $\hat{\mathbf{Y}} = \mathbf{XB}$ (Legendre & Legendre 2012) to quantify the variation in **Y** that is related to **X**. Permutation tests can be performed overall, with the null hypothesis that **X** does not explain more variation in **Y** than expected by chance, or for individual canonical axes (i.e., eigenvectors of a PCA of the fitted values). Note that significance tests are not performed on individual predictors, which makes RDA relatively robust towards colinearity in predictors. A PCA of the residuals **U** may reveal further genetic structure not explained by **X**, e.g. by mapping the scores of the first PCA axis in space (residual analysis).

Link-level analysis has been used to assess whether spatial genetic structure can be related to isolation-by-distance (Wright 1943; Epperson 2003), isolation-by-resistance (McRae 2006; Spear et al. 2010), or isolation-by-barrier (IBB) (Vignieri 2005). As these questions are similar to those addressed in evolutionary biology and spatial genetics, landscape genetics uses the same suite of link-based methods as in spatial genetics (Epperson 2003; Sokal & Wartenberg 1983; Guillot et al. 2009): Mantel test (Mantel 1967), partial Mantel test (Smouse et al. 1986), and multiple regression based on distance matrices (MRM) (Legendre & Legendre 2012; Smouse et al. 1986; Lichstein 2007).

Mantel tests have been widely used to assess the association between two distance matrices $\mathbf{D_X}$ and $\mathbf{D_Y}$. The lower (or upper) triangle of each distance matrix of size $n \times n$ is extracted as a vector of length $N = n(n - 1)/2$, so that each link is included only once. In the following, we will use "vector $\mathbf{D_Y}$" to refer to a vector of unique pairwise distances and "matrix $\mathbf{D_Y}$" to refer to the corresponding distance matrix. As an example, a sample of size $n = 100$ would result in a distance vector $\mathbf{D_Y}$ of length $N = 4950$, thus heavily inflating sample size. The Mantel statistic quantifies the linear ($r_M$) or rank correlation ($rho_M$) between the distance vectors $\mathbf{D_Y}$ and $\mathbf{D_X}$. However, the $N = 4950$ pair-wise distance values in a distance vector are not independent of each other. In our example, each of the $n = 100$ observed values is compared to all $n - 1 = 99$ other values. As a consequence, the statistical significance of $r_M$ or $rho_M$ (Mantel test) must be assessed with a permutation test where rows and columns of the $n \times n$ matrix $\mathbf{D_Y}$ are permuted together before extracting distance vector $\mathbf{D_Y}$. Note that it is sufficient to permute either matrix $\mathbf{D_Y}$ or matrix $\mathbf{D_X}$.

In a partial Mantel test, the Mantel statistic is calculated after partialling out the linear effect of a third distance vector $\mathbf{D_{X2}}$, thus assessing the residual correlation between distance vectors $\mathbf{D_Y}$ and $\mathbf{D_{X1}}$ after accounting for $\mathbf{D_{X2}}$. Causal modeling with distance matrices uses all Mantel and partial Mantel statistics calculated between pairs of distance vectors to discriminate between competing hypotheses (Legendre & Trousselier 1988; Cushman et al. 2006).

The main difference between a Mantel test of two distance vectors $\mathbf{D_Y}$ and $\mathbf{D_X}$ and a simple linear regression of vectors $\mathbf{D_Y}$ on $\mathbf{D_X}$ is the postulation of a directed relationship (see above), which is appropriate if we want to explain genetic structure by landscape predictors. In addition, multiple regression with distance matrices (MRM) allows the simultaneous consideration of multiple explanatory distance vectors $\mathbf{D_X}$ as in a multiple regression. However, as in the Mantel test, the residuals in MRM are not independent of each other, which violates an important assumption of linear regression. To test the slope coefficients $\mathbf{b}$, $\mathbf{D_Y}$ needs to be permuted as described above. If there are no further problems, such as spatial autocorrelation in the residuals, and the goal is significance testing of the regression model and its parameters, this permutation test should be valid.

However, the interpretation of the regression model is rather different if we use vector $\mathbf{D_Y}$ instead of $\mathbf{Y}$, and tests based on distance vectors $\mathbf{D_Y}$ and $\mathbf{D_X}$ are notoriously less powerful than those based on $\mathbf{Y}$ and $\mathbf{X}$ (Legendre & Fortin 2010). This is true both for correlation-type Mantel tests and regression-type MRM. Despite the fact that simple and partial Mantel tests and their extension to MRM are the most commonly applied statistics in landscape genetics (Storfer et al. 2010), multiple studies have identified various shortcomings of Mantel-based approaches (e.g., Balkenhol et al. 2009a; Cushman & Landguth 2010; Legendre & Fortin 2010; Graves et al. 2013; Guillot & Rousset 2013). Throughout this chapter, we highlight several of these issues and discuss alternative statistical approaches.

### 5.2.4  Model selection

Model selection (Johnson & Omland 2004) refers to the problem of selecting, from a set of candidate models, either a single best model or models weighed according to how well they fit the data and perform weighted averaging. The problem is that (i) more than one model may be statistically significant, (ii) the parameter estimate of the regression slope (and its statistical significance) for a given landscape predictor may depend on which other predictors are included in the model (unless predictors are uncorrelated), and (iii) everything else being equal, we expect a model with more predictors to fit the data better than a model with fewer predictors. In fact, any regression model of $n$ observations of response $\mathbf{Y}$ with $n$ predictors will explain 100% of the variation in $\mathbf{Y}$, even if the predictors were sampled at random. Several methods have been proposed to penalize models for the number of predictors, including adjusted $R^2$, Akaike information criterion (AIC, or AICc with small-sample correction), and Bayes information criterion (BIC), where the best model would have the highest adjusted $R^2$ or the lowest AIC, AICc, or BIC (Johnson & Omland 2004).

In MRM, the errors in $\mathbf{D_U}$ are not independent. In statistical hypothesis testing, this problem can be addressed with an appropriate permutation test (see above), but the pairwise nature of the data complicates residual analysis and model selection. Indeed, AIC and similar indices as listed above should not be applied to MRM models (Van Strien et al. 2012), as model rankings are highly unreliable unless sample size is corrected and, when correcting for sample size, their power to detect meaningful predictors is very low (Franckoviak et al. submitted). This is a pressing problem and more research is needed to show how the correlation structure among the errors $\mathbf{D_U}$ may be explicitly modeled to allow the use of AIC or similar measures in model selection. Clarke et al. (2002) proposed maximum-*likelihood* population effects (MLPEs) to explicitly model the dependence among pairwise observations. As MLPE models are fitted by residual maximum-likelihood (REML) methods, AIC, AICc, and BIC are not applicable, but a marginal $R^2$ statistic may be used to identify the most parsimonious model (Orelien & Edwards 2008; Van Strien et al. 2012).

### 5.2.5  How to put space into the multivariate regression model

Thus far, we have assumed that residuals of a regression model do not show any spatial autocorrelation. However, as spatial autocorrelation in the residuals violates the assumption of independent errors, we need to test for spatial autocorrelation in the residuals and remove it before interpreting a regression model. The methods for testing and accounting for spatial autocorrelation are similar and will be discussed jointly in this section. We will only present a subset of methods and refer to the ongoing debate about valid regression-type analysis of spatial ecological data (e.g., Beale et al. 2010; Kühn & Dormann 2012). Note that while generalized linear mixed models (GLMMs) (Zuur et al. 2009; Gałecki & Burzykowski 2013) and generalized least squares (GLS) regression as a special case of a GLMM are gaining importance in ecological data analysis, their application to multivariate genetic data has not been fully developed yet and their coverage is beyond the scope of this chapter.

### 5.2.6 Multivariate linear regression with OLS

The basic regression model used in ordinary least-squares (OLS) regression is spatially implicit and the observations are assumed to be spatially independent. Thus, although observations were taken at specific sampling locations, this spatial information is not considered in the analysis. Consequently, each observation receives the same weight in the estimation of regression coefficients **B**, which are assumed to be constant across the study area. The predicted values $\hat{\mathbf{Y}}$ are modeled as a linear combination **XB** of predictors in **X** (Fig. 5.2A).

### 5.2.7 Spatial weights matrix W

Spatial analysis is based on the expectation that nearby observations are on average more similar than distant ones. The first step in building such spatial relationships into the regression model is to define a spatial weights matrix **W** (Fig. 5.2B). Based on the spatial coordinates, we first define for each sampling unit $i$ whether other sampling units $j$ are its neighbors. Then we assign weights $w_{ij}$ to each pair, either as binary weights with $w_{ij} = 1$ for neighbors and $w_{ij} = 0$ for non-neighbors, or as a decreasing function of distance, often up to a threshold distance beyond which all weights are $w_{ij} = 0$. Finally, we may want to adjust for the number of neighbors $j$ of each observation $i$ so that the weights $w_{ij}$ of all neighbors of $i$ sum to one.

### 5.2.8 Spatial regression

Different types of spatial regression (Fig. 5.2C) associate the spatial weights matrix **W** with different terms of the regression model, which implies different assumptions about the origin of the spatial autocorrelation. (i) Spatial lag models (e.g., conditional autoregressive models, CAR) add a weighted mean of the response **Y** at neighboring locations as a predictor to the model. The assumption is that the presence or abundance of an allele at location $i$ depends on its presence or abundance in the neighborhood, which is a reasonable assumption under IBD. (ii) Spatial error models (e.g., simultaneous autoregressive models, SAR) add a weighted mean of the error **U** at neighboring locations as a predictor to the model. This is indicated if we attribute spatial autocorrelation to an unmeasured nuisance factor. Spatial regression methods are commonly applied to a univariate response **y**, although multivariate CAR models have been proposed (Gelfand & Vounatsou 2003). More research is needed to show their applicability to model IBD in landscape genetic data.

If the process is **non-stationary** so that the spatial autocorrelation structure varies across the study area (Box 5.1), one should consider statistics that can measure the degree of spatial structure at local scales, such as local indicators of spatial association (LISA; Anselin 1995; Sokal et al. 1998). Geographically weighted regression (GWR) (Fotheringham 2002) can detect and address situations where the relationship between **X** and **Y** is non-stationary, so that the slope coefficients **B** vary across the study area. Such statistical analyses have been used in ecology (e.g., Fortin & Melles 2009; Windle et al. 2012) and evolution (e.g., Ochoa-Ochoa et al. 2014) but they are not yet commonly used in landscape genetics. A recent example used GWR to demonstrate that different models of genetic connectivity for Rocky Mountain tailed frogs (*Ascaphus montanus*) were supported in privately and publicly managed forests despite close spatial proximity of the two types of land ownerships (Spear & Storfer 2010).

### 5.2.9 Spatial eigenvectors

***Spatial eigenvector*** methods (Griffith 2000; Borcard & Legendre 2002; Griffith & Peres-Neto 2006; Dray et al. 2006, 2012; Jombart et al. 2008, 2009; Dray 2011), such as ***Moran's eigenvector map***s (MEM) (Dray et al. 2006; Dray 2011), are based on eigenanalysis (Box 5.4) of the spatial weights matrix **W** (Fig. 5.2D). In the case of MEM, **W** is made symmetric and column and row means are removed before eigenanalysis. If matrix **W** is of full rank one of the $n$ eigenvectors will have zero variance (and an eigenvalue close to zero) and will be discarded. Each of the remaining $n-1$ spatial eigenvectors describes a periodic spatial pattern. If sampling locations form a regular transect, the spatial eigenvectors will represent sine-type patterns (similar to a Fourier decomposition), whereas two-dimensional and irregular sampling designs will result in more complex patterns resembling two-dimensional sine waves (Box 5.3). The $(n-1)$ spatial eigenvectors, which form the columns of matrix **V**, are orthogonal and uncorrelated among themselves. Just as we can describe a set of 2 points by a single line with two parameters, the variation in $n$ observations can be fully described by any set of $n$ variables or by $n-1$ variables if the mean has been removed. This means that the variation in **Y** can be completely modeled by a linear combination of the variables in **V** (i.e., a regression of **Y** on **V** will have **U** = 0 and $R^2 = 1$). While this full model is not of interest in itself, the matrix $\mathbf{R}_{YV}$ of correlations between the columns in **Y** (alleles) and the columns in **V** (spatial eigenvectors) provides a spatial decomposition of the genetic variation at all spatial scales that can be used to quantify multivariate spatial autocorrelation with Moran's $I$ or to model significant spatial genetic variation as spatial predictors that can be included in the regression model (spatial filtering, see Section 5.2.11).

### 5.2.10  Multivariate Moran's I

When using MEM, the spatial eigenvectors in **V** are sorted by the spatial scale they represent, so that the first spatial eigenvector ($k = 1$) represents a single sine wave spanning the maximum extent of the study area. Each subsequent spatial eigenvector represents a smaller-scale spatial pattern and the spatial scale of each spatial eigenvector can be quantified by Moran's $I$, a measure of spatial autocorrelation (Box 5.1). Conveniently, the eigenvalue $\lambda_k$ associated with each spatial eigenvector $k$ is proportional to its Moran's $I_k$, up to a constant $a$, which corresponds to the inverse of the average sum of weights per observation (Fig. 5.2E). Moreover, a weighted mean of the eigenvalues $\lambda$, weighted by the squared correlations $\mathbf{R}^2_{yV}$ of **V** with a response variable **y** (i.e., frequency of a specific allele), results in Moran's $I_y$ of allele **y** (Dray 2011). Multivariate Moran's $I_Y$ can then be found as the mean of Moran's $I_y$ of all alleles, weighted by column weights **c** (Fig. 5.2E; Wagner 2013).

When averaging across alleles and loci, the coding of allele frequencies will affect their relative weight (Smouse and Peakall 1999). As each allele is treated as a variable, it makes sense to weigh the contribution of each allele to Moran's $I$ of the locus by the inverse of the number of alleles per locus, so that the total weight of each locus equals one, as recommended by Jombart et al. (2008). Adjusting for differences in allele frequencies within a locus seems less important (Smouse and Peakall 1999). To average across loci, we further divide weights by the number of loci so that the column weights **c** sum to one.

While MEM derives the matrix of eigenvectors **V** from the sampling design alone, spatial principal component analysis (sPCA) (Jombart et al. 2008) derives **V** from a combination of the sampling design and the observed data. This has the advantage that spatial eigenvectors are sorted by their contribution to multivariate Moran's $I$ of the data. However, sPCA does not provide an additive decomposition of $I_Y$. Furthermore, the matrix **V** of spatial eigenvectors derived for a response matrix **Y** of allele frequencies and for a predictor matrix **X** may differ, which will require more prior knowledge about the key spatial scales of interest in the analysis of the relationship between genetic variation and landscape predictors. sPCA is a new explicitly spatial ordination technique that deals with spatial structures multivariate data and its merits still need to be evaluated.

### 5.2.11  Spatial filtering

Another approach to include space in multivariate models is spatial filtering, which uses spatial eigenvectors to control for significant spatial autocorrelation at any spatial scale. Matrix **V** as defined above can be used to decompose the spatial variation in the response **Y** in the matrix of fitted values **Ŷ** or in the matrix of residuals **U**. We can thus ask what is the overall spatial structure in the genetic data, which spatial patterns are explained by landscape predictors, and which remain unexplained? Significant spatial eigenvectors for fitted values indicate shared patterns, whereas significant spatial eigenvectors for residuals indicate unexplained patterns that may be related either to dispersal processes or to unmeasured landscape factors.

For testing the statistical significance of spatial eigenvectors, the method by Jombart et al. (2009) should be preferred over forward selection (Blanchet et al. 2008), as the former takes into account the relationship between spatial eigenvectors (Wagner 2013). The test statistic is defined as the maximum (multivariate) variance explained by a single spatial eigenvector. For each spatial eigenvector, its observed multivariate $R^2$ is compared to the distribution of the test statistic obtained from a permutation test, where observations are permuted randomly (Jombart et al. 2009; Wagner 2013).

In a regression framework (Fig. 5.2F), significant spatial eigenvectors for **Y** or for **U** are added as predictors to the regression model to account for significant spatial structure in the genetic data when testing the association between **X** and **Y** (spatial filtering; Legendre & Legendre 2012, Griffith & Peres-Neto 2006; Peres-Neto & Legendre 2010). Spatial filtering with MEM has been used to account for unmeasured environmental variation when identifying loci that are potentially under selection (Manel et al. 2010). However, the statistical validity of such an approach remains to be thoroughly tested, and the existing evidence from non-genetic simulation studies suggests that spatial filtering with MEM and related methods may lead to biased parameter estimates and inflated type I error rates (Dormann et al. 2007; Bini et al. 2009; Beale et al. 2010).

## 5.3 HOW TO MODEL ISOLATION-BY-DISTANCE

Isolation-by-distance plays a key role in landscape genetics. First, analyzing landscape effects on genetic variation is only warranted if there is spatial structure in the genetic data (i.e., if the population is not panmictic). Second, the effects of matrix resistance (IBR) should not be tested against a null model of panmixis but against a null model of IBD, thus testing for effects of landscape features on rates of gene flow beyond the effect of IBD. Third, if gene flow is spatially restricted (by IBD or IBR), this creates spatial autocorrelation in the genetic data that needs to be accounted for when testing the association between **X** and **Y** (e.g., in the identification of outlier loci that may indicate selection). Figure 5.3 summarizes approaches for testing and accounting for IBD in the frameworks of spatial regression with conditional autoregressive modeling CAR, spatial filtering with MEM, and regression of distance matrices (MRM).

### 5.3.1 IBD and spatial regression with CAR

In this section, the symbol **Z** will be used to refer to a "space only" model (IBD) and **X** to other landscape predictors. Isolation-by-distance can be modeled as a stationary *isotropic* spatial process, thus assuming that the interaction between sampling locations depends on distance alone (Fig. 5.3A). Spatial regression with a conditional autoregressive (CAR) model is appropriate for modeling gene flow as it explicitly models the interaction between allele frequencies of neighboring sampling locations (i.e., it assumes that allele frequency at location $i$ depends on the frequency of the same allele at neighboring locations $j$; Fig. 5.3). In essence, $\mathbf{WYB_W}$ calculates a weighted mean of allele frequencies at neighboring locations, where **W** defines which locations are neighbors and what is their relative weight (i.e., the spatial covariance structure). $\mathbf{B_W}$ defines the contribution of the weighted mean $\mathbf{Z} = \mathbf{WY}$ to the fitted values $\hat{\mathbf{Y}}$ (i.e., the strength of positive spatial autocorrelation defined by **W**). The spatial weights matrix **W** remains constant, whereas $\mathbf{B_W}$ contains a separate autoregression coefficient for each variable in **Y**.

A simple CAR model without additional terms can be used to test for IBD as a spatial process defined by **W** against a null hypothesis of panmixis. When testing for landscape effects (representing IBR or selection), partial regression (conditioning by the CAR term $\mathbf{WYB_W}$) can be used to incorporate IBD into the null model. Thus, first an IBD model defined by $\mathbf{Z} = \mathbf{WY}$ is fitted separately to **Y** and to **X**, and the regression of **Y** on **X** is carried out on the residuals of both **Y** and **X**. In Fig. 5.3, the conditioning by **Z** is indicated by subscript **X|Z**.

Residual analysis (see above) should be performed before interpreting the model. If the spatial regression model is correctly specified and the spatial process (gene flow) is stationary and isotropic, then the residuals **U** should not show any further spatial structure. In multivariate regression, this can be tested with multivariate Moran's $I$ (see above). If the test indicates significant spatial autocorrelation in the residuals, it may be useful to perform a PCA of **U**, which is implicitly done in RDA, and plot the scores of the first few axes (unconstrained RDA axes) in space for visual interpretation of unexplained spatial patterns.

### 5.3.2 IBD and spatial filtering with MEM

If the spatial scale of gene flow is unknown, or if we expect it may not be stationary and isotropic (e.g., IBR leading to variation in rates of gene flow beyond mere distance effects), the assumptions of spatial regression with a CAR model may be too restrictive. Hence we may want to define a flexible model of gene flow using MEM (Fig. 5.3B). In essence, spatial filtering with MEM uses a subset $\mathbf{Z} = \mathbf{V}_K$ of spatial eigenvectors in $\mathbf{V}$ to model significant spatial structure of any shape and at any spatial scale in the response $\mathbf{Y}$. $\mathbf{Z}$ may thus serve as a proxy for unmeasured landscape factors or unspecified spatial biotic processes.

Using MEM to test for a significant spatial structure in $\mathbf{Y}$ at any scale (i.e., against a null hypothesis that none of the spatial eigenvectors are significant – panmixis) is often not informative. Rejecting this null hypothesis does not inform us whether the deviation is due to IBD or IBR. Spatial filtering with MEM may be most applicable when we want to test for evidence of selection (e.g., identifying outlier loci) while accounting for spatial autocorrelation due to gene flow, whether due to IBD or IBR, without using a predefined model.

Once $\mathbf{Z} = \mathbf{V}_K$ has been defined, it may be useful to perform partial regression conditioned by $\mathbf{Z}$ instead of treating both $\mathbf{X}$ and $\mathbf{Z}$ equally as predictors. This makes sense conceptually (space itself is not a meaningful predictor) and avoids problems of interpretation, as we can then distinguish between conditioned variance (variance explained by $\mathbf{Z}$), constrained variance (variance explained by $\mathbf{X}$ after conditioning for $\mathbf{Z}$), and residual variance (Wagner 2013).

As a caveat, Beale et al. (2010) found that spatial filtering with MEM resulted in biased parameter estimates and inflated type I error rates. While spatial eigenvectors are promising as a flexible tool for modeling spatial structure in landscape genetic data, more research is needed on how they can be applied in hypothesis testing, accounting for the special nature of spatial eigenvectors (Gilbert & Bennett 2010; Wagner 2013).

### 5.3.3   IBD and multiple regression of distance matrices (MRM)

In MRM, a vector $\mathbf{D}_Z$ of pairwise geographic distances is commonly used to model IBD (Fig. 5.3C). The null model of panmixis is rejected if a simple regression of vectors $\mathbf{D}_Y$ on $\mathbf{D}_Z$ is significant, based on a permutation test where the rows and columns of matrix $\mathbf{D}_Y$ are permuted simultaneously (see above). This is equivalent to a Mantel test between vectors $\mathbf{D}_Y$ and $\mathbf{D}_Z$.

Partial regression of distance matrices can be used to test landscape effects against a null model of IBD by testing whether landscape effects defined in vector $\mathbf{D}_X$ (e.g., IBR) are statistically significant after accounting

for vector $\mathbf{D}_Z$. If $\mathbf{D}_X$ is a single vector (which may reflect a complex hypothesis about the resistance values of multiple land cover types), this is equivalent to a partial Mantel test between vectors $\mathbf{D}_Y$ and $\mathbf{D}_X$, conditioning for vector $\mathbf{D}_Z$. In essence, the residuals of a regression of vector $\mathbf{D}_Y$ on $\mathbf{D}_Z$ are regressed on the residuals of a regression of vector $\mathbf{D}_X$ on $\mathbf{D}_Z$. If there are multiple vectors $\mathbf{D}_X$, each representing a different landscape element (e.g., length of a single cover type along the transect between two sampling locations), they may be added together in the same model. In that case, after conditioning for vector $\mathbf{D}_Z$, the regression coefficients for the $\mathbf{D}_X$ vectors are tested as if each was added to the model last (i.e., accounting for all other predictors in the model). In contrast, the effect of vector $\mathbf{D}_Z$ is tested without accounting for vectors $\mathbf{D}_X$ because IBD as defined in vector $\mathbf{D}_Z$ is part of the null model for testing vectors $\mathbf{D}_X$. In many landscape genetic studies, IBD will be an appropriate null model, e.g. for testing IBR, whereas IBR would rarely be an appropriate null model for testing IBD. A notable exception may be the simultaneous testing for a barrier effect (IBB) and IBD, as a complete barrier to gene flow may occur with or without IBD on either side. Note that the causal modeling framework (Cushman et al. 2006) involves testing all possible partial Mantel correlations to rule out alternative hypotheses.

There are several issues with MRM, relating mainly to the shape of the relationship, the validity and statistical power of significance tests, and model selection. While regression assumes a linear relationship and constant variance around the regression line (homoscedasticity), these assumptions may not hold for genetic distances, so that a plot of vector $\mathbf{D}_Y$ against vector $\mathbf{D}_Z$ may show a non-linear relationship or an increase of variance with distance. For instance, Rousset (1997) proposed that $F_{ST}/(1 - F_{ST})$ is approximately linearly related to geographic distance in a one-dimensional stepping stone model, but linearly related to the natural logarithm of geographic distance in a two-dimensional stepping stone model. Under equilibrium conditions in a two-dimensional stepping stone model, Hutchison and Templeton (1999) expected a monotonic increase of pairwise $F_{ST}$ values and an increase of their spread with geographic distance (i.e., non-constant variance), due to a shift in the relative importance of the homogenizing effect of gene flow at short distances and the divergent effect of drift at large distances. Lack of regional equilibrium may result in a scatter plot where pairwise $F_{ST}$ increases at short distances and levels off at larger distances, thus introducing non-linearity (Hutchison & Templeton 1999). In a simulation study, Graves et al. (2013) found that after 300 non-overlapping generations, the relationship between pairwise genetic distances $D_{ps}$ and the matrix of cost distances used to simulate gene flow (i.e., the known truth) was asymptotic rather than linear and could not be linearized by common transformations. Furthermore, measures of genetic distance differ in their sensitivity to population genetic processes (e.g., Whitlock 2011; Raeymaekers et al. 2012; see Chapter 3) and population size or divergence time may be more important for explaining population genetic structure than gene flow (Marko & Hart 2011).

Compared to regression of node-based data, tests based on distance matrices (Mantel test, partial Mantel test, and MRM) have considerably lower statistical power (Legendre & Fortin 2010) and thus higher type II error rates, so that existing landscape effects are less likely to be detected in a regression based on distance matrices. On the other hand, if there is positive spatial autocorrelation in both the response and predictor matrices (e.g., due to IBD), type I error rates may be considerably inflated (Guillot & Rousset 2013), so that spurious effects are more likely to become statistically significant. Moreover, recent findings suggest that accounting for IBD by a vector $\mathbf{D_Z}$ of geographic distances does not sufficiently remove spatial autocorrelation in partial analysis (Guillot & Rousset 2013). Goldberg and Waits (2010) proposed a method to identify and remove observations that are non-independent due to spatial autocorrelation, though this does not account for the issue of inflated sample size.

The pairwise nature of distance data impedes the checking of assumptions and conditions with residual analysis. As noted above, the residuals $\mathbf{U}$ are not independent of each other and information-theoretic indices commonly used for model selection (AIC, AICc, or BIC) are not applicable to distance matrices (Van Strien et al. 2012; Franckoviak et al. submitted). Finally, the identification of influential points is hampered by the problem that one unusual observation will affect $n-1$ pairwise distance values, so that the influential observations may best be identified with leave-one-out jackknife methods.

## 5.4  FUTURE DIRECTIONS

For landscape genetics of adaptive variation (e.g., detection of outlier loci; see Chapter 9), multivariate regression provides a natural framework for modeling the individual response of alleles to at-site conditions, such as soil, vegetation, or bioclimatic variables related to selection. However, valid inference needs to account for spatial autocorrelation induced by gene flow. Spatial filtering with MEM is useful for partialling out spatial structure without specifying a particular process of IBD or IBR, as long as the majority of alleles reflects the same process of gene flow. However, the statistical validity of spatial filtering needs to be thoroughly tested and alternatives may need to be developed. Future directions may involve studying selection and gene flow at the same time with multivariate spatial regression (e.g., the CAR model). In contrast to MEM, this involves explicitly modeling the spatial process of IBD or IBR at the appropriate spatial scale(s). The spatial process may be averaged over a majority of alleles with similar parameters (i.e., exclude potential outlier loci that may show a different spatial autocorrelation structure) to obtain a quantification of gene flow, which can then be used to account for gene flow when testing the response to at-site variables.

The study of gene flow focuses on between-site characteristics, which remains a challenge for model selection and valid statistical inference. Solutions are emerging along three avenues:

**i** Remain in link-level analysis and explicitly model the error structure in MRM (Clarke et al. 2002; Van Strien et al. 2012).

**ii** Adopt a neighborhood-level approach, where between-site characteristics are integrated in a connectivity measure such as an incidence function model to obtain a connectivity value for each sampling location (e.g., Keyghobadi et al. 2005; James et al. 2011; Rico et al. 2014). These connectivity values can then be used as predictors of genetic variation, where the response is a measure of genetic diversity or differentiation, a matrix of allele frequencies, or a set of PCoA scores (dbRDA).

**iii** In node-level analysis of allele frequencies $\mathbf{Y}$, use one or multiple distance matrices $\mathbf{D_X}$ of landscape predictors to model the covariance structure of the errors $\mathbf{U}$ (but see Guillot et al. 2014 on valid covariance models). While Bradburd et al. (2013) present an implementation for binary SNP data in a Bayesian framework, the approach could be extended to multinomial logistic regression to accommodate codominant markers such as microsatellites and could potentially be implemented in the framework of generalized linear mixed models (GLMM) (Zuur et al. 2009; Gałecki & Burzykowski 2013).

## ACKNOWLEDGMENTS

This chapter was supported by NSERC Discovery Grants to HHW and MJF.

## REFERENCES

Anselin, L. (1995) Local indicators of spatial association – LISA. *Geographical Analysis* **27**, 93–115.

Balkenhol, N., Waits, L.P., & Dezzani, R.J. (2009a) Statistical approaches in landscape genetics: an evaluation of methods for linking landscape and genetic data. *Ecography* **32**, 818–30.

Balkenhol, N., Gugerli, F., Cushman, S.A., Waits, L.P., Coulon, A., Arntzen, J.W., Holderegger, R., & Wagner, H.H. (2009b) 'Identifying future research needs in landscape genetics: where to from here?', *Landscape Ecology* **24**, 455–63.

Beale, C. M., Lennon, J. J., Yearsley, J. M., Brewer, M. J. and Elston, D. A. (2010), Regression analysis of spatial data. *Ecology Letters* **13**, 246–64.

Bini, L.M., Diniz-Filho, J.A.F., Rangel, T.F.L.V.B., Akre, T.S.B., Albaladejo, R.G., Albuquerque, F.S., Aparicio, A., Araujo, M.B., Baselga, A., Beck, J., Bellocq, M.I., Bohning-Gaese, K., Borges, P.A.V., Castro-Parga, I., Chey, V.K., Chown, S.L., de Marco, P., Dobkin, D.S., Ferrer-Castan, D., Field, R., Filloy, J., Fleishman, E., Gomez, J.F., Hortal, J., Iverson, J.B., Kerr, J.T., Kissling, W.D., Kitching, I.J., Leon-Cortes, J.L., Lobo, J.M., Montoya, D., Morales-Castilla, I., Moreno, J.C., Oberdorff, T., Olalla-Tarraga, M.A., Pausas, J.G., Qian, H., Rahbek, C., Rodriguez, M.A., RUeda, M., Ruggiero, A., Sackmann, P., Sanders, N.J., Terribile, L.C., Vetaas, O.R., & Hawkins, B.A. (2009) Coefficient shifts in geographical ecology: an empirical evaluation of spatial and non-spatial regression. *Ecography* **32**, 193–204.

Blanchet, F.G., Legendre, P., & Borcard, D. (2008) Forward selection of explanatory variables. *Ecology* **89**, 2623–32.

Bolker, B.M. (2008) *Ecological Models and Data in R*. Princeton University Press, Princeton, NJ.

Borcard, D. & Legendre, P. (2002) All-scale spatial analysis of ecological data by means of principal coordinates of neighbour matrices. *Ecological Modeling* **153**, 51–68.

Bradburd, G.S., Ralph, P.L., & Coop, G.M. (2013) Disentangling the effects of geographic and ecological isolation on genetic differentiation. *Evolution* **67**, 3258–73.

Clarke, R.T., Rothery, P., & Raybould, A.F. (2002) Confidence limits for regression relationships between distance matrices: estimating gene flow with distance. *Journal of Agricultural, Biological and Environmental Statistics* **7**, 361–72.

Cushman, S.A. & Landguth, E.L. (2010) Spurious correlations and inference in landscape genetics. *Molecular Ecology* **19**, 3592–602.

Cushman, S.A., McKelvey, K.S., Hayden, J., & Schwartz, M.K. (2006) Gene flow in complex landscapes: testing multiple hypotheses with causal modeling. *The American Naturalist* **168**, 486–99.

Dale, M.R.T. & Fortin, M.-J. (2010) From graphs to spatial graphs. *Annual Review of Ecology, Evolution and Systematics* **41**, 21–38.

de Jong, P., Sprenger, C., & van Veen, F. (1984) On extreme values of Moran's $I$ and Geary's $c$. *Geographical Analysis* **16**, 17–24.

Dormann, C.F., McPherson, M., Araújo, J.B., Bivand, M., Bolliger, R., Carl, J., Davies, G., Hirzel, R., Jetz, A., Kissling, D.W., Kühn, I., Ohlemüller, R., Peres-Neto, P., Reineking, B., Schröder, B., Schurr, M.F., & Wilson, R. (2007) Methods to account for spatial autocorrelation in the analysis of species distributional data: a review. *Ecography* **30**, 609–28.

Dormann, C.F., Elith, J., Bacher, S., Buchmann, C., Carl, G., Carré, G., Marquéz, J.R.G., Gruber, B., Lafourcade, B., Leitão, P.J., Münkemüller, T., McClean, C., Osborne, P.E., Reineking, B., Schröder, B., Skidmore, A.K., Zurell, D., & Lautenbach, S. (2013) Collinearity: a review of methods to deal with it and a simulation study evaluating their performance. *Ecography* **36**, 27–46.

Dray, S. (2011) A new perspective about Moran's coefficient: spatial autocorrelation as a linear regression problem. *Geographical Analysis* **43**, 127–41.

Dray, S., Legendre, P., & Peres-Neto, P.R. (2006) Spatial modeling: a comprehensive framework for principal coordinate analysis of neighbour matrices (PCNM). *Ecological Modeling* **196**, 483–93.

Dray, S., et al. (2012) Community ecology in the age of multivariate multiscale spatial analysis. *Ecological Monographs* **82**, 257–75.

Dutilleul, P., Stockwell, J.D., Frigon, D., & Legendre, P. (2000) The Mantel Test versus Pearson's Correlation Analysis: assessment of the differences for biological and environmental studies. *Journal of Agricultural, Biological and Environmental Statistics* **5**, 131–50.

Dyer, R.J., Nason, J.D., & Garrick, R.C. (2010) Landscape modeling of gene flow: improved power using conditional genetic distance derived from the topology of population networks. *Molecular Ecology* **19**, 3746–59.

Epperson, B.K. (2003) *Geographical Genetics*. Princeton University Press, Princeton, NJ.

Florax, R.J.G.M. & Rey, S. (1995) The impacts of misspecified spatial interaction in linear regression models. In: Anselin, L. & Florax, R.J.G.M. (eds.), *New Directions in Spatial Econometrics*. Springer, Berlin, Heidelberg.

Foll, M. & Gaggiotti, O.E. (2006) Identifying the environmental factors that determine the genetic structure of populations. *Genetics* **174**, 875–91.

Fortin, M.-J. & Dale, M.R.T. (2005) *Spatial Analysis: A Guide for Ecologists*. Cambridge University Press, New York.

Fortin, M.-J. & Melles, S.J. (2009) Avian spatial responses to forest spatial heterogeneity at the landscape level: conceptual and statistical challenges. In: Miao, S., Carstenn, S., &

Nungesser, M. (eds.), *Real World Ecology: Large-Scale and Long-Term Case Studies and Methods.* Springer, New York.

Fotheringham, A.S. (2002) In: Brunsdon, C. & Charlton, M. (eds.), *Geographically Weighted Regression: The Analysis of Spatially Varying Relationships.* John Wiley & Sons, Ltd, Chichester.

Franckoviak, R.P., Jarvis, K., Acuna, I., Landguth, E.L., Fortin, M.-J., & Wagner, H.H. (submitted) Model selection with multiple regression on distance matrices leads to incorrect inferences. *Molecular Ecology Resources.*

François, O. & Durand, E. (2010) Spatially explicit Bayesian clustering models in population genetics. *Molecular Ecology Resources* **10**, 773–84.

Gałecki, A.T. and Burzykowski, T. (2013) *Linear Mixed-Effects Models Using R: A Step-by-Step Approach.* Springer, New York.

Gelfand, A.E. & Vounatsou, P. (2003) Proper multivariate conditional autoregressive models for spatial data analysis. *Biostatistics* **4**, 11–15.

Gilbert, B. & Bennett, J.R. (2010) Partitioning variation in ecological communities: do the numbers add up? *Journal of Applied Ecology* **475**, 1071–82.

Goldberg, C.S. & Waits, L.P. (2010) Comparative landscape genetics of two pond-breeding amphibian species in a highly modified agricultural landscape. *Molecular Ecology* **19**, 3650–63.

Graves, T.A., Beier, P., & Royle, J.A. (2013) Current approaches using genetic distances produce poor estimates of landscape resistance to interindividual dispersal. *Molecular Ecology* **22**, 3888–903.

Griffith, D.A. (1996) Some guidelines for specifying the geographic weights matrix contained in spatial statistical models. In Arlinghaus, S.L. (ed.), *Practical Handbook of Spatial Statistics.* CRC Press, Boca Raton, FL.

Griffith, D.A. (2000) A linear regression solution to the spatial autocorrelation problem. *Journal of Geographical Systems* **2**, 141–56.

Griffith, D.A. & Peres-Neto, P.R. (2006) Spatial modeling in ecology: the flexibility of eigenfunction spatial analyses. *Ecology* **87**, 2603–13.

Guillot, G. & Rousset, F. (2013) Dismantling the Mantel tests. *Methods in Ecology and Evolution* **4**, 336–44.

Guillot, G., Mortier, F. & Estoup, A. (2005) Geneland: a computer package for landscape genetics. *Molecular Ecology Notes* **5**, 712–15.

Guillot, G., Leblois, R., Coulon, A., & Frantz, A.C. (2009) Statistical methods in spatial genetics. *Molecular Ecology* **18**, 4734–56.

Guillot, G., Schilling, R.L., Porcu, E. & Bevilacqua, M. (2014) Validity of covariance models for the analysis of geographical variation. *Methods in Ecology and Evolution* **5**, 329–35.

Hanski, I. (1994) A practical model of metapopulation dynamics. *Journal of Animal Ecology* **63**, 151–62.

Hutchison, D.W. & Templeton, A.R. (1999) Correlation of pairwise genetic and geographic distance measures: inferring the relative influences of gene flow and drift on the distribution of genetic variability. *Evolution* **53**, 1898–914.

James, P.M.A., Coltman, D.W., Murray, B.W., Hamelin, R.C., & Sperling, F.A.H. (2011) Spatial genetic structure of a symbiotic beetle-fungal system: toward multi-taxa integrated landscape genetics. *PLoS One* **6**, e25359.

Jay, F. (2011) *PoPS: Prediction of Population Genetic Structure – Program Documentation and Tutorial.* University Joseph Fourier, Grenoble, France.

Johnson, J.B. & Omland, K.S. (2004) Model selection in ecology and evolution. *Trends in Ecology and Evolution* **19**, 101–8.

Jombart, T., Devillard, S., Dufour, A., & Pontier, D. (2008) Revealing cryptic spatial patterns in genetic variability by a new multivariate method. *Heredity* **101**, 92–103.

Jombart, T., Dray, S., & Dufour, A.-B. (2009) Finding essential scales of spatial variation in ecological data: a multivariate approach. *Ecography* **32**, 161–8.

Keyghobadi, N., Roland, J., Matter, S.F., & Strobeck, C. (2005) Among- and within-patch components of genetic diversity respond at different rates to habitat fragmentation: an empirical demonstration. *Proceedings of the Royal Society B: Biological Sciences* **272**, 553–60.

Kühn, I. & Dormann, C.F. (2012) Less than eight (and a half) misconceptions of spatial analysis. *Journal of Biogeography* **39**, 995–8.

Legendre, P. (1993) Spatial autocorrelation: trouble or new paradigm? *Ecology* **74**, 1659–73.

Legendre, P. & Fortin, M.-J. (2010) Comparison of the Mantel test and alternative approaches for detecting complex multivariate relationships in the spatial analysis of genetic data. *Molecular Ecology Resources* **10**, 831–44.

Legendre, P. & Legendre, L. (2012) *Numerical Ecology,* 3rd edition. Elsevier Science & Technology Books, San Diego, CA.

Legendre, P. & Trousselier, M. (1988) Aquatic heterotrophic bacteria: modeling in the presence of spatial autocorrelation. *Limnology and Oceanography* **33**, 1055–67.

Lichstein, J.W. (2007) Multiple regression on distance matrices: a multivariate spatial analysis tool. *Plant Ecology* **188**, 117–31.

Manel, S., Joost, S., Epperson, B.K., Holderegger, R., Storfer, A., Rosenberg, M.S., Scribner, K.T., Bonin, A., & Fortin, M.-J. (2010) Perspectives on the use of landscape genetics to detect genetic adaptive variation in the field. *Molecular Ecology* **19**, 3760–72.

Manel, S., Gugerli, F., Thuiller, W., Alvarez, N., Legendre, P., Holderegger, R., Gielly, L., & Taberlet, P. (2012) Broad-scale adaptive genetic variation in alpine plants is driven by temperature and precipitation. *Molecular Ecology* **21**, 3729–38.

Mantel, N. (1967) The detection of disease clustering and a generalized regression approach. *Cancer Research* **27**, 209–20.

McRae, B.H. (2006) Isolation by resistance. *Evolution* **60**, 1551–61.

Moilanen, A. & Nieminen, M. (2002) Simple connectivity measures in spatial ecology. *Ecology* **83**, 1131–45.

Moran, P.A.P. (1950) Notes on continuous stochastic phenomena. *Biometrika* **37**, 17–23.

Muirhead, J.R., Gray, D.K., Kelly, D.W., Ellis, S.M., Heath, D.D., & MacIsaac, H.J. (2008) Identifying the source of species invasions: sampling intensity vs. genetic diversity. *Molecular Ecology* **17**, 1020–35.

Murphy, M.A., Dezzani, R.J., Pilliod, D.S., & Storfer, A. (2010) Landscape genetics of high mountain frog metapopulations. *Molecular Ecology* **19**, 3634–49.

Ochoa-Ochoa, L.M., Campbell, J.A., & Flores-Villela, O.A. (2014) Patterns of richness and endemism of the Mexican herpetofauna, a matter of spatial scale? *Biological Journal of the Linnean Society* **111**, 305–16.

Orelien, J.G. & Edwards, L.J. (2008) Fixed-effect variable selection in linear mixed models using statistics. *Computational Statistics and Data Analysis* **52**, 1896.

Peres-Neto, P.R. (2006) A unified strategy for estimating and controlling spatial, temporal and phyogenetic autocorrelation in ecological models. *Oecologia Brasiliensis* **10**, 105–19.

Peres-Neto, P.R. & Legendre, P. (2010) Estimating and controlling for spatial structure in the study of ecological communities. *Global Ecology and Biogeography* **19**, 174–84.

Raeymaekers, J.A.M., Lens, L., van den Broeck, F., van Dongen, S., & Volckaert, F.A.M. (2012) Quantifying population structure on short timescales. *Molecular Ecology* **21**, 3458–73.

Rico, Y., Boehmer, H.J., & Wagner, H.H. (2014) Effect of rotational shepherding on demographic and genetic connectivity of calcareous grassland plants. *Conservation Biology* **28**, 467–77.

Rousset, F. (1997) Genetic differentiation and estimation of gene flow from F-statistics under isolation by distance. *Genetics* **145**, 1219–28.

Ryman, N., Palm, S. André, C., Carvalho, G.R., Dahlgren, T.G., Jorde, P.E., Laikre, L., Larsson, L.C., Palmé, A., & Ruzzante, D.E. (2006) Power for detecting genetic divergence: differences between statistical methods and marker loci. *Molecular Ecology* **15**, 2031–45.

Safner, T., Miller, M.P., McRae, B.H., Fortin, M.-J., & Manel, S. (2011) Comparison of Bayesian clustering and edge detection methods for inferring boundaries in landscape genetics. *International Journal of Molecular Sciences* **12**, 865–89.

Saura, S. & Rubio, L. (2010) A common currency for the different ways in which patches and links can contribute to habitat availability and connectivity in the landscape. *Ecography* **33**, 523–37.

Schoville, S.D., Bonin, A., François, O., Lobreaux, S., Melodelima, C., & Manel, S. (2012) Adaptive genetic variation on the landscape: methods and cases. *Annual Review of Ecology, Evolution and Systematics* **43**, 23–43.

Selkoe, K.A. & Toonen, R.J. (2006) Microsatellites for ecologists: a practical guide to using and evaluating microsatellite markers. *Ecology Letters* **9**, 615–29.

Smouse, P.E. & Peakall, R. (1999) Spatial autocorrelation analysis of individual multiallele and multilocus genetic structure. *Heredity* **82**, 561–73.

Smouse, P., Long, J.C., & Sokal, R.R. (1986) Multiple regression and correlation extension of the Mantel test of matrix correspondence. *Systematic Zoology* **35**, 627–32.

Sokal, R.R. & Wartenberg, D.E. (1983) A test of spatial autocorrelation analysis using an isolation-by-distance model. *Genetics* **105**, 21–37.

Sokal, R.R., Oden, N.L., & Thomson, B.A. (1998) Local spatial autocorrelation in a biological model. *Geographical Analysis* **30**, 331–54.

Spear, S.F. & Storfer, A. (2010) Anthropogenic and natural disturbance lead to differing patterns of gene flow in the Rocky Mountain tailed frog, *Ascaphus montanus*. *Biological Conservation* **143**, 778–86.

Spear, S.F., Balkenhol, N., Fortin, M.-J., McRae, B.H., & Scribner, K.T. (2010) Use of resistance surfaces for landscape genetic studies: considerations for parameterization and analysis. *Molecular Ecology* **19**, 3576–91.

Storfer, A., Murphy, M.A., Spear, S.F., Holderegger, R., & Waits, L.P. (2010) Landscape genetics: Where are we now? *Molecular Ecology* **19**, 3496–514.

Van Strien, M.J., Keller, D., & Holderegger, R. (2012) A new analytical approach to landscape genetic modeling: least-cost transect analysis and linear mixed models. *Molecular Ecology* **21**, 4010–23.

Vignieri, S.N. (2005) Streams over mountains: influence of Riparian connectivity on gene flow in the Pacific jumping mouse (*Zapus trinotatus*). *Molecular Ecology* **14**, 1925–37.

Wagner, H.H. (2013) Rethinking the linear regression model for spatial ecological data. *Ecology* **94**, 2381–91.

Wagner, H.H. & Fortin, M.J. (2013) A conceptual framework for the spatial analysis of landscape genetic data. *Conservation Genetics* **14**, 253–61.

Whitlock, M.C. (2011) $G'_{ST}$ and D do not replace $F_{ST.}$ *Molecular Ecology* **20**, 1083–91.

Windle, M.J.S., Rose, G.A., Devillers, R., & Fortin, M.J. (2012) Spatio-temporal variations in invertebrate–cod–environment relationships on the Newfoundland-Labrador Shelf, vol. 1995–2009. *Marine Ecology Progress Series* **469**, 263–78.

Wright, S. (1943) Isolation-by-distance. *Genetics* **28**, 114–38.

Zuur, A.F., Ieno, E.N., Walker, N.J., Saveliev, A.A., & Smith, G.M. (2009) *Mixed Effects Models and Extensions in Ecology with R.* Springer, New York.

# PART 2

# METHODS

## Chapter 6

# SIMULATION MODELING IN LANDSCAPE GENETICS

*Erin Landguth,*[1] *Samuel A. Cushman,*[2] *and Niko Balkenhol*[3]

[1]*Division of Biological Sciences, University of Montana, USA*
[2]*Forest and Woodlands Ecosystems Program, Rocky Mountain Research Station, United States Forest Service, USA*
[3]*Department of Wildlife Sciences, University of Göttingen, Germany*

## 6.1  INTRODUCTION

In this chapter, we briefly introduce the general concepts and definitions related to simulation modeling and explain its potential for scientific research. We then focus specifically on the current utility and future potential of simulation modeling in landscape genetics. For this, we explain vital differences between landscape genetic simulations and population genetic simulations, and highlight simulation studies that have used landscape genetic simulations to (a) evaluate analytical approaches, (b) develop landscape genetic theory, or (c) enhance our interpretation of empirical findings. We also present an overview of available landscape genetic simulation software, provide basic guidelines for conducting simulation studies, and identify future research avenues related to landscape genetic simulation modeling. As we illustrate, simulation modeling has already substantially contributed to landscape genetics and increased consideration of simulation modeling can tremendously advance future developments in the field.

## 6.2  A BRIEF OVERVIEW OF MODELS AND SIMULATIONS

In essence, a model is a simplified version of reality. The simplification is achieved by focusing only on certain key characteristics and properties of selected systems, while disregarding all other characteristics and properties. For example, as explained in the first three chapters of this book, many different factors influence genetic variation in heterogeneous environments. Attempts to consider all of these factors simultaneously make it extremely challenging to understand how changes in a single factor will affect genetic diversity and structure in a particular landscape. Thus, instead of including everything that can possibly impact spatial genetic variation, researchers create models that only include certain factors that are of particular interest, or those that are assumed to be biologically most important. Indeed, a crucial aspect of modeling is to identify ***essential processes*** and ***essential parameters*** for accurately representing the system for the specific research question.

*Landscape Genetics: Concepts, Methods, Applications*, First Edition. Edited by Niko Balkenhol, Samuel A. Cushman, Andrew T. Storfer, and Lisette P. Waits.

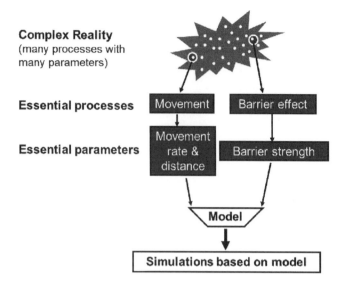

**Complex Reality**
(many processes with
many parameters)

**Essential processes**

**Essential parameters**

**Fig. 6.1** Schematic of the simplification process required for modeling and subsequent simulations. Out of complex reality, only few processes of interest are chosen. For example, if the aim of a simulation study is to assess the consequences of dispersal barriers to gene flow, essential processes would be movement behavior (i.e., the process of individuals moving across the landscape) and the barrier effect (i.e., the process of altering individual movements across the barrier). The essential processes are modeled via essential parameters, for example, the rate and distance of individual movements and the strength of the barrier (e.g., whether it is a complete barrier to movement or allows a certain proportion of individuals to successfully cross). Varying the essential parameters makes it possible to quantify how different barrier strengths affect gene flow in species with different movement characteristics. Note that essential processes and parameters are determined by the goal of the simulation study, and deciding which processes and parameters to include is a very important aspect when designing a simulation study.

Identifying these essential components of a model also means that researchers have to make **simplifying assumptions**, because not all processes and parameters influencing a system can be included in a model.

For example, a researcher might deem the process of dispersal to be especially important for a model that is supposed to represent gene flow across a fragmented landscape, while mutation may be deemed relatively unimportant. Thus, the model would include certain relevant parameters, such as dispersal rate and distance, but would not include other parameters, such as mutation rate or type. It is important to realize that our decisions about essential model components and simplifying assumptions are really just hypotheses that may or may not be valid. As such, the components used to build a model should be well justified in terms of biological understanding of the study organism. Representing the same study system with different essential components and simplifying assumptions may lead to quite different conclusions about how the system works. The final model should strike a good balance between model simplicity and realism. A model that is too complex will not make it any easier to understand reality or explain underlying generating mechanisms, while a model that is too simple will not lead to very meaningful insights about reality and the emergence of empirical patterns.

Models are also the basis for simulations. The Latin word *simulatus* (pp. of *simulare*) literally means "to make

like, to imitate, to copy, or to feign", and the purpose of a simulation is to mimic the behavior of a system of interest, based on how that system is represented in the underlying model (Fig. 6.1). Importantly, researchers can control the essential processes and parameters in their simulation model by including or excluding certain processes and by changing the parameter settings for the processes included in the model. Thus, simulation modeling can involve many different treatments and replicate measurements. This level of control is difficult or impossible to achieve with most ecological and genetic data sets gathered in the field, and simulations essentially present an alternative and complement to actual field experiments (see also Chapter 4).

## 6.3   GENERAL BENEFITS OF SIMULATION MODELING

The quasi-experimental framework of simulation models offers several important benefits for research in general. For example, the repeated simulation of a system with varying parameters allows researchers to assess how much confidence we can put in the conclusions derived from simulation results. This is important not only for making accurate scientific inferences but also for deriving reliable recommendations for practical applications. With empirical data, the range of parameter values and assumptions that can be tested is usually much more limited, potentially leading to weaker or more uncertain inferences.

Simulations can also mimic perfect sampling conditions, which can lead to stronger inferences. As explained in Chapter 4, study design can substantially alter the outcome of research projects, and simulations make it possible to circumvent variability induced by incomplete sampling of populations, imperfect spatial distribution of samples, or inadequate representation of environmental heterogeneity.

Importantly, simulation modeling can be used to predict how a system or its behavior will change if certain processes or parameters are altered. This is particularly relevant for predicting the effects of environmental change on a system or for evaluating the likely outcomes of various management scenarios. Moreover, simulation models can also be parameterized with historical data and thus help to address research questions relating to the past, for example, by testing whether historic conditions in the past actually could have resulted in data patterns observed today. Finally, by using explicit simulation models, researchers can allow others to directly replicate study results or to modify and extend previous studies. Thus, simulation modeling can increase the reproducibility of science if software or programming code, as well as researcher-defined parameters, are made publically available.

In sum, simulation modeling has high value for science and society, as it can be used to predict and explain, guide data collection, illuminate core dynamics of a system, discover new questions, bound outcomes to plausible ranges, quantify uncertainties, offer crisis options in near-real time, demonstrate trade-offs and suggest efficiencies, challenge the robustness of prevailing theory through perturbations, expose prevailing wisdom as incompatible with available data, train students and practitioners, educate the general public, reveal the apparently simple to be complex, and vice versa (Epstein 2007). Consequently, simulations are increasingly used in scientific research, have provided many important findings in various disciplines (e.g., Grimm et al. 2005), and are increasingly accepted by empiricists (Jeltsch et al. 2013).

A detailed treatment of general simulation modeling is beyond the scope of this chapter, but an excellent and comprehensive introduction can be found in Grimm and Railsback (2005, 2012). In the remainder of the chapter, we will specifically focus on simulation modeling in landscape genetics, and then will elaborate on certain aspects that distinguish these kinds of simulation models from other population genetic simulations.

## 6.4  LANDSCAPE GENETIC SIMULATION MODELING

Simulations have been used in genetic research for many years (e.g., Kimura and Otha 1974), and the availability of software for simulating genetic data is increasing steadily (reviewed in Hoban et al. 2012). However, it is important to realize several differences between landscape genetic simulations and "classic" population genetic simulations.

First, many population genetic simulation approaches generate genetic data only at the population level, meaning that the output of these simulations is typically in the form of summary statistics for each simulated population, for example, in terms of population-specific allele frequencies or inbreeding coefficients. In contrast, many landscape genetic simulations actually produce genetic data for every individual, even if these individuals are grouped into populations. Thus, landscape genetic simulations generally rely on *individual-based models (IBMs)*, which are often called *agent-based models (ABMs)* in the non-ecological literature. IBMs are classes of computational models for simulating the actions and interactions of autonomous individuals, with a key aspect being the variability among these individuals. Thus, individuals differ in their attributes (e.g., males versus females, different age classes) and these attributes influence the actions of individuals (e.g., different dispersal characteristics for males versus females), as well as their reactions to each other (e.g., mating schemes) or to other simulation settings (e.g., varying propensity to cross simulated barriers for males versus females).

These interindividual differences increase the realism of individual-based models, but also increase their complexity. This means that results obtained from IBMs can be very informative and meaningful, but only if the added complexity is adequately addressed through uncertainty and sensitivity analyses (see below). Since most landscape genetic simulations are based on IBMs, they also do not necessarily depend on classic population genetic theory, such as Hardy–Weinberg or linkage equilibrium, which often form the basis for more traditional population genetic simulations. Indeed, the individual-based nature of most landscape genetic simulations makes it possible to simulate genetic data and underlying processes without the need to define any discrete populations, which may be difficult to define for highly mobile species that are not restricted to discrete habitat patches or for very fine-scale analyses (i.e., analyses within, rather than among, populations).

A second major characteristic of landscape genetic simulations is the fact that they are always based on ***spatially-explicit*** models. These models are defined by placing individuals or groups of individuals (i.e., populations) on one- or two-dimensional regular lattices, or in irregular (*x*-, *y*-) coordinate space. Specific rules in the simulation model then define how individuals move and interact across space, for example, by defining the distances that individuals can move away from their birth location or the distance within which they can find a mating partner. Clearly, population genetic simulations are often also spatially-explicit (e.g., Balloux 2001), but while simulating population genetic data without space is possible, this is not the case for landscape genetic simulations.

In addition to space, another vital characteristic of landscape genetic simulations is the direct incorporation of environmental heterogeneity into the underlying model. This usually requires that in addition to the locations of individuals or populations in space, there also needs to be a spatial representation of the environment that individuals are placed in. Importantly, this environment is variable (i.e., non-homogeneous) in space, and potentially also in time, and it directly affects some or all of the essential processes included in the model. Thus, the rules that govern the actions and reactions of simulated individuals not only depend on pure space but also on the user-defined environmental heterogeneity included in the model. The fact that essential processes are directly affected by environmental heterogeneity is the key distinguishing feature between population genetic and landscape genetic simulation modeling.

In sum, the landscape genetic simulation modeling framework is generally individual-based, always spatially-explicit, and always incorporates environmental heterogeneity as a major influence on the actions and reactions of simulated organisms.

## 6.5  EXAMPLES OF SIMULATION MODELING IN LANDSCAPE GENETICS

An increasing number of studies have used landscape genetic simulations as defined above to study the effects of past and present environmental heterogeneity on population genetic processes, such as gene flow, drift, and selection (Table 6.1).

More specifically, landscape genetic simulations have been used primarily to investigate three categories of research questions. First, simulations have a particular value for evaluating the strengths and weaknesses of different *analytical* approaches and for determining how to best quantify landscape–genetic relationships. Second, simulations provide unmatched ability to evaluate *theoretical* questions related to how and why landscapes influence genetic diversity and structure. Third, simulations have been very fruitfully applied in guiding *empirical* studies of gene flow. Below, we briefly review several examples of simulation studies addressing each of these three main types of research questions.

## 6.5.1  Analytical evaluations: (When) Do methods work? How can we best quantify landscape-genetic relationships?

Evaluating the strengths and weaknesses of different statistical approaches in landscape genetics is among the most active research areas at present, and simulation modeling has contributed tremendously to advancing this important task. Balkenhol et al. (2009) was the first formal simulation study to compare multiple statistical approaches commonly used in landscape genetics. They simulated several different landscape genetic scenarios, comparing the statistical power, type I error rates, and the overall ability to lead researchers to accurate conclusions about landscape–genetic relationships among 11 statistical methods. Their results showed that commonly applied techniques (e.g., Mantel and partial Mantel tests) have high type I error rates and that different methods generally show only moderate levels of agreement. Based on these findings, they suggested combining different multivariate methods with different strengths and weaknesses for reliable landscape genetic analyses.

Cushman and Landguth (2010a) evaluated the specific performance of causal modeling with partial Mantel tests using simulation modeling. They found that simple correlational analyses between genetic data and proposed explanatory models indeed produce strong spurious correlations, which lead to incorrect inferences, but that the causal modeling approach was extremely effective at rejecting incorrect explanations,

**Table 6.1** Examples of simulation modeling in landscape genetics.

| Type | Research question | Reference |
|---|---|---|
| **Analytical:** *(When) Do methods work? How can we best quantify landscape–genetic relationships?* | Strengths and weaknesses of different statistical approaches | Balkenhol et al. (2009) |
| | Performance of causal modeling with partial Mantel tests | Cushman & Landguth (2010a); Cushman et al. (2013b) |
| | Sample design | Oyler-McCance et al. (2013) |
| | Sample size, number of alleles and loci | Landguth et al. (2012a); Prunier et al. (2013) |
| | Effects of spatial and temporal scale of analysis | Jaquiery et al. (2011) |
| | Influence of grain, extent and thematic resolution | Cushman & Landguth (2010b) |
| | Adaptive versus neutral markers; ability to detect barrier resistance | Landguth & Balkenhol (2012) |
| **Theoretical:** *How why does landscape heterogeneity influence genetics?* | Time lag before barrier signatures to emerge in spatial genetic data | Landguth et al. (2010a) |
| | Interactions between migration rates and landscape resistance barrier strengths | Landguth et al. (2010b) |
| | Limiting factors on strength of detectability, such as habitat area and fragmentation | Cushman et al. (2013a) |
| | Influence of landscape characteristics and home-range size | Graves et al. (2012) |
| **Empirical:** *Using simulation to elucidate, evaluate, and explain empirical observations* | Common vole and anthropogenic fragmentation | Gauffre et al. (2008) |
| | Mountain goats and landscape connectivity | Shirk et al. (2012) |
| | American marten and climate change | Wasserman et al. (2011, 2013) |

correctly identifying the true causal process, and rejecting highly correlated, but false, alternative models.

More recently, Graves et al. (2013) evaluated the causal modeling framework based on the partial Mantel test for identifying landscape resistance to dispersal. In their simulations, Graves et al. (2013) found that the causal modeling framework was not able to detect the correct underlying resistance model, and recommended to directly model the processes underlying dispersal and gene flow, thus indirectly calling for a greater use of simulation modeling in landscape genetics.

Finally, Cushman et al. (2013b) used simulation modeling to evaluate the effectiveness of several forms of the causal modeling framework to support the correct model and reject the null hypotheses. Their work found that partial Mantel tests have very low type II

error rates, but elevated type I error rates, consistent with the findings of Balkenhol et al. (2009) and Graves et al. (2013). Furthermore, Cushman et al. (2013b) found that the frequency of these errors is directly related to the degree of correlation between the true and alternative resistance model and proposed an improvement based on the ratio of relative support of the causal modeling diagnostic tests, instead of formal hypothesis testing. Recently, Castillo et al. (2014) used simulation modeling to evaluate the performance of this relative support approach to causal modeling with partial Mantel tests and found that it correctly identified the landscape variables driving gene flow among a large collection of highly correlated alternative models, although it seemed to consistently underestimate the magnitude of resistance.

In sum, even though simulation studies have not fully resolved ongoing debates about statistical approaches for landscape genetics, they clearly have helped to identify shortcomings of common analytical techniques, and have improved the reliability of landscape genetic inferences. Note that many papers suggesting novel analytical approaches in landscape genetics also incorporate simulations to illustrate the utility of the presented methods (e.g., Shirk & Cushman 2011; Wang 2012; Bradburd et al. 2013).

Aside from testing analytical methods, simulations have also been used effectively in several evaluations of the effects of sample design (see Chapter 4), effects of different genetic distance measures and genetic markers (e.g., Dyer et al. 2010), and effects of the spatial and temporal scale of analysis on landscape genetic results (e.g., Jaquirey et al. 2011). For example, Cushman and Landguth (2010b) used an individual-based, spatially-explicit simulation model to evaluate how changes to the grain, extent, and thematic resolution of landscape models affect the nature and strength of observed landscape genetic pattern–process relationships. They found that changes to the thematic resolution of resistance models produced very large effects on observed relationships, while changes to the grain and extent of analysis have smaller, but statistically significant, effects on landscape genetic analyses.

Blair et al. (2012) used a subset of the data simulated by Landguth et al. (2010a) and extended the original analyses by including additional individual-based methods. They found that Bayesian clustering methods (see Chapter 7) performed best overall for barrier detection, while boundary detection methods performed poorly. This result is congruent with findings of Safner et al. (2011), who also used simulations to investigate the ability of different methods for genetically detecting barriers.

### 6.5.2 Theoretical developments: How/why does landscape heterogeneity influence genetics?

As shown in Chapter 3, classic population genetic theory is based on idealized mathematical relationships that trade realism for mathematical tractability. There are likely major obstacles to developing a coherent mathematical theory that can explain the interactions of spatial complexity and temporal dynamics on genetic processes. However, simulations enable direct investigation of these questions, even in the absence of mathematical equations describing the processes, which enables rapid exploration and elaboration on theoretically very important questions in landscape genetics.

Thus, simulation modeling has a particularly important role in testing existing theories, developing new ideas, and identifying underlying mechanisms in landscape genetics. As noted in Chapter 3, spatial complexity and temporal variation violate critical assumptions of classical population genetic theory. Modifications of established mathematical theory to understand population genetic structure in spatially complex and temporally varying landscapes is extremely difficult. As a result, simulation modeling is of immense value in exploring how the pattern–process relationships change as these assumptions are relaxed. One of the foundational issues in developing landscape genetic theory is translating population genetics theory based on panmictic and discretely bounded populations to individual-based, spatially-explicit processes of dispersal, mating, and gene flow. For example, Landguth et al. (2010b) designed a simulation study to link spatially-explicit landscape genetics to classical population genetics theory. Using classical Wright–Fisher models and spatially-explicit, individual-based, landscape genetic models to simulate gene flow via dispersal and mating, they developed a mathematical formula that predicts the relationship between barrier strength (i.e., permeability) and the migration rate ($m$) across the barrier. In the process, they showed that relaxing some of the assumptions of the Wright–Fisher model can substantially change population substructure and argue that individual-based, spatially-explicit modeling provides a general framework to investigate how interactions between movement and landscape resistance drive population genetic patterns and connectivity across complex landscapes.

A second important question for both theoretical development and practical application of landscape genetics pertains to the effects of temporal lags in population response. Understanding how spatial genetic patterns are influenced by landscape change is crucial for developing a robust spatially-explicit landscape genetic theory, for reliable interpretation of empirical observations, as well as for development of management recommendations. Landguth et al. (2010b) used spatially-explicit, individual-based simulations to quantify the number of generations for new landscape barrier signatures to become detectable and for old signatures to disappear after barrier removal. They found that the lag time for the signal of a new barrier to become established is short using individual-based analyses, but much longer using population-based methods. These results suggest that individual-based approaches may have a higher power to detect effects of contemporary landscape features on genetic structure and connectivity. However, following the removal of a barrier formerly dividing a population, they found that individual-based and population-based approaches perform similarly. Both types of approaches are able to detect historical discontinuities from more than 100 generations ago.

A third type of theoretical development that has been addressed using landscape genetic simulations relates to the relationship between landscape structure and the strength and detectability of genetic differentiation caused by that structure. Cushman et al. (2013a) explored the impacts of limiting factors on landscape genetic processes using simulation modeling. Specifically, they quantified the effects of habitat area, fragmentation, and the contrast in resistance between habitat and non-habitat on the apparent strength and statistical detectability of landscape–genetic relationships. Their findings suggest that landscape genetic effects are often not detectable when habitat is highly connected or the contrast in resistance between habitat and non-habitat is low (i.e., the landscape structure does not limit gene flow). They also produced regression equations that can predict whether or not isolation-by-resistance (IBR) will be detected independently of isolation-by-distance as a function of (a) habitat fragmentation and (b) contrast in resistance between habitat and non-habitat. The approach could therefore provide a quantitative means to predict the strength and detectability of landscape–genetic relationships given a wide range of landscape configurations and relative resistances.

Finally, simulation modeling can also help to formally link landscape effects to key evolutionary processes through individual movement and natural selection, and thus has a high potential for adaptive landscape genetics (see Chapter 9). Jones et al. (2013) used a landscape genomics approach to reveal adaptive genetic variation along a cline and through simulations determined the sensitivity and error rates among different methods. Simulation modeling offers great promise for developing robust spatially-explicit evolutionary landscape genetic theory and offers an opportunity to understand the range of conditions and causes that can generate complex patterns. A few software packages for simulating both neutral and adaptive genetic data are already available (see Table 6.2).

### 6.5.3 Empirical applications: using simulation to elucidate, evaluate, and explain empirical observations

Simulations complement empirical research in several important ways. Empirical landscape genetic research uses inductive inference to associate patterns of observed genetic variation in real populations with hypotheses about what factors in the landscape may be responsible for creating that variation. In empirical studies, the true driving processes are never known, but only inferred. Simulation modeling makes it possible to validate these inferences by using the processes identified to be important for the empirical patterns as input for the simulations. If the inferred underlying processes have indeed formed the genetic patterns observed in the empirical data, then it should be possible to recreate these patterns through simulations that are based on these processes. Essentially, such *pattern-oriented* or *pattern-process modeling* evaluates whether an underlying process inferred through empirical induction can produce the patterns observed in the data and how well it can do so (e.g., Shirk et al. 2012).

Gauffre et al. (2008) is one of the first studies that coupled empirical data and simulation modeling to test the effects of landscape structure on genetic differentiation. They investigated the population genetic structure and gene flow pattern for the common vole (*Microtus arvalis*) in a heterogeneous landscape. They were surprised to find one panmictic population, especially because a large motorway that traversed the study area was expected to act as a dispersal barrier. Using computer simulations, they demonstrated that recent anthropogenic barriers to effective dispersal are difficult to detect through population-based analysis of species with large effective population sizes.

Similarly, Shirk et al. (2012) used simulations to validate a landscape resistance model previously developed by Shirk et al. (2010) for mountain goats (*Oreamnos americanus*) inhabiting a fragmented landscape in the Cascade Range, Washington. The simulations confirmed that the model identified in Shirk et al. (2010) was sufficient to explain the observed genetic structure in the mountain goat population, and that alternative explanations were not able to account for those patterns of genetic differentiation. The strong relationship

**Table 6.2** List of spatially-explicit landscape genetics simulation software packages and specific features. Type – type of simulation program: ***forward simulator*** (F) or ***backward simulator*** (B). Level – population-based (P) or individual-based (I). Land – incorporation of explicit environmental variables (E), cost distance matrix calculated from environmental variables (CD), landscape heterogeneity (LH). Note that links to these programs can be found on the website accompanying this book: www.landscapegenetics.info.

| Program name | Type | Level | Land | Example Use for Landscape Genetics | Reference |
|---|---|---|---|---|---|
| AquaSplatche | B | P | E | – | Neuenschwander (2006) |
| Splatche | B | P | E | Identifying the environmental factors that determine the genetic structure of populations. Foll and Gaggiotti (2006) | Ray et al. (2010) |
| CDPOP | F | P, I | CD, E | Landscape effects on gene flow for a climate-sensitive montane species, the American pika. Castillo et al. (2014) | Landguth & Cushman (2010) |
| CDFISH | F | P, I | CD | Combining demographic and genetic factors to assess population vulnerability in stream species. Landguth et al. (2014) | Landguth et al. (2012b) |
| EcoGenetics | F | P | LH | – | http://www2.unil.ch/ biomapper/ecogenetics/ index.html |
| SimAdapt | F | P, I | LH | – | Rebaudo et al. (2013) |
| QuantiNemo | F | P | LH | – | Neuenschwander et al. (2008) |

between the empirical and simulated patterns of genetic isolation provides independent validation of the Shirk et al. (2010) resistance model and demonstrates the utility of this pattern–process modeling approach in supporting landscape genetic inferences.

One of the greatest potential strengths of simulation modeling is in adapting the results of empirical studies to consider genetic connectivity over larger spatial extents or to reflect responses to future scenarios of landscape change. Wasserman et al. (2010) used individual-based analyses to discover that gene flow in American marten (*Martes americana*) in northern Idaho was primarily related to elevation and that alternative hypotheses involving isolation-by-distance (IBD), geographical barriers, effects of canopy closure, roads, tree size class, and an empirical habitat model were not supported. These results were extended by Wasserman et al. (2011, 2013) through simulation modeling to evaluate the influences of five potential future climate scenarios, each involving different levels of warming.

They found that even moderate warming scenarios resulted in very large reductions in population connectivity, and thus more isolated and smaller genetic neighborhoods. These, in turn, resulted in substantial loss of allelic richness and reductions in expected heterozygosity.

While this short overview of simulation studies is by no means exhaustive, it clearly illustrates the important contributions and future potential of simulation modeling for landscape genetics. In the next section, we present some general guidelines for researchers interested in conducting landscape genetic simulation modeling.

## 6.6 DESIGNING AND CHOOSING LANDSCAPE GENETIC SIMULATION MODELS

Simulation modeling studies in landscape genetics have recently begun to proliferate, as individual-based, spatially-explicit genetic modeling tools become more

(a)  (b)  (c)  (d)  (e)

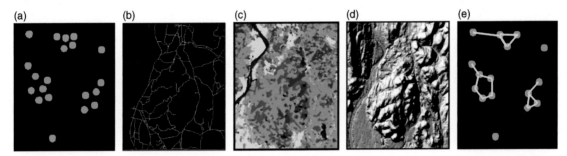

**Fig. 2.2** Conceptual models of landscape structure. The structure of a landscape can be represented in various ways. For example, the distribution of point elements (a) might be a suitable landscape model for a system in which the location of entities is the only factor that is important. Conversely, a linear network model (b) might be appropriate when the question involves connectivity of a hydrological network or the influences of a road network on fragmenting terrestrial habitats. A landscape mosaic model (c) could be chosen when the research goal is to assess the effects of different categorical land cover types on gene flow or selection. A gradient model (d) would be appropriate when gene flow or selection processes are affected by continuously varying attributes of a landscape such as elevation, density of vegetation or human population density. Landscape genetics studies sometimes represent spatial locations of genetic samples, with or without graph edges connecting them (e).

*Landscape Genetics: Concepts, Methods, Applications*, First Edition. Edited by Niko Balkenhol, Samuel A. Cushman, Andrew T. Storfer, and Lisette P. Waits.
© 2016 John Wiley & Sons, Ltd. Published 2016 by John Wiley & Sons, Ltd.

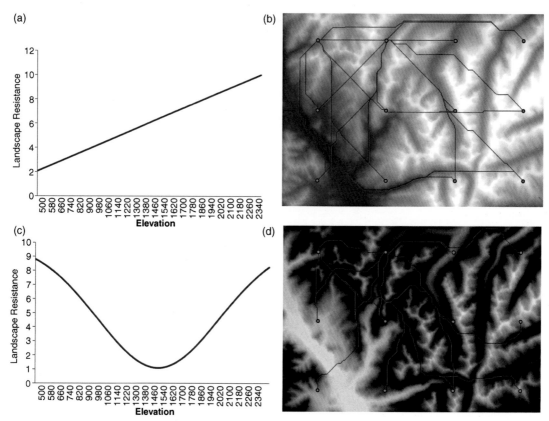

**Fig. 2.4** The thematic resolution is the resolution and functional form at which each factor included in the thematic content is represented. For example, the effect of elevation on a population process could be represented in various ways. For example, resistance to gene flow could increase linearly with elevation (a). In this case landscape resistance would be low in the valley areas (dark) and high on ridges (white) (b). The least-cost routes (red lines) among a network of individuals (green dots) would minimize cumulative cost by preferentially following low-elevation paths (b). Conversely, resistance to gene flow might be lowest at middle elevations (c). In this case resistance to gene flow would be lowest at intermediate elevations and higher in the deepest valleys and on the highest ridges (c), and least-cost routes connecting a network of individuals would preferentially follow paths that avoid low and high elevations (red paths in d). Notably, even though (b) and (d) both have the same thematic content (elevation) and address the same question (gene flow), they produce very different predictions of the pattern and degree of connectivity among individuals.

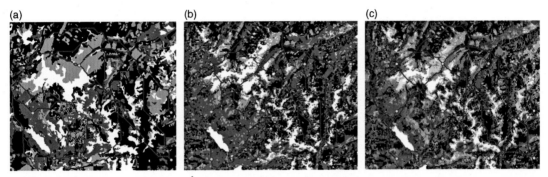

**Fig. 2.5** In a mosaic model of landscape structure and when the chosen thematic content is forest cover, there are a number of ways to represent the thematic resolution of forest in a landscape. For example, one could represent the effect of forest cover on a population process as four classes of different forest types (a), or as three classes of different successional stages (b), or as a combination of the four cover types and three seral stages (c). A network of individuals is shown as yellow dots and the least-cost paths connecting them are shown as red lines. The location of these least-cost routes and the relative cost of movement among pairs of points differ notably between these three thematic resolutions of forest cover.

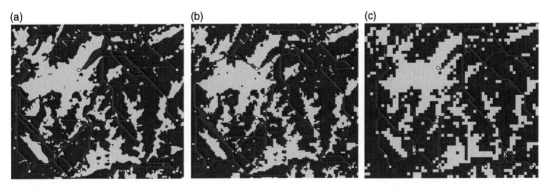

**Fig. 2.7** Effect of changing the grain on predicted connectivity patterns. This figure shows how changing the grain of a two-class mosaic of forest and non-forest affects predicted connectivity among a network of individuals. Individuals are shown as yellow points and the least-cost paths connecting all combinations of individuals are shown as red lines. The routes and relative cost among pairs of individuals change as the grain of the landscape map changes.

(a)   (b)   (c)

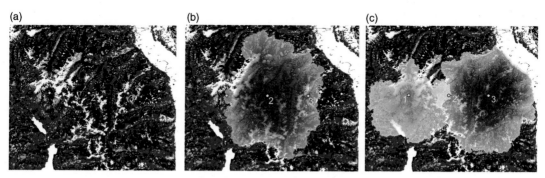

**Fig. 2.10** Depiction of genetic neighborhoods in a continuously distributed population on a resistance surface. The gray-scale background is a landscape resistance model in which gene flow is facilitated by a mature closed canopy forest; the resistance increases as the forest cover becomes open and is highest in non-forest cover types. The three numbered yellow dots represent the locations of three individuals taken from a continuous population that is distributed across the map. The blue patch surrounding point 2 in (b) is a dispersal kernel that originates on the location of individual 2 and extends outward to a maximum of 20,000 cost units. If this species had a genetic neighborhood extent of 20,000 cost units, this blue patch would indicate the extent of the genetic neighborhood centered on the location of individual 2. Both individuals 1 and 3 are within this genetic neighborhood (e.g., they are both covered by the blue dispersal kernel originating at individual 2). However, given that the dispersal kernel value is higher (darker blue) at the location of individual 3 than at the location of individual 1 (which is right on the edge of the kernel), one would predict higher genetic similarity between individuals 2 and 3 than between individuals 2 and 1, if gene flow is governed by the landscape features depicted in the resistance surface. In (c) the genetic neighborhood extent (20,000 cost units) is shown surrounding each of individuals 1 and 3. These two neighborhood kernels both overlap individual 2, but they do not overlap each other. That is, the genetic neighborhood extent surrounding individual 1 does not include individual 3 and vice versa. This shows the non-transitive nature of continuously distributed populations governed by isolation-by-distance or resistance. The extent of a local population is a function of each location and there are often no discrete boundaries between genetic neighborhoods.

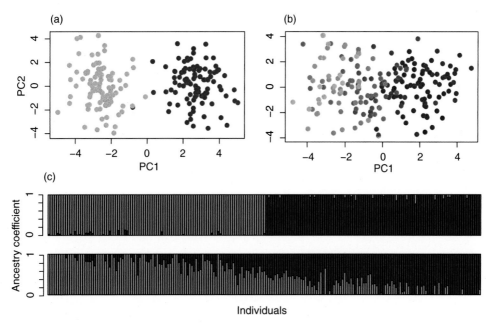

**Fig. 7.1** Results for PCA and STRUCTURE algorithms for simulated data without and with admixture. (a) and (b) PC plots for data from a two-population model without and with admixture. (c) STRUCTURE bar-plot representations of ancestry coefficients for the same data (top: no admixture, bottom: recent admixture).

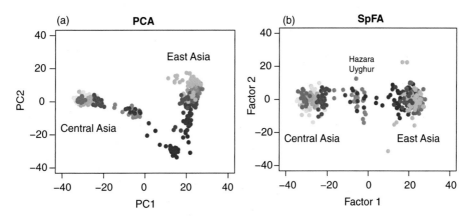

**Fig. 7.2** Principal component analysis and spatial factor analysis for human data (individuals from 27 Asian populations). (a) The PC1–PC2 plot shows a "horseshoe" structure as expected under an equilibrium IBD model. (b) The spatial factor analysis axes include correction for spatial autocorrelation, and exhibit two main clusters in Central and East Asia and two admixed populations (Hazara and Uyghur).

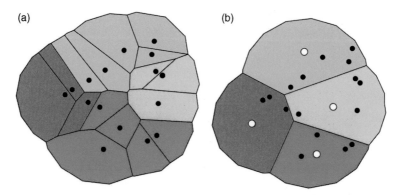

**Fig. Box 7.2** Use of spatial information in the prior distribution of MBC algorithms. (a) Dirichlet cells are centered on individuals in TESS and (b) Dirichlet cells are centered on population "territories" in GENELAND (white cells represent the territory centers).

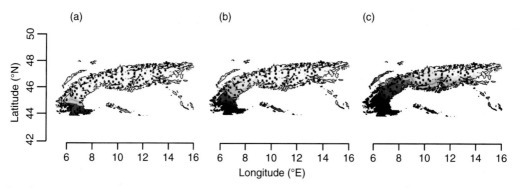

**Fig. Box 7.3** Predictions of contact zone movements in the alpine plant species *Gypsophila repens* under scenarios of climate change with increasing temperature levels. The predictions were obtained from the POPS model with $K = 2$ for mean annual temperatures rising by (a) 0 °C (current population structure), (b) 2 °C and (c) 4 °C. The grey color intensities represent the amounts of genetic ancestry in the warm south-western cluster. The dark line represents the contact zone between warm and cold populations, defined as 50% of ancestry in each cluster.

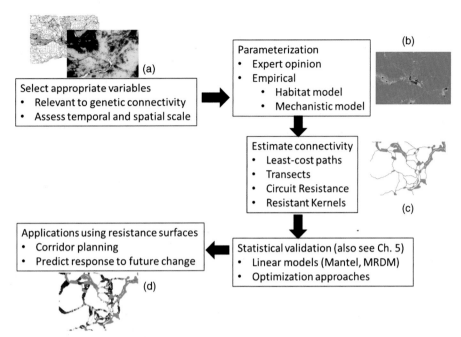

**Fig. 8.1** Framework for incorporating resistance surfaces into landscape genetic studies. The accompanying figures depict a scenario for each step on the same landscape. Inset (a) represents two spatial layers used as input for a resistance surface (in this case, roads and land cover). Inset (b) represents a resistance surface reflecting resistance due to roads and land cover, with warmer colors indicating higher resistance. Inset (c) shows a network of least-cost paths (connecting core areas in green) for connectivity across the resistance surface. Finally, inset (d) is a network of connectivity corridors based on the least-cost paths, with corridor width and color representing the degree of connectivity expected along the corridor (greater resistance along narrow paths with warm colors).

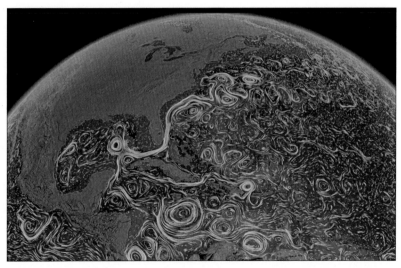

**Fig. 13.1** "Perpetual Ocean," a visualization of the world's surface ocean currents. Width of lines indicates speed of flow. The brightest features are large-scale stable and more stationary mean flow, which tends to be laminar (e.g., the Gulf Stream running around Florida and up the east coast of North America). Only large-scale eddies are depicted here, as they are fairly stable and slow moving (even when fast-spinning). Small-scale fleeting eddies are important transport features that are not visible here. The many thin, dash-size lines indicate areas of the ocean with slow flow, which can create areas "cut off" from mixing, such as the panhandle of Florida. (Picture from NASA/Reuters. See animation at http://www.nasa.gov/topics/earth/features/perpetual-ocean.html.)

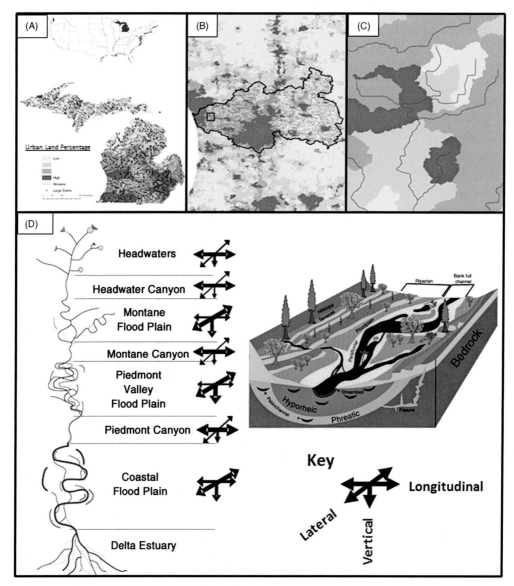

**Fig. 13.4** (A–C) Hierarchical representation of freshwater landscapes from regional to local scales including superficial (subsurface) sources of connectivity for aquatic organisms. (Figure provided by Dana Infante and Kyle Herreman.) (D) Landscape map highlighting hydrogeomorphologic features and habitat types, and with inset showing hydrodynamic features that make up flow. (From Stanford et al. 2005, Stanford 2006.)

widely available (see Table 6.2). As explained in Chapter 2, different landscape models can be used to represent real landscapes (e.g., habitat-matrix and gradient models, categorical and continuous landscape data). Thus, deciding on a landscape model that adequately represents environmental heterogeneity and its effects on the processes of interest is one of the first and most important steps when conducting landscape genetic simulation modeling. Similarly, researchers need to decide whether they are interested in genetic variation and spatial coordinates of individual organisms, or groups of individuals, and whether they want to model neutral genetic data or markers under selection.

After making decisions about what to include in the model and how to present different processes and variables, one either has to choose among available software options for the actual simulations (see below) or write new, tailored code. Either way, it is important to assure that the code used for the simulations has been *verified* and *validated* before *evaluating* its utility for certain research questions (see Rykiel 1996). **Verification** is essentially the process of error-checking and debugging the programming code. **Validation**, on the other hand, is the process of ensuring that the simulations actually accomplish their intended purpose. For example, it is often possible to define a parameter set for which specific results can be expected. Simulations that do not lead to expected results under "unambiguous" parameter settings may have an inadequate underlying model structure or may simply be coded incorrectly. Validation is a very important step in simulation modeling because it can be used to test whether decisions about essential features and simplifying assumptions are valid. Furthermore, if the programming architecture allows, landscape genetic simulators should be validated against the rich theory and equations from population genetics. For example, baseline Wright–Fisher assumptions could be stipulated and the theoretical rate of decay of heterozygosity could be matched against the individual-based simulations (e.g., Landguth and Cushman 2010; Rebaudo et al. 2013). Finally, **model evaluation** (also termed model credibility; Rykiel 1996) means that researchers assess how useful a model is for their specific research purpose. The utility of any given model depends on model structure (i.e., essential processes and parameters; simplifying assumptions), required input data and parameters that may or may not be available, speed of simulations, and type, tractability, and interpretability of simulation output.

### 6.6.1 Software for landscape genetics simulations

Because essential model features and simplifying assumptions will vary from study to study, it is not surprising that available programs for landscape genetic simulations focus on different processes and represent environmental heterogeneity in different ways. In Table 6.2, we summarize programs that can conduct landscape genetic simulations and highlight their specific utilities. We focus on IBMs that are spatially-explicit and allow inclusion of environmental variables affecting a key process. For a general review of genetic simulation software, see Hoban et al. (2012) or the Genetic Simulation Resources catalogue provided by the National Institute of Health (https://popmodels.cancercontrol.cancer.gov/gsr/).

Each program models environment differently. For example, AQUASPLATCHE and SPLATCHE use "vegetation" and "roughness" data sets to create a "resistance" that reduces effective migration rates by altering the survival of migrants. In contrast, CDPOP and CDFISH simulate dispersal and mating movement of individuals as a function of a user-defined resistance matrix that can be built from any number of environmental variables. In addition, CDPOP allows environmental surface inputs specific to genotypes under spatial selection. Thus, selection is implemented through differential survival of offspring as a function of the relative fitness of the offspring's genotype at the location on that surface where the dispersing individual settles. Similarly, SIMADAPT allows users to input a layer of landscape resistance, which determines the movement of simulated individuals, and also a layer of habitat quality, which influences local population dynamics (i.e., growth rates) and also the reproductive fitness of individuals. The software outputs both neutral genetic data and data under selection. ECOGENETICS uses landscape characteristics (patch size and topology) and metapopulation dynamics in its simulation approach. QUANTINEMO models population fusion/fission and thus incorporates landscape heterogeneity and habitat configuration and change through time to investigate the effects of selection, mutation, recombination, and drift on quantitative traits with varying architectures in structured populations.

Thus, different software can and should be be chosen to address specific research questions in landscape genetics.

### 6.6.2 Practical guidelines for conducting landscape genetic simulations

For those who want to conduct their own landscape genetic simulations or want to evaluate published studies, we recommend considering the following guidelines. First and foremost, evaluate carefully the model structure, its essential variables, and simplifying assumptions. Do you agree with this representation of reality? Are all essential processes and variables included that are of interest for the research question being addressed? Can you accept the simplifications? Remember that since a model is a simplified version of reality, it is in itself based on assumptions that may or not be correct. Thus, it is quite possible that two different simulation studies will lead to conflicting answers about a certain research question, simply because the underlying models vary. For example, Safner et al. (2011) and Blair et al. (2012) both used simulations to evaluate the utility of various analytical approaches for detecting barriers to gene flow. While the two studies came to similar conclusions overall, the time that methods needed to detect genetic barriers varied greatly among the two studies, ranging from several hundred generations in Safner et al. (2011) to less than 10 generations in Blair et al. (2012). This difference is likely due to the different simulation approaches that were used in the two studies, which involved a different model structure and implementation of processes. As stated by Box and Draper (1987) *"All models are wrong, but some are useful"*, and deciding whether the simplifications implemented in a model are acceptable is a major component of conducting and evaluating simulation studies.

Second, before investing a huge amount of time in setting up and running complex simulations with a model structure that fits your research needs, make sure to ascertain that the simulation model has been verified and validated. For this, it is very helpful to initially simulate small, simple data sets to learn how the model works and what output it provides. Just because a model promises to simulate data in a certain way, does not always mean that expected results will be obtained. If this is the case, it is vital to assess whether detected discrepancies between obtained and expected results are due to an incorrect model structure (i.e., assumptions about how the system works are incorrect to begin with) or because the model structure is not correctly programmed.

Third, keep it simple! Since simulations essentially represent multifactorial experiments, each additional process or variable adds another factor level to the experiment, and this increases the workload exponentially because different parameter settings of each variable usually need to be matched up with the parameter settings of all other parameter settings. Simulation experiments involving too many factorial levels do not only increase computer-processing time but are also much less tractable. In the end, a simulation involving too many processes, variables, or parameter alterations may not be much help in understanding landscape genetic complexity. This point is also very important to convey to reviewers and editors who evaluate simulation studies submitted to scientific journals. It may seem easy to add another parameter to an existing model or to obtain model results for a greater number of runs, but the time and analytical effort to do so should not be underestimated. More importantly, simplifying assumptions of a model essentially represent hypotheses about how a system works, and of course these hypotheses might be wrong. If one disagrees with a certain model structure or implementation, it should be seen as motivation for another simulation study that uses a different model based on different assumptions. This is no different from empirical research, where certain research topics are discussed for long periods of time based on conflicting scientific evidence. It takes many such empirical studies before their results can be synthesized and potentially lead to an agreement on a certain subject. Thus, it is unrealistic to think that a single simulation study can address all aspects of a given research question simultaneously or that results from a single simulation study will lead to a final and ultimate solution for a scientific problem. Indeed, the strength of simulation modeling is that processes and parameters can be controlled for, so that investigating different aspects of a research question is possible through multiple successive studies that disentangle a system in a step-by-step manner.

Fourth, we recommend conducting **sensitivity** and **uncertainty analyses**, and potentially also **robustness analyses**. Sensitivity analysis means that values of input parameters are varied by small increments for one variable at a time, while keeping all other variable values constant. These analyses help researchers assess the relative importance of individual variables in influencing model outcomes and predictions. On the other hand, uncertainty analysis makes it possible to quantify how likely certain study results will be achieved, given a certain range of input parameters. For example, confidence intervals can be calculated from multiple simulations that use the same parameter sets but involve one or several stochastic components. Finally, robustness analyses help to determine how much simulation results depend on essential features or simplifying assumptions. For this, several models for the same system need to be constructed that differ slightly in their simplifying assumptions. Report the results of these analyses and carefully assess how much confidence one can put into your results and conclusions.

Fifth, if software for your specific purpose is not available and you write your own code, make it available to the scientific community, verify, validate, and explain it sufficiently so that others can understand, evaluate, and improve it.

Finally, regardless of whether you use an existing software or write your own code, make sure to report exactly how your model is structured, what processes and variables it includes, and what parameter values you used. We strongly advocate the **ODD protocol** originally developed by Grimm et al. (2006) for describing and formulating (individual-based) models (see also Grimm et al. 2010; Grimm and Railsback 2005, 2012). ODD stands for Overview, Design concepts, and Details, and following this protocol ensures that other researchers can follow both the general model structure and the specifics of model parameterization and implementation. Since the ODD protocol can substantially help in clarifying the model structure and experimental design required for specific research questions, we actually recommend starting simulation projects with the ODD protocol.

## 6.7  THE FUTURE OF LANDSCAPE GENETIC SIMULATION MODELING

In sum, simulation modeling provides many interesting opportunities for landscape genetics and has already influenced the field substantially. However, simulation modeling also requires a certain amount of expertise in order to adequately apply and evaluate simulation models. Just like any other field study or experiment, simulation modeling needs well-thought-out designs for accurate and reliable inferences. Furthermore, it is vital to remember that no simulation model can or should be as complex as reality, so that their results are strictly valid only within the roam of the modeled system, with all its simplifications. Thus, results from

a certain model cannot always be generalized to other study systems, and two models representing the same system may lead to different conclusions.

In the future, we expect to see many more applications of simulation modeling in landscape genetics. Improved availability of simulation software, more widespread programming skills of young scientists, and faster and more powerful computer processors will greatly enhance landscape genetic simulations, which will play a major part in shaping the future development in the field. We envision that such future simulations will encompass much greater ecological realism, including individual variation in life-history strategies and behavior, including mate choice, landscape-dependent emi- and immigration decisions, or interindividual competition (e.g., for mates or space). Future models in landscape genetics will also include different modes of natural and sexual selection, interactions among multiple species, seasonal variation in the environment, and overlapping generations. With the help of such simulation approaches, landscape genetics will allow us to tackle many important and exciting research questions in ecology and evolution. In particular, we expect that simulation modeling will play a key role for the future development of adaptive landscape genetics (Chapter 9), because the many complex mechanisms linking neutral and adaptive processes in heterogeneous environments can only fully be assessed through quasi-experimental simulation studies. Simulation has and will continue to play an invaluable role in evaluating alternative landscape definitions, sampling designs and the power and effectiveness of emerging analytical techniques. We hope that researchers will expand their use of simulations in conjuction with field studies and controlled experiments to benefit from this synergy (e.g. Cushman et al. 2014).

## REFERENCES

Balloux, F. (2001) EASYPOP (Version 1.7): a computer program for population genetic simulations. *Journal of Heredity* **92**, 301–2.

Balkenhol, N., Waits, L.P., & Dezzani, R.J. (2009) Statistical approaches in landscape genetics: an evaluation of methods for linking landscape and genetic data. *Ecography* **32**, 818–30.

Blair, C., Weigel, D.E., Balazik, M., Keeley, A.T.H., Walker, F.M., Landguth, E.L., Cushman, S.A., Murphy, M., Waits, L.P., & Balkenhol, N. (2012) A simulation-based evaluation of methods for inferring linear barriers to gene flow. *Molecular Ecology Resources* **12**, 822–33.

Box, G.E.P. & Draper, N.R. (1987) *Empirical Model-Building and Response Surfaces*, John Wiley & Sons, Ltd, Oxford.

Bradburd, G., Ralph, P.L., & Coop, G.M. (2013) Disentangling the effects of geographic and ecological isolation on genetic differentiation. *Evolution* **67**, 3258–73.

Castillo, J.A., Epps, C.W., Davis, A.R., & Cushman, S.A. (2014) Landscape effects on gene flow for a climate-sensitive montane species, the American pika. *Molecular Ecology* **23**, 843–56.

Cushman, S.A. (2014) 'Grand challenges in evolutionary and population genetics: the importance of integrating epigenetics, genomics, modeling and experimentation.' *Frontiers in Genetics* **5**, 197.

Cushman, S.A. & Landguth, E.L. (2010a) Spurious correlations and inferences in landscape genetics. *Molecular Ecology* **19**, 3592–602.

Cushman, S.A. & Landguth, E.L. (2010b) Scaling landscape genetics. *Landscape Ecology* **25**, 967–79.

Cushman, S.A., Shirk, A.J., & Landguth, E.L. (2013a) Landscape genetics and limiting factors. *Conservation Genetics* **14**, 263–74.

Cushman, S.A., Wasserman, T.N., Landguth, E.L., & Shirk, A. (2013b) Re-evaluating casual modeling with Mantel tests in landscape genetics. *Diversity* **5**, 51–72.

Dyer, R.J., Nason, J.D., & Garrick, R.C. (2010) Landscape modeling of gene flow: improved power using conditional genetic distance derived from the topology of population networks. *Molecular Ecology* **19**, 3746–59.

Epstein, J.M. (2007) Remarks on the role of modeling in infectious disease mitigation and containment. In: Lemon, S.M., Hamburg, M.A., Sparling, P.F., Choffnes, E.R., & Mack, A. (eds.), *Ethical and Legal Considerations in Mitigating Pandemic Disease: Workshop Summary: Forum on Microbial Threats*. National Academies Press.

Foll, M. & Gaggiotti, O. (2006) Identifying the environmental factors that determine the genetic structure of populations. *Genetics* **174**, 875–91.

Gauffre, B., Estoup, A., Bretangnolle, V., & Cosson, J.F. (2008) Spatial genetic structure of a small rodent in a heterogeneous landscape. *Molecular Ecology* **17**, 4619–29.

Graves, T.A., Wasserman, T.N., Ribeiro, M., Landguth, E.L., Spear, S.F., Balkenhol, N., Higgins, C.B., Fortin, M.-J., Cushman, S.A., & Waits, L.P. (2012) The influence of landscape characteristics and home-range size on the quantification of landscape-genetics relationships. *Landscape Ecology* **27**, 253–66.

Graves, T.A., Beier, P., & Royle, A. (2013) Current approaches using genetic distances produce poor estimates of landscape resistance to interindividual dispersal. *Molecular Ecology* **22**, 3888–903.

Grimm, V. and Railsback, S.F. (2005) *Individual-Based Modeling in Ecology*. Princeton University Press, Oxford.

Grimm, V. and Railsback, S.F. (2012) Designing, formulating and communicating agent-based models. In: Heppenstall, A.J., Crooks, A.T., See, L.M., & Batty, M. (eds.), *Agent-Based Models of Geographical Systems*, Springer.

Grimm, V., Revilla, E., Berger, U., Jeltsch, F., Mooij, W.M., Railsback, S.F., Thulke, H.-H., Weiner, J., Wiegand, T., DeAngelis, D.L. (2005) Pattern-oriented modeling of agent-based complex systems: lessons from ecology. *Science* **310**, 987–91.

Grimm, V., Berger, U., Bastiansen, F., Eliassen, S., Ginot, V., Giske, J., Goss-Custard, J., Grand, T., Heinz, S., Huse, G., Huth, A., Jepsen, J.U., Jørgensen, C., Mooij, W.M., Müller, B., Pe'er, G., Piou, C., Railsback, S.F., Robbins, A.M., Robbins, M.M., Rossmanith, E., Rüger, N., Strand, E., Souissi, S., Stillman, R.A., Vabø, R., Visser, U., & DeAngelis, D.L. (2006) A standard protocol for describing individual-based and agent-based models. *Ecological Modelling* **198**, 115–26.

Grimm, V., Berger, U., DeAngelis, D.L., Polhill, G., Giske, J., Railsback, S.F. (2010) The ODD protocol: a review and first update. *Ecological Modelling* **221**, 2760–8.

Guillot, G. & Rousset, F. (2013) Dismantling the Mantel tests. *Methods in Ecology and Evolution* **4**, 336–44.

Hoban, S., Bertorelle, G. & Gaggiotti, O.E. (2012) Computer simulations: tools for population and evolutionary genetics. *Nature Review Genetics* **13**, 110–22.

Jaquiery, J., Broquet, T., Hirzel, A.H., Yearsley, J., & Perrin, N. (2011) Inferring landscape effects on dispersal from genetic distances: how far can we go? *Molecular Ecology* **20**, 692–705.

Jeltsch, F., Bonte, D., Pe'er, G., Reineking, B., Leimgruber, P., Balkenhol, N., Schröder, B., Buchmann, M.C., Mueller, T., Blaum, N., Zurell, D., Böhning-Gaese, K., Wiegand, T., Eccard, A.J., Hofer, H., Reeg, J., Eggers, U., & Bauer, S. (2013) Integrating movement ecology with biodiversity research – exploring new avenues to address spatiotemporal biodiversity dynamics. *Movement Ecology* **1**, 1–13.

Jones, M.R., Forester, B.R., Teufel, A.I., Adams, R.V., Anstett, D.N., Goodrich, B.A., Landguth, E.L., Joost, S., & Manel, S. (2013) Integrating spatially-explicit approaches to detect adaptive loci in a landscape genomics context. *Evolution* **677**, 3455–68.

Kimura, M. & Ohta, T. (1974) On some principles governing molecular evolution. *Proceedings of the National Academy of Sciences of the United States of America* **71**, 2848–52.

Landguth, E.L. & Balkenhol, N. (2012) Relative sensitivity of neutral versus adaptive genetic data for assessing population differentiation. *Conservation Genetics* **13**, pp. 1421–26.

Landguth, E.L. & Cushman, S.A. (2010) CDPOP: a spatially explicit cost distance population genetics program. *Molecular Ecology Resources* **10**, 156–61.

Landguth, E.L., Cushman, S.A., Schwartz, M.K., Murphy, M., McKelvey, K.S., & Luikart, G. (2010a) Quantifying the lag time to detect barriers in landscape genetics. *Molecular Ecology* **19**, 4179–91.

Landguth, E.L., Cushman, S.A., Murphy, M., & Luikart, G. (2010b) Relationships between migration rates and landscape resistance assessed using individual-based simulations. *Molecular Ecology* **10**, 854–62.

Landguth, E.L., Fedy, B., Garey, A., Mumma, M., Emel, S., Oyler-McCance, S., Cushman, S.A., Wagner, H.H., & Fortin, M.-J. (2012a) Effects of sample size, number of markers and allelic richness on the detection of spatial genetic pattern. *Molecular Ecology Resources* **12**, 276–84.

Landguth, E.L., Muhlfeld, C.C., & Luikart, G. (2012b) CDFISH: an individual-based, spatially-explicit, landscape genetics simulator for aquatic species in complex riverscapes. *Conservation Genetics Resources* **4**, 133–6.

Landguth, E.L., Muhlfeld, C.C., Waples, R.S., Jones, L., Lowe, W.H., Whited, D., Lucotch, J., Neville, H., & Luikart, G. (2014) Combining demographic and genetic factors to assess population vulnerability in stream species. *Ecological Applications* **24**, 1505–24.

Neuenschwander, S. (2006) AQUASPLATCHE: a program to simulate genetic diversity in populations living in linear habitats. *Molecular Ecology Notes* **6**, 583–5.

Neuenschwander, S., Hospital, F., Guillaume, F., & Goudet, J. (2008) quantiNEMO: an individual-based program to simulate quantitative traits with explicit genetic architecture in a dynamic metapopulation. *Bioinformatics* **24**, 1552–3.

Oyler-McCance, S.J., Fedy, B.C., & Landguth, E.L. (2013) Sample design effects in landscape genetics. *Conservation Genetics* **14**, 275–85.

Prunier, J.G., Kaufmann, B., Fenet, S., Picard, D., Pompanon, F., Joly, P., & Lena, J.P. (2013) Optimizing the trade-off between spatial and genetic sampling efforts in patchy populations: towards a better assessment of functional connectivity using an individual-based sampling scheme. *Molecular Ecology* **22**, 5516–30.

Ray, N. & Excoffier, L. (2010) A first step toward inferring levels of long distance dispersal during past expansions. *Molecular Ecology Resources* **10**, 902–14.

Ray, N., Currat, M., Foll, M., & Excoffier, L. (2010) SPLATCHE2: a spatially-explicit simulation framework for complex demography, genetic admixture and recombination. *Bioinformatics* **26**, 2993–4.

Rebaudo, R., Le Rouzic, A., Dupas, S., Silvain, J.-F., Harry, M., & Dangles, O. (2013) SimAdapt: an individidual-based genetic model for simulating landscape management impacts on populations. *Methods in Ecology and Evolution* **4**, 595–600.

Rykiel, E.J. (1996) Testing ecological models: the meaning of validation. *Ecological Modeling* **90**, 229–44.

Safner, T., Miller, M.P., McRae, B.H., Fortin, M.-J., & Manel, S. (2011) Comparison of Bayesian clustering and degree detection methods for inferring boundaries in landscape genetics. *International Journal of Molecular Sciences* **12**, 865–89.

Shirk, A.J. & Cushman, S.A. (2011) sGD software for estimating spatially explicit indices of genetic diversity. *Molecular Ecology Resources* **11**, 923–34.

Shirk, A.J., Wallin, D.O., Cushman, S.A., Rice, C.G., & Warheit, K.I. (2010) Inferring landscape effects on gene flow: a new model selection framework. *Molecular Ecology* **19**, 3603–19.

Shirk, A.J., Cushman, S.A., & Landguth, E.L. (2012) Simulating pattern–process relationships to validate landscape genetic models. *International Journal of Ecology* **2012**, 539109.

Wan, I. (2012) Examining the full effects of landscape heterogeneity on spatial genetic variation: a multiple matrix regression approach for quantifying geographic and ecological speciation. *Evolution* **67**, 3402–11.

Wasserman, T.N., Cushman, S.A., Schwartz, M.K., & Wallin, D.O. (2010) Spatial scaling and multi-model inference in landscape genetics: *Martes americana* in northern Idaho. *Landscape Ecology* **25**, 1601–12.

Wasserman, T.N., Cushman, S.A., Shirk, A.S., Landguth, E.L., & Little, J.S. (2011) Simulating the effects of climate change on population connectivity of American marten (*Martes americana*) in the northern Rocky Mountains, USA. *Landscape Ecology* **27**, 211–25.

Wasserman, T.N., Cushman, S.A., Little, J.S., Shirk, A.J., & Landguth, E.L. (2013) Population connectivity and genetic diversity of American marten (*Martes americana*) in the United States northern Rocky Mountains in a climate change context. *Conservation Genetics* **14**, 529–41.

*Chapter 7*

# CLUSTERING AND ASSIGNMENT METHODS IN LANDSCAPE GENETICS

*Olivier François[1] and Lisette P. Waits[2]*

[1]*Grenoble INP, Université Grenoble-Alpes, France*
[2]*Fish and Wildlife Sciences, University of Idaho, USA*

## 7.1 INTRODUCTION

Clustering and assignment methods that provide synthetic representations of population genetic structure and evaluate membership of individuals to a fixed number of genetic clusters are essential tools of landscape genetic researchers (Manel et al. 2003). Historically, predefined populations constituted the basis of traditional approaches for investigating population genetic structure with fixation indices (Wright 1951; Weir & Cockerham 1984; Hartl & Clark 1997; Holsinger & Weir 2009; see Chapter 3). Individuals were assigned to predefined genetic clusters on the basis of a **likelihood** function that computes the probability of their genotype for each cluster (Paetkau et al. 1995). The distinction between traditional population genetic approaches and modern clustering methods comes from the a *priori* definition of genetic clusters in traditional approaches. In modern clustering methods, genetic clusters are not predefined units (Pritchard et al. 2000; Manel et al. 2005). In these methods,

clusters are inferred from a set of individual multilocus genotypes by estimating genetic ancestry from unobserved source populations. The inferred clusters can then be used for assigning group membership to genotypes. In this chapter, we review statistical methods that ascertain population structure and estimate genetic ancestry without the use of predefined populations. We put an emphasis on methods that incorporate spatial and environmental heterogeneity data in their outputs. We summarize 10 methods that synthesize concepts from population genetics, landscape ecology, and landscape genetics, and that integrate geographic and ecological information on population samples (Table 7.1).

The chapter is structured as follows. The next section will describe two general approaches called ***exploratory data analysis*** (EDA) and ***model-based clustering*** (MBC). The general principle of EDA methods is to summarize multilocus genotype data sets within a reduced number of variables that are then used to delineate genetic clusters without any

---

*Landscape Genetics: Concepts, Methods, Applications*, First Edition. Edited by Niko Balkenhol, Samuel A. Cushman, Andrew T. Storfer, and Lisette P. Waits.
© 2016 John Wiley & Sons, Ltd. Published 2016 by John Wiley & Sons, Ltd.

**Table 7.1** Summary of 10 statistical software packages and their underlying models for estimating population structure in landscape genetics.

| Methods | Admixture | Features | Algorithm | Choice of $K$ | References |
|---|---|---|---|---|---|
| PCA | Yes | Axes of greatest variation | SVD (fast) | Tracy–Widom Cross-validation | Patterson et al. (2006) |
| sNMF | Yes | Mean-square estimates of ancestry proportions | Numerical optimization (fast) | Cross-validation | Frichot et al. (2013) |
| sPCA | – | Axes of greatest spatial autocorrelation | Eigenanalysis (fast) | – | Jombart et al. (2008) |
| spFA | Yes | Correction for IBD effects | SVD (fast) | Cross-validation | Frichot et al. (2012) |
| CCA | – | Correlation between genetic and environmental gradients | Regression methods (fast) | – | ter Brack (1986) |
| STRUCTURE | Yes | Bayesian estimation of ancestry proportions | MCMC (slow) | Estimation of model evidence | Pritchard et al. (2000) Falush et al. (2003) |
| TESS | Yes | Spatial estimation of ancestry proportions (Bayesian) | MCMC (slow) | Deviance information criterion | Chen et al. (2007) Durand et al. (2009) |
| BAPS | Yes | Spatial estimation of membership coefficients (Bayesian) | Numerical optimization (fast) | Split and merge algorithm | Corander et al. (2003) Corander et al. (2008) |
| GENELAND | – | Spatial estimation of membership coefficients | MCMC (slow) | Reversible jump algorithm | Guillot et al. (2005) |
| POPS | Yes | Correlation between ancestry and environmental gradients | MCMC (slow) | Deviance information criterion | Jay et al. (2011) Jay et al. (2012) |

Abbreviations: PCA, principal component analysis; sNMF, sparse non-negative matrix factorization; sPCA, spatial component analysis; spFA, spatial factor analysis; CCA, canonical correspondence analysis; SVD, singular value decomposition; MCMC, Markov chain Monte Carlo.

assumptions about the processes that generate the data. In contrast, MBC methods are based on population genetic models of the distribution of allele frequencies in structured populations. A classical MBC method is implemented in the computer program STRUCTURE, for which a popular algorithm underlies a simplified divergence model from a unique ancestral population (Pritchard et al. 2000).

In landscape genetics, landscapes are often regarded as highly heterogeneous areas that form the templates on which spatial patterns influence ecological and evolutionary processes (Chapter 2). A central focus is on incorporating spatial and environmental heterogeneity into population genetic analysis, and we survey the abilities of EDA and MBC methods to achieve this

objective. In the final part of the chapter, we will present some remaining challenges and future opportunities for clustering and assignment methods.

## 7.2 EXPLORATORY DATA ANALYSIS AND MODEL-BASED CLUSTERING FOR POPULATION STRUCTURE ANALYSIS

In clustering analysis, groupings can be analyzed based on *exploratory data analysis* (EDA) or *model-based clustering* (MBC) methods. Using multilocus genotypic data, EDA and MBC methods can estimate individual ancestry coefficients or predict membership of individuals to genetic clusters (Patterson et al. 2006; Pritchard et al.

2000). Both approaches provide useful visual representations of population genetic structure.

## 7.2.1    Exploratory data analysis

EDA methods summarize multilocus genotype data sets in terms of a reduced number of components that can be used to interpret the data. Many EDA methods apply to population and landscape genetics including, for example, ordination and multidimensional scaling techniques (McGarigal et al. 2000). The most representative method used for exploratory data analysis is a dimension reduction technique called *principal component analysis* (PCA) (Joliffe 1986; also see Chapter 5 for more information on PCAs). PCA has become a popular tool for exploring genetic data since its earliest use (Cavalli-Sforza

et al. 1994). Technically PCA seeks orthogonal projections of multidimensional data, called principal axes or components (PCs), along which the data show the highest variance. The first principal component (PC1) is the projection with the largest variance. Subsequent PCs represent summaries of the data, which are uncorrelated with the earlier PCs and account for the residual variance in the data. While the first two PCs are often used to explore the structure of variation in the sampled populations, it is generally useful to determine a number of significantly representative components, and this goal can be achieved with Tracy–Widom testing procedures (Patterson et al. 2006). Recent works on PCA include modifications of this method, which lead to clustering results similar to those obtained with MBC algorithms (Engelhardt and Stephens 2010; Frichot et al. 2014; see Box 7.1).

---

### Box 7.1   Theoretical connections between EDA and MBC algorithms

To better interpret the outputs of PCA and Bayesian clustering algorithms, an active domain of research consists of seeking theoretical connections between exploratory and model-based algorithms (Patterson et al. 2006; Engelhardt and Stephens 2010). The connection arises as most EDA and MBC algorithms attempt to represent the genotypic matrix as a product of two matrices. Those matrices are called scores and loadings in PCA. In the admixture model of STRUCTURE, they correspond to the ancestry matrix (**Q**) and ancestral allele frequency matrix (**F**). With this perspective, the inference problem solved by STRUCTURE may be viewed as a version of PCA with non-negative scores and loadings. The matrix factorization perspective also provides means to compute accurate least squares estimates of the **Q**-matrix. Least squares estimates can be computed using algorithms similar to PCA, and are much faster than STRUCTURE. For example, an efficient algorithm that can compute least squares estimates of ancestry coefficients is non-negative matrix factorization (NMF) (Lee & Seung 1999, 2001). NMF algorithms require only a few minutes of computing resources where STRUCTURE could take hours (Frichot et al. 2014; Parry & Wang 2013). When we applied NMF algorithms to our simulations of recent contact between two divergent populations, the ancestry coefficients strongly resembled STRUCTURE estimates (Pearson's correlation between estimates was greater than 0.97).

Theoretical connections between EDA and MBC can also be exploited to suggest a value for the number of clusters in MBC programs. Several approaches have been proposed to estimate the number of cluster, $K$, in STRUCTURE and the number of significant components in PCA, or the number of factors in *factor analysis*. Originally, Pritchard et al. (2000) suggested choosing the value of $K$ that maximizes the likelihood function, $\Pr(X|K)$, but acknowledged that it might oversplit genetic groups and should be interpreted cautiously. The current way of choosing $K$ in Bayesian clustering models is by applying a rule-of-thumb proposed by Evanno et al. (2005); which is based on an approximation of model evidence that considers the rate of change in the likelihood function. See Gilbert et al. (2012) for recommendations on the use of the Evanno method. An alternative to the Evanno method is the use of the deviance information criterion, a standard index for statistical model selection using Monte Carlo algorithms (Spiegelhalter et al. 2002; François & Durand 2010; Gao et al. 2011). Another way to choose $K$ is by applying Tracy–Widom tests for the significance of PC scores in PCA (Patterson et al. 2006). Patterson et al. (2006) proposed to choose $K$ clusters in STRUCTURE when they obtain $K - 1$ significant tests. With fast algorithms like NMF, a more accurate way of choosing $K$ is by using a cross-validation approach where the data set is divided into smaller subsets, each of them being used in turn for learning model parameters and for validating their predictions (Alexander & Lange 2011; Wold 1978; Eastment & Krzanowski 1982; Frichot et al. 2014).

### 7.2.2 Model-based clustering approaches

While EDA methods avoid formulating hypotheses about the biological processes that underlie the observed data, MBC methods incorporate population genetic models in their analysis. MBC approaches are computationally more intensive than EDA methods and their development is stimulated by the availability of powerful computing devices. MBC approaches infer clusters of individuals based on multilocus genotypes, and assign individuals to the identified clusters. In addition, many clustering models compute individual ancestry coefficients, considering alleles as proportionally inherited from two or more parental populations. The parental populations are usually thought of as being the relics of some ancestral populations. The ancestry coefficients represent the respective contributions of the ancestral populations to the observed sample. Because they can predict cluster membership without predefined populations, MBC approaches are thus useful when parental populations are not available to the study (Manel et al. 2005). In addition, MBC methods help to delineate population boundaries when those boundaries are uncertain (Pritchard et al. 2000; Cegelski et al. 2003).

The most commonly used MBC method is implemented in the computer program STRUCTURE (Pritchard et al. 2000). The STRUCTURE program includes models with or without admixture. Models without admixture assume that the sample consists of $K$ diverging genetic clusters. Individuals are then probabilistically assigned to one of the $K$ clusters and the resulting probabilities are called membership coefficients. Admixture models suppose that the genetic data originate from the mixing of $K$ putative parental populations that may be unavailable to the study. The ancestry coefficients correspond to the admixture proportions and can be computed for each individual in the sample. These coefficients are stored in a matrix, **Q**, in which each element, $q_{ik}$, represents the proportion of individual $i$'s genome that originates from the parental population $k$. An interpretation for the "clusters" in admixture models is as unobserved source populations that had diverged in the past, had reached equilibrium, and had been brought into contact again at a later date. In admixture models, individuals are not assigned to a cluster of origin, only alleles are, but the ancestry coefficients can be used to assign individuals to clusters.

Mathematically, Bayesian MBC algorithms like STRUCTURE describe the joint probability distribution of the genotypic and ancestry coefficient matrices. The joint probability distribution decomposes into the product of two terms: the likelihood and the prior distribution, which summarizes a priori information about the parameters. Posterior estimates for the parameters of interest are computed by updating the prior distribution based on the genetic data. The parameters are estimated by using Markov chain Monte Carlo (MCMC) algorithms (Gilks et al. 1996). MCMC methods use stochastic simulation algorithms to produce samples from the posterior distribution and their run-times are much slower than running a PCA. Applications of MBC analysis are abundant in the literature. Classical examples encompass the status of wildcat *Felis sylvestris* populations in Scotland and their hybridization with domestic cats (Beaumont et al. 2001), the delineation of population units for the wolverine (*Gulo gulo*) in Montana (Cegelski et al. 2003), or the colonization of the Swiss Alps by the Valais shrew (*Sorex antinorii*) (Yannic et al. 2008).

### 7.2.3 Visualization of PCA and STRUCTURE results

Visualization of EDA and MBC outputs is an important step that facilitates interpretation of results. A classic way of displaying PCA outputs is by using PC plots, which represent individual scores along the first and second principal components (Fig. 7.1A and B). PC1–PC2 plots may sometimes be restrictive, as describing population structure may require more than two principal components. Using the geographic origin of samples, an alternative approach is to represent synthetic maps (Cavalli-Sforza et al. 1994). Synthetic maps are useful geographic representations of genetic variation through space that interpolate PC scores to create continuous contours for each PC (see an illustration of synthetic maps in Manel et al. 2003). Regarding MBC outputs, a classic way to display ancestry coefficients stored in **Q**-matrices is by using bar-plot representations. Using bar-plot representations, each individual is represented by $K$ colored segments of length equal to their ancestry coefficients (Rosenberg et al. 2002; see Fig. 7.1C). One drawback of bar-plot representations is that they display a one-dimensional description of population structure. In this case, interpolation of ancestry coefficients on geographic maps can lead to useful two-dimensional representations of population genetic structure. An example of geographic interpolation of ancestry coefficients can be found for European populations of the plant species *Arabidopsis thaliana* (François et al. 2008).

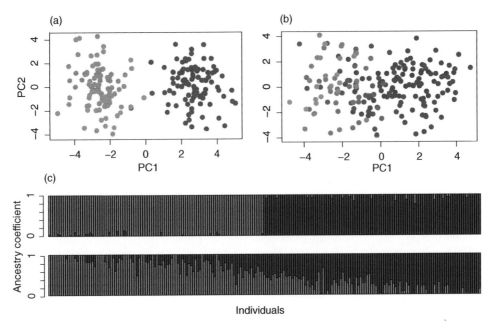

**Fig. 7.1** Results for PCA and STRUCTURE algorithms for simulated data without and with admixture. (a) and (b) PC plots for data from a two-population model without and with admixture. (c) STRUCTURE bar-plot representations of ancestry coefficients for the same data (top: no admixture, bottom: recent admixture). (For a color version of this figure, please refer to the color plates section.)

### 7.2.4  Simulated examples

To illustrate the use of PCA and MBC algorithms, we performed some simulations of genotypic data under two classic models of discrete and continuous populations, and examined the results produced by these methods. A first data set was simulated according to Wright's two-island model where evolution occurs under migration-drift equilibrium. Under this scenario, two populations of equal effective sizes, $N_e$, exchange an average of $mN_e$ migrants per generation, where $m$ is the migration rate of individuals. Using a coalescent simulation approach, we set $m = 2/N_e$ and generated genotypes at 50 diploid loci for samples of $n = 100$ individuals taken from each island. For the obtained genotypic data, PCA clearly separated the two samples along its first principal axis, PC1 (Fig. 7.1A). As expected under a two-population model, PC2 did not provide additional evidence of population structure. Put together, the two axes revealed an apparent population genetic structure consisting of two genetic clusters that matched with the samples from each island perfectly. When the admixture model of the Bayesian program STRUCTURE was used, the estimated ancestry

coefficients confirmed the results obtained from PCA. Individuals from each genetic cluster did not share any ancestry with individuals from the other cluster, suggesting that no admixed individuals were present in the sample (Fig. 7.1C).

Next, another data set was created to mimic recent contact and admixture between two divergent populations. In those data, the level of differentiation between the two ancestral populations, measured by $F_{ST}$, was around 0.05. Each individual genotype shared ancestry in each island population. As predicted for secondary contact zones, levels of genetic admixture were assumed to vary continuously along a geographic gradient. Here, variation of genetic admixture occurred along a longitudinal gradient. In this situation, PCA captured population genetic structure within PC1 (Fig. 7.1B). The PC1 scores, measuring the projection of the genetic data on the first principal axis, displayed continuous variation and exhibited high correlation with the simulated ancestry proportions ($R$-squared $= 0.77$). As expected from an approach that models admixture explicitly, the correlation between the estimated ancestry coefficients and their true value was higher with STRUCTURE than with PCA ($R$-squared $= 0.97$; Fig. 7.1C) and the admixture cline was well recovered by STRUCTURE.

## 7.3  SPATIALLY-EXPLICIT METHODS IN LANDSCAPE GENETICS

Geography is often reported to be a major determinant of population genetic variation in landscape genetic studies (Manel et al. 2003; Storfer et al. 2007). Recent developments of EDA and MBC algorithms include geographic information in the inference of genetic clusters. The goal of this category of methods is to improve the assignment of individuals to genetic clusters and account for isolation-by-distance (IBD) effects in statistical estimates of ancestry coefficients. EDA and MBC methods that consider spatial information reveal spatial genetic patterns more efficiently than methods that do not include this information. In this section, we describe landscape genetic EDA and MBC methods that make explicit use of geographic information.

## 7.4  SPATIAL EDA METHODS: SPATIAL PCA AND SPATIAL FACTOR ANALYSIS

Let us first review EDA methods that consider both geographic and genetic data. Spatial principal component analysis (sPCA) is a modification of the PCA algorithm in which principal axes maximize spatial autocorrelation instead of correlation (Jombart et al.

2008). The sPCA method is a particular case of Moran's eigenvector maps (Dray et al. 2006). In sPCA, there are positive and negative eigenvalues. Components associated with positive eigenvalues have positive autocorrelation and can be interpreted as describing global structures. Components associated with negative eigenvalues are assumed to describe local structures. For example, Jombart et al. (2008) used sPCA to discuss the identification of management units for conservation purposes in the endangered brown bear (*Ursus arctos*) (Waits et al. 2000; Manel et al. 2004). Their study detected differentiation among northern individuals along a gradient that correlates with the IBD gradients and found that the conservation units defined in previous studies corresponded to non-negative eigenvalues.

Spatial factor analysis (spFA) is a recent EDA approach that infers population genetic structure from spatial genetic data and accounts for statistical artifacts generated by IBD (Frichot et al. 2012). The spFA method is based on a covariance model for spatially correlated genotypes. For spatially-explicit coalescent simulations where genetic discontinuities were masked by IBD patterns, spFA identified discontinuities better than PCA or sPCA (Frichot et al. 2012). In addition, spFA algorithms are based on approximations of the genotypic matrix that reduce the dimension of the genetic data. The spFA algorithms are then appropriate for analyzing large genotypic data sets, including thousands of genetic polymorphisms. In Fig. 7.2, we applied PCA and spFA to a worldwide sample of genomic DNA from 388 individuals in 27 Asian populations (Frichot et al. 2012). The data set

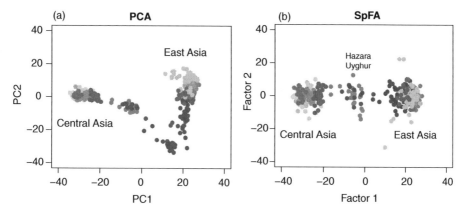

**Fig. 7.2** Principal component analysis and spatial factor analysis for human data (individuals from 27 Asian populations). (a) The PC1–PC2 plot shows a "horseshoe" structure as expected under an equilibrium IBD model. (b) The spatial factor analysis axes include correction for spatial autocorrelation, and exhibit two main clusters in Central and East Asia and two admixed populations (Hazara and Uyghur). (For a color version of this figure, please refer to the color plates section.)

used a panel of 10,664 SNPs from the Harvard Human Genome Diversity Project (Patterson et al. 2012). Samples from Central Asia were represented with light colors, whereas populations from East Asia were represented with dark colors. The PC plot exhibited a horseshoe effect, revealing the existence of IBD patterns in the data. PCA described a continuum of populations without observable genetic discontinuities (Fig. 7.2A). spFA was applied to the data using a scale parameter that corrected for the effects of IBD in the first factors (Fig. 7.2B). The results provided evidence of a major discontinuity separating two human genetic clusters, one in Central Asia and one in East Asia. Uyghur and Hazara populations were projected in an intermediate position, suggesting shared ancestry from ancestral gene pools located in Central and East Asia.

## 7.5  SPATIAL MBC METHODS

The spatial MBC models rely on geographically explicit prior distributions and adopt the same likelihood framework as non-spatial MBC methods (details are given in Box 7.2). The spatial MBC models are implemented in three main software packages: GENELAND (Guillot et al. 2005), TESS (Chen et al. 2007; Durand et al. 2009), and BAPS (Corander et al. 2008). Spatial models can be applied with or without admixture.

Regarding models without admixture, BAPS and TESS are based on Markov random *fields* that are spatial models of correlation between cluster memberships (François et al. 2006; Corander et al. 2008). GENELAND uses a prior distribution based on a colored tessellation model explained in Box 7.2 (Guillot et al. 2005). The original

---

**Box 7.2**  Incorporation of geographic data in TESS, BAPS, and GENELAND

The spatial models implemented in TESS and BAPS define the neighborhood of each individual based on a Voronoi tessellation of the study area (François et al. 2006; Corander et al. 2008). In a Voronoi tessellation, each individual sampling site is surrounded by a cell made of points that are closer to the given site than to any other sampling site. Two sampling sites are neighbors if their cells share a common edge. In TESS and BAPS, neighboring individuals are more likely to be co-assigned to a cluster than individuals that are far apart. In addition, the correlation between cluster labels decreases with the distance between sampling sites, as expected under spatially restricted dispersal and IBD models (Kimura & Weiss 1964). BAPS and TESS are based on correlation models and attempt to minimize the Walhund (1928) effect by incorporating spatial dependencies in their model. In GENELAND, the prior distribution on cluster labels is based on a different probabilistic model (Guillot et al. 2005). GENELAND detects genetic boundaries by considering that these boundaries separate $K$ random mating subpopulations. In GENELAND, Voronoi cells are associated with "territories", each of which groups several individuals within a unique cell. The geographic locations of the cells as well as their number are considered as parameters of the model and are estimated using an MCMC algorithm. The differences between TESS and GENELAND models are illustrated in the figure.

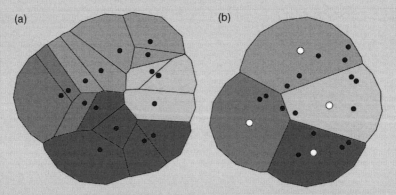

Use of spatial information in the prior distribution of MBC algorithms. (a) Dirichlet cells are centered on individuals in TESS and (b) Dirichlet cells are centered on population "territories" in GENELAND (white cells represent the territory centers). (For a color version of this figure, please refer to the color plates section.)

A relevant question is when GENELAND or TESS and BAPS algorithms could produce interesting insights on landscape genetic data. Based on simulation studies, the GENELAND model seems appropriate to detect recent linear geographic boundaries to gene flow (Blair et al. 2012; Safner et al. 2011). From the landscape genetic perspective, those barriers could be the consequence of recent landscape fragmentation and habitat destruction. For example, it is useful to study barrier effects of roads on local genetic diversity and structure using molecular approaches (Balkenhol & Waits 2009). In contrast, the TESS model addresses situations where barriers exhibit complex geographic shapes and may be permeable to migrants (Chen et al. 2007). In addition, TESS can also analyze population fusion events in addition to divergence events, which can be useful for detecting recent contact between previously isolated populations.

GENELAND model did not include admixture, but a recent update of the software implemented a model of admixture similar to TESS. For TESS, the admixture model does not require that the source populations have been sampled and the prior distribution on ancestry coefficients is spatially-explicit, decomposing the geographic variation of ancestry coefficients into regional and local scales. Spatial models of admixture are often more accurate than models without admixture, and they are useful for interpreting population structure when ecological or habitat changes create large-scale population movements (François & Durand 2010; Debout et al. 2011).

## 7.6 HABITAT AND ENVIRONMENTAL HETEROGENEITY MODELS

### 7.6.1 Going beyond geography

A particular feature of landscape genetic studies is a focus on landscape heterogeneity, which is characterized by a variety of habitat patches or gradients with different resources and levels of permeability (see Chapters 2 and 8). The non-uniformity of habitat can result in niche variations where some aspects of genetic variation and the variation of morphological traits are determined by the variety of habitats used by the populations. Association of genetic variation with habitat and ecological variables is a central question of landscape genetics (Manel et al. 2003; Storfer et al. 2007; Aitken et al. 2008; Balkenhol et al. 2009). For example, Richardson et al. (2009) analyzed correlations between climate and population genetic patterns in the western white pine (*Pinus monticola*). Their results suggested that divergent climatic selection has influenced population genetic structure and phenotypic traits associated with tree growth. Environmental and climatic conditions can influence the timing of phenological events in plant species, such as flowering, which is an obvious target for divergent selection (Stanton and Galen 1997; Stinson 2004; Jump and Penuela 2005; Franks et al. 2007). Measuring correlations between genetic variation and environmental conditions is thus an important step in the analysis and explanation of population genetic structure (Manel et al. 2003; Storfer et al. 2007; Jay et al. 2012; Schoville et al. 2012). In this section, we survey EDA and MBC methods that include models of habitat and habitat suitability and that evaluate correlations between genetic variation and environmental gradients.

### 7.6.2 Canonical correspondence analysis and redundancy analysis

An important class of EDA methods, called *constrained ordination*, can be employed to incorporate the effects of environmental gradients into landscape genetics studies. The methods include *canonical correspondence analysis* (CCA) and *redundancy analysis* (RDA). Originally introduced by ter Braak (1986); CCA was developed as a multivariate method to elucidate the relationships between biological assemblages of species and their environment. The CCA method is widely used in landscape ecology (Birks 1996; Cushman & McGarigal 2002; see Chapter 2). To apply constrained ordination, landscape geneticists can replace species abundance data by intraspecific diversity or genetic variation (Angers et al. 1999; Manel et al. 2003; Parisod & Christin 2008; Sork et al. 2010). For instance, Parisod and Christin (2008) used multivariate analyses to evaluate fine-scale ecological heterogeneity within continuous populations of *Biscutella laevigata*, a high mountain flowering plant. CCA identified the factors responsible for the partitioning of the genetic variance and revealed a composite effect of IBD, phenological divergence, and local adaptation to habitats characterized by different solar radiation regimes. CCA and RDA can also be designed to extract environmental gradients from landscape genetic data sets. These gradients are the basis for evaluating the differential habitat preferences (assuming niche variation) of populations via an ordination diagram (Sork et al. 2010). The constrained ordination methods thus provide a general framework for estimation and statistical testing of the effects of environmental variables on genetic variation and genetic diversity.

### 7.6.3 Bayesian MBC algorithms using environmental data

Bayesian methods that use predefined populations can identify environmental factors that correlate with population subdivision (Foll & Gaggiotti 2006). Similarly, MBC algorithms were recently extended to include environmental variables (Jay et al. 2011; 2012). Implemented in the computer program POPS, MBC algorithms can evaluate population structure and can simultaneously infer ancestry coefficients based on associations with environmental variables. Similar approaches – separating habitat and spatial covariates – are commonly adopted in landscape ecology (Lichstein et al. 2002). The POPS model measures the effects of environmental variables on individual ancestry while correcting for geographic trends and spatial auto-correlation. Box 7.3 gives example of application of POPS software to the alpine plant *Gypsophyla repens* (Jay et al. 2012).

---

**Box 7.3** **Predicting impact of climate change on population genetic structure**

Using ancestry models implemented in POPS, Jay et al. (2012) forecasted the magnitudes of displacement of contact zones between plant populations potentially adapted to warmer environments and other populations in 20 alpine plant species. For this they considered global temperature increase scenarios according to the IPPC predictions (IPPC 2007). In the European Alps, ADM predicted a global trend of movement in a north-east direction for 17 of the 20 plant species. A related approach was applied to California valley oaks, where regional distribution models were fitted to genetically differentiated groups with distinct responses to climate change (Sork et al. 2010).

The figure displays ADM projections obtained from the POPS algorithm for the alpine plant species *Gypsophyla repens*. The ADM maps were computed on the basis of 94 amplified fragment length polymorphisms genotyped in 319 individuals from 107 sampling sites in the European Alps (Jay et al. 2012). Climatic predictors used annual average of minimum and maximum daily temperatures (years 1980–1989) and spring and summer seasonal precipitations (years 1980–1989) measured at each site. For a temperature increase of 2 °C, contact zones were predicted to move in a north-east direction by approximately 81 km and by 210 km for an increase of 4 °C.

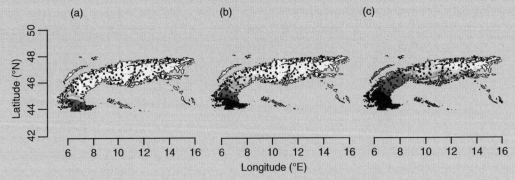

Predictions of contact zone movements in the alpine plant species *Gypsophila repens* under scenarios of climate change with increasing temperature levels. The predictions were obtained from the POPS model with $K = 2$ for mean annual temperatures rising by (a) 0 °C (current population structure), (b) 2 °C and (c) 4 °C. The grey color intensities represent the amounts of genetic ancestry in the warm south-western cluster. The dark line represents the contact zone between warm and cold populations, defined as 50% of ancestry in each cluster. (For a color version of this figure, please refer to the color plates section.)

### 7.6.4  Ancestry distribution models

Based on geographic and environmental data, EDA and MBC algorithms can provide routines to project ancestry coefficients on geographic maps under scenarios of environmental change, leading to *ancestry distribution models* (ADM). The ADM approach shares an analogy to species distribution models (SDM), which are statistical models correlating species abundance data to environmental predictors (Guisan & Zimmermann 2000). Approaches based on SDM are useful for inferring the realized niche of a species and to identify locations that satisfy the species habitat requirements by projecting the inferred niche on a geographic map. Analyzing data at a finer scale than SDMs, ADMs build projections of intraspecific genetic variation based on correlations between environmental variables and individual ancestry. ADMs can be used to estimate changes in population genetic structure and the magnitudes of pole-ward displacement for areas of mixed ancestry separating pairs of genetic clusters (Storfer et al. 2007; Thomassen et al. 2010). For example, forecasts from regression models can be done by varying climatic variables under various scenarios, considering temperatures globally increasing from 2 °C to 4 °C in the next century (IPCC 2007).

## 7.7  DISCUSSION

### 7.7.1  From landscape ecology and population genetics to landscape genetic methods

The development of EDA and MBC methods has been among the most active areas of methodological research in landscape ecology, population genetics, and landscape genetics (Manel et al. 2003; 2005; Storfer et al. 2007, 2010). Since the inception of landscape genetics, researchers have fostered the development of approaches that incorporate landscape heterogeneity in population genetic models. Geography and habitat modeling have been integrated in MBC approaches and geographic mapping of genetic variation is routinely used in landscape genetic studies. Ancestry distribution models, which summarize genetic variation at the intraspecific level in geographic space, enable predictions of change in population genetic structure under scenarios of environmental modification.

### 7.7.2  Interpretations of EDA and MBC outputs

EDA and MBC approaches have been developed from very distinct perspectives: EDA methods focus on the description and the explanation of the data, while MBC approaches enable statistical predictions. A shared objective of EDA and MBC methods is the description of ancestral relationships among individuals using landscape genetic data. From an evolutionary standpoint, genetic clusters are often interpreted as genetically divergent, discrete groups of individuals that can, for example, arise when gene flow is impeded by physical obstacles and particular landscape or environmental features. In many cases, habitat fragmentation and species range contraction induce genetic drift that can cause local genetic divergence and lead to genetically clustered populations. For most species, population demography results, however, in phases of expansion, and interpretation of discrete populations is complicated by the existence of ancient events and continuous genetic variation through geographic space.

### 7.7.3  Ancient events

Though landscape genetic studies have focused more intensively on trying to understand recent and current landscape effects on genetic variation, the discipline is also concerned with untangling these effects from more ancient events (Balkenhol et al. 2009; Anderson et al. 2010; Landguth et al. 2010). For example, changes in local conditions, such as climate, might have triggered population movements toward more favorable areas and resulted in modifications of species distributions. Populations isolated for a long time might be brought into contact in a certain area, leading to genetic admixture of different gene pools in contact zones (Barton & Hewitt 1985; Hewitt 2000). EDA and MBC methods have enough flexibility to detect genomic signatures of both fission and fusion events when they have occurred during the history of a species. In this case, individuals can trace ancestry to more than one genetic cluster, and genetic clusters are better interpreted as ancestral populations that are not necessarily represented in actual samples.

### 7.7.4  Continuous variation

Continuous genetic variation commonly materializes through the observation of genetic gradients and patterns of IBD. The concept of IBD was introduced by S. Wright to describe the accumulation of local genetic differences under spatially restricted dispersal (Wright 1943). In species where interbreeding occurs at small distances and where there are continuously distributed populations, the theory predicts that genetic differentiation will increase with geographic distance and generate spatial autocorrelation in allele frequencies. IBD is often idealized within models in which regularly spaced subpopulations can exchange migrants only locally (stepping stone models; Kimura & Weiss 1964). Several recent studies have reported distortions caused by IBD in patterns inferred from PCA or from Bayesian clustering methods (Novembre & Stephens 2008; Frantz et al. 2009; Durand et al. 2009; Schwartz & McKelvey 2009; François et al. 2010). In PCA those distortions generate a "horseshoe effect" that may bias interpretations of population genetic structure (Fig. 7.2). The MBC program STRUCTURE may discern multiple clusters where there is only a single large area with IBD (Frantz et al. 2009; Schwartz & McKelvey 2009). The problems generated by IBD effects arise because of the presence of spatial autocorrelation in the genetic data, which is not modeled by standard EDA or MBC approaches. The problems can be alleviated by using approaches presented in this chapter that correct for IBD effects (spFA, BAPS, or TESS) (Meirmans 2012; Frichot et al. 2012).

### 7.7.5  Strengths and weaknesses of EDA and MBC methods

Model assumptions are sometimes critical in MBC and the number of genetic clusters detected by MBC (or EDA) algorithms does not necessarily correspond to the number of biologically meaningful populations in the sample (Pritchard et al. 2000; Kalinowski 2011; François & Durand 2010). In addition, EDA and MBC results may be inaccurate when population sampling is geographically irregular, when key populations remain unsampled, or when sampling efforts are unevenly distributed among populations (Rosenberg et al. 2005; Schwartz & McKelvey 2009; McVean 2009). For EDA and MBC methods, the ability to detect population structure generally depends on the degree of genetic differentiation among groups, on the degree of spatial overlap of genetic groups (for spatial methods), on total sample size, and on the number of markers (Latch et al. 2006; Chen et al 2007; Patterson et al. 2006; Fogelqvist et al. 2010). Predictions of ancestry distribution models based on environmental variables also depend on several critical assumptions and simplifications, and ADM are more accurate when environmental change occurs within short timescales (Jay et al. 2012). Results obtained with EDA and MBC methods therefore need to be interpreted with caution, and the users of these methods should be aware of the above limitations.

The advantage of using EDA methods is that these approaches scale with the dimension of data, tolerating hundreds of thousands of markers and thousands of individuals. With PCA, it is however sometimes difficult to summarize the information contained in more than two components and researchers sometimes restrict the description of the data to PC1–PC2 plots. Using MBC approaches can overcome this issue as MBC algorithms produce graphical representations that allow users to visualize $K$ clusters simultaneously. Recently developed EDA approaches like NMF, however, provide direct estimates of ancestry coefficients, which can be visualized through bar-plot representations. These new approaches can compute ancestry coefficients within run-times that are much faster than STRUCTURE (Parry & Wang 2013; Frichot et al. 2014).

The advantage of using MBC methods is that they include prior distributions based on environmental and geographic data, and improve the description of the genetic data. The use of informative prior distributions can be particularly beneficial when the number of markers is moderate (tens of markers) and when detailed information about geography and habitat is available (Chen et al. 2007). A last advantage of MBC algorithms is to provide an efficient statistical framework for the automatic input of missing data. In conclusion, no method can be universally better than the other methods. The choice of an approach could be based on the ability of each approach to capture the essential features in the particular data studied, which could be done using information criteria and model-checking in MBC or cross-validation in EDA.

### 7.7.6  Divergent selection and population structure

The majority of applications of EDA and MBC methods in landscape genetics have focused on analyses of neutral markers to infer a population genetic structure. The increasing ease of generating large genomic data sets using next-generation sequencing methods is transforming population and landscape genetics (see Chapter 9). We are transitioning from analyses of 10–20 neutral microsatellite loci or a few 100 AFLP markers to data sets containing thousands of SNPs that include both neutral and adaptive loci. This will provide new opportunities for understanding the role of divergent selection and adaptation in shaping population genetic structure as well as new challenges. For example, conservation geneticists are developing a new framework for defining conservation units that consider the different types of information provided by neutral and adaptive markers (Funk et al. 2012).

When using genomics data, it will be important to distinguish between neutral and adaptive loci since they reflect the outcome of different genetic processes. Neutral loci are dominated by the processes of gene flow and drift and reflect demographic history, while adaptive loci are dominated by the process of natural selection (Chapter 3). Thus, the inadvertent inclusion of adaptive loci in genotypic matrices could bias interpretations of neutral population structure. Conversely, the presence of population structure could inflate the number of false positive tests for detection of loci undergoing selective pressure due to local adaptation (Frichot et al. 2013; see Chapter 9). Most methods to detect allele frequencies that correlate with environmental gradients require an accurate assessment of population structure (Schoville et al. 2012). One great challenge for the future of landscape genetics is to design methods to infer population genetic structure while separating neutral from adaptive genetic variation. This will become a necessary preliminary step when analyzing genomic data sets that will inevitably include alleles under selection (Schoville et al. 2012).

## REFERENCES

Aitken, S.N., Yeaman, S., Holliday, J.A., Wang, T., & Curtis-McLane, S. (2008) Adaptation, migration or extirpation: climate change outcomes for tree populations. *Evolutionary Applications* **1**, 95–111.

Alexander, D.H. & Lange, K. (2011) Enhancements to the ADMIXTURE algorithm for individual ancestry estimation. *BMC Bioinformatics* **12**, 246.

Anderson, C.D., Epperson, B.K., Fortin M.-J., Holderegger, R., James, R.M.A., Rosenberg, M.S., Scribner, K.T., & Spear, S. (2010) Considering spatial and temporal scale in landscape-genetic studies of gene flow. *Molecular Ecology* **19**, 3565–75.

Angers, B., Magnan, P., Plante, M., & Bernatchez, L. (1999) Canonical correspondence analysis for estimating spatial and environmental effects on microsatellite gene diversity in brook charr (*Salvelinus fontinalis*). *Molecular Ecology* **8**, 1043–53.

Balkenhol, N. & Waits, L.P. (2009) Molecular road ecology: exploring the potential of genetics for investigating transportation impacts on wildlife. *Molecular Ecology* **18**, 4151–64.

Balkenhol N., Gugerli F., Cushman S.A., Waits, L.P., Coulon, A., Arntzen, J.W., Holderegger, R., Wagner, H.W., & Participants of the Landscape Genetics Research Agenda Workshop 2007 (2009) Identifying future research needs in landscape genetics: where to from here? *Landscape Ecology* **24**, 455–63.

Barton, N.H. & Hewitt, G.M. (1985) Analysis of hybrid zones. *Annual Review of Ecology and Systematics* **16**, 113–48.

Beaumont, M., Barratt, E.M., Gottelli, D., Kitchener, A.C., Daniels, M.J., Pritchard, J.K., & Bruford, M.W. (2001) Genetic diversity and introgression in the Scottish wildcat. *Molecular Ecology* **10**, 319–33.

Birks, H.J.B. (1996) Statistical approaches to interpreting diversity patterns in the Norwegian mountain flora. *Ecography* **19**, 332–40.

Blair, C., Weigel, D.E., Balazik, M., Keeley, A.T.H., Walker, F.M., Landguth, E.L., Cushman, S.A., Murphy, M., Waits, L.P., & Balkenhol, N. (2012) A simulation-based evaluation of methods for inferring linear barriers to gene flow. *Molecular Ecology Resources* **12**, 822–33.

Cavalli-Sforza, L.L., Menozzi, P., & Piazza, A. (1994) *The History and Geography of Human Genes*. Princeton University Press, Princeton, NJ.

Cegelski, C., Waits, L.P., & Anderson, N.J. (2003) Assessing population structure and gene flow in Montana wolverines (*Gulo gulo*) using assignment-based approaches. *Molecular Ecology* **12**, 2907–18.

Chen, C., Durand, E., Forbes, F., & François, O. (2007) Bayesian clustering algorithms ascertaining spatial population structure: a new computer program and a comparison study. *Molecular Ecology Notes* **7**, 747–56.

Corander, J., Sirén, J., & Arjas, E. (2008) Bayesian spatial modeling of genetic population structure. *Computational Statistics* **23**, 111–29.

Corander, J., Waldmann, P., & Sillanpää, M.J. (2003) Bayesian analysis of genetic differentiation between populations. *Genetics* **163**, 367–74.

Cushman, S.A. & McGarigal K. (2002) Hierarchical, multi-scale decomposition of species–environment relationships. *Landscape Ecology* **17**, 637–46.

Debout, G.D.G., Doucet, J.-L., & Hardy O.J. (2011) Population history and gene dispersal inferred from spatial genetic structure of a Central African timber tree, *Distemonanthus benthamianus* (Caesalpinioideae). *Heredity*, **106**, 88–99.

Dray, S., Legendre, P., & Peres-Neto, P. (2006) Spatial modeling: a comprehensive framework for principal coordinate analysis of neighbours matrices (PCNM). *Ecological Modeling* **196**, 483–93.

Durand, E., Jay, F., Gaggiotti, O.E., & François, O. (2009) Spatial inference of admixture proportions and secondary contact zones. *Molecular Biology and Evolution* **26**, 1963–73.

Eastment, H.T. & Krzanowski, W.J. (1982) Cross-validatory choice of the number of components from a principal component analysis. *Technometrics* **24**, 73–7.

Engelhardt, B.E. & Stephens, M. (2010) Analysis of population structure: a unifying framework and novel methods based on sparse factor analysis. *PLoS Genetics* **6**, 12.

Evanno, G., Regnaut, S., & Goudet, J. (2005) Detecting the number of clusters of individuals using the software STRUCTURE: a simulation study. *Molecular Ecology* **14**, 2611–20.

Falush, D., Stephens, M., & Pritchard, J.K. (2003) Inference of population structure using multilocus genotype data: linked loci and correlated allele frequencies. *Genetics* **164**, 1567–87.

Fogelqvist, J., Niittyvuopio, A., Ågren, J., Savolainen, O., & Lascoux, M. (2010) Cryptic population genetic structure: the number of inferred clusters depends on sample size. *Molecular Ecology Resources* **10**, 314–23.

Foll, M. & Gaggiotti, O.E. (2006) Identifying the environmental factors that determine the genetic structure of populations. *Genetics* **174**, 875–91.

François, O. & Durand, E. (2010) Spatially explicit Bayesian clustering models in population genetics. *Molecular Ecology Resources* **10**, 773–84.

François, O., Ancelet, S., & Guillot, G. (2006) Bayesian clustering using hidden Markov random fields in spatial population genetics. *Genetics*, **174**, 805–16.

François, O., Blum, M.G.B., Jakobsson, M., & Rosenberg, N.A. (2008) Demographic history of European populations of *Arabidopsis thaliana*. *PLoS Genetics* **4**, e1000075.

François, O., Currat, M., Ray, N., Han, E., Excoffier, L., & Novembre, J. (2010) Principal component analysis under population genetic models of range expansion and admixture. *Molecular Biology and Evolution* **27**, 1257–68.

Franks, S.J., Sim, S., & Weis, A.E. (2007) Rapid evolution of flowering time by an annual plant in response to a climate fluctuation. *Proceedings of the National Academy of Sciences of the United States of America* **104**, 1278–82.

Frantz, A.C., Cellina, S., Krier, A., Schley, L., & Burke, T. (2009) Using spatial Bayesian methods to determine the genetic structure of a continuously distributed population: clusters or isolation-by-distance? *Journal of Applied Ecology* **46**, 493–505.

Frichot, E., Schoville, S.D., Bouchard, G., & François, O. (2012) Correcting principal component maps for effects of spatial autocorrelation in population genetic data. *Frontiers in Genetics* **3**, 254.

Frichot, E., Schoville, S.D., Bouchard, G., & François, O. (2013) Testing for associations between loci and environmental gradients using latent factor mixed models. *Molecular Biology and Evolution* **30**, 1687–99.

Frichot, E., Mathieu, F., Trouillon, T., Bouchard, G., & François, O. (2014) Fast and efficient estimation of individual ancestry coefficients. *Genetics* **196**, 973–83.

Funk, W.C., McKay, J.K., Hohenlohe, P.A., & Allendorf, F.W. (2012) Harnessing genomics for delineating conservation units. *Trends in Ecology and Evolution* **27**, 489–96.

Gao, H., Bryc, K., & Bustamante, C.D. (2011) On identifying the optimal number of population clusters via the Deviance Information Criterion. *PLoS One* **6**, e21014.

Gilbert, K.J., Andrew, R.L., Bock, D.G., Franklin, M.T., Kane, N.C., Moore, J.-S., Moyers, B.T., Renaut, S., Rennison, D.J., Veen, T., & Vines, T.H. (2012) Recommendations for utilizing and reporting population genetic analyses: the reproducibility of genetic clustering using the program structure. *Molecular Ecology* **21**, 4925–30.

Gilks, W.R., Richardson, S., & Spiegelhalter, D.J. (1996) *Markov Chain Monte Carlo in Practice.* Chapman and Hall, New York.

Guillot, G., Estoup, A., Mortier, F., & Cosson, J.F. (2005) A spatial statistical model for landscape genetics. *Genetics* **170**, 1261–80.

Guisan, A. & Zimmermann, N.E. (2000) Predictive habitat distribution models in ecology. *Ecological Modeling* **135**, 147–86.

Hartl, D.L. & Clark, A.G. (1997) *Principles of Population Genetics*, 3rd edition. Sinauer Associates Inc., Sunderland, MA.

Hewitt, G. (2000) The genetic legacy of the quaternary ice ages. *Nature* **405**, 907–13.

Holsinger, K.E. & Weir, B.S. (2009) Genetics in geographically structured populations: defining, estimating and interpreting FST. *Nature Reviews Genetics* **10**, 639–50.

Intergovernmental Panel on Climate Change (IPCC) (2007) *Climate Change 2007 – The Physical Science Basis: Working Group I Contribution to the Fourth Assessment Report of the IPCC*, vol. **4**, Cambridge University Press, Cambridge, UK, and New York, USA.

Jay, F., François, O., & Blum, M.G.B. (2011) Predictions of native American population structure using linguistic covariates in a hidden regression framework. *PLoS One* **6**, e16227.

Jay, F., Manel, S., Alvarez, N., Durand, E.Y., Thuiller, W., Holderegger, R., Taberlet, P., & François, O. (2012) Forecasting changes in population genetic structure of alpine plants in response to global warming. *Molecular Ecology* **21**, 2354–68.

Jolliffe, I.T. (1986) *Principal Component Analysis.* Springer Verlag, New York.

Jombart, T., Devillard, S., Dufour, A.B., & Pontier, D. (2008) Revealing cryptic spatial patterns in genetic variability by a new multivariate method. *Heredity* **101**, 92–103.

Jump, A.S. & Penuelas, J. (2005) Running to stand still: adaptation and the response of plants to rapid climate change. *Ecology Letters* **8**, 1010–20.

Kalinowski, S.T. (2011) The computer program structure does not reliably identify the main genetic clusters within species: simulations and implications for human population structure. *Heredity* **106**, 625–32.

Kimura, M. & Weiss, G.H. (1964) The stepping stone model of population structure and the decrease of genetic correlation with distance. *Genetics* **49**, 561–76.

Latch, E.K., Dharmarajan, G., Glaubitz, J.C., & Rhodes, O.E. Jr., (2006) Relative performance of Bayesian clustering software for inferring population substructure and individual assignment at low levels of population differentiation. *Conservation Genetics* **7**, 295–302.

Landguth, E.L., Cushman, S.A., Schwartz, M.K., McKelvey, K.S., Murphy, M., & Luikart, G. (2010) Quantifying the lag time to detect barriers in landscape genetics. *Molecular Ecology* **19**, 4179–91.

Lee, D.D. & Seung, H.S. (1999) Learning the parts of objects by non-negative matrix factorization. *Nature* **401**, 788–91.

Lee, D.D. & Seung, H.S. (2001) Algorithms for non-negative matrix factorization. *Advances in Neural Information Processing Systems* **13**, 556–62.

Lichstein, J.W., Simons, T.R., Shriner, S.A., & Franzreb, K.E. (2002) Spatial autocorrelation and autoregressive models in ecology. *Ecological Monographs* **72**, 445–63.

Manel, S., Schwartz, M.K., Luikart, G., & Taberlet, P. (2003) Landscape genetics: combining landscape ecology and population genetics. *Trends in Ecology and Evolution* **18**, 189–97.

Manel, S., Bellemain, E., Swenson, J., & François, O. (2004) Assumed and inferred spatial structure of populations: the Scandinavian brown bears revisited. *Molecular Ecology* **13**, 1327–31.

Manel, S., Gaggiotti, O.E., & Waples, R.S. (2005) Assignment methods: matching biological questions with appropriate techniques. *Trends in Ecology and Evolution* **20**, 136–42.

McGarigal, K., Cushman, S.A., & Stafford, S.G. (2000) *Multivariate Statistics for Wildlife and Ecology Research*. Springer-Verlag, New York.

McVean, G. (2009) A genealogical interpretation of principal components analysis. *PLoS Genetics* **5**.

Meirmans, P.G. (2012) The trouble with isolation-by-distance. *Molecular Ecology* **21**, 2839–46.

Novembre, J. & Stephens, M. (2008) Interpreting principal component analyses of spatial population genetic variation. *Nature Genetics* **40**, 646–9.

Paetkau, D., Calvert, W., Stirling, I., & Strobeck C. (1995) Microsatellite analysis of population structure in Canadian polar bears. *Molecular Ecology* **4**, 347–54.

Parisod, C. & Christin, P.-A. (2008) Genome-wide association to fine-scale ecological heterogeneity within a continuous population of *Biscutella laevigata* (Brassicaceae). *New Phytologist* **178**, 436–47.

Parry, R.M. & Wang, M.D. (2013) A fast least-squares algorithm for population inference. *BMC Bioinformatics* **14**.

Patterson, N., Price, A., & Reich, D. (2006) Population structure and eigenanalysis. *PLoS Genetics* **2**, e190.

Patterson, N., Moorjani, P., Luo, Y., Mallick, S., Rohland, N., Zhan, Y., Genschoreck, T., Webster, T., & Reich, D. (2012) Ancient admixture in human history. *Genetics* **192**, 1065–93.

Pritchard, J.K., Stephens, M., & Donnelly, P. (2000) Inference of population structure using multilocus genotype data. *Genetics* **155**, 945–59.

Richardson, B., Rehfeldt, G., & Kim, M. (2009) Congruent climate related genecological responses from molecular markers and quantitative traits for western white pine (*Pinus monticola*). *International Journal of Plant Sciences* **170**, 1120–31.

Rosenberg, N.A., Pritchard, J.K., Weber, J.L., Cann, H.M., Kidd, K.K., Zhivotovsky, L.A., & Feldman, M.W. (2002) The genetic structure of human populations. *Science* **298**, 2381–5.

Rosenberg, N.A., Mahajan, S., Ramachandran, S., Zhao, C., Pritchard, J.K., & Feldman, M.W. (2005) Clines, clusters and the effect of study design on the inference of human population structure. *PLoS Genetics* **1**, e70.

Safner, T., Miller, M.P., McRae, B.H., Fortin, M.J., & Manel, S. (2011) Comparison of Bayesian clustering and edge detection methods for inferring boundaries in landscape genetics. *International Journal of Molecular Sciences* **12**, 865–89.

Schoville, S.D., Bonin, A., François, O., Lobreaux, S., Melodelima, C., & Manel S. (2012) Adaptive genetic variation on the landscape: methods and cases. *Annual Review of Ecology, Evolution and Systematics* **43**, 23–43.

Schwartz, M.K. & McKelvey, K.S. (2009) Why sampling scheme matters: the effect of sampling scheme on landscape genetic results. *Conservation Genetics* **10**, 441–52.

Sork, V.L., Davis, F.W., Westfall, R., Flint, A.L., Ikegami, M., Wang, H., & Grivet, D. (2010) Gene movement and genetic association with regional climate gradients in California valley oak (*Quercus lobata* Née). *Molecular Ecology* **19**, 3806–23.

Spiegelhalter, D.J., Best, N.G., Carlin, B.P., & van der Linde, A. (2002) Bayesian measures of model complexity and fit. *Journal of the Royal Statistical Society: Series B – Statistical Methodology* **64**, 583–639.

Stanton, M.L. & Galen, C. (1997) Life on the edge: adaptation versus environmentally mediated gene flow in the snow buttercup, *Ranunculus adoneus*. *American Naturalist* **150**, 143–78.

Stinson, K.A. (2004) Natural selection favors rapid reproductive phenology in *Potentilla pulcherrima* k (Rosaceae) at opposite ends of a subalpine snowmelt gradient. *American Journal of Botany* **91**, pp. 531–539.

Storfer, A., Murphy, M.A., Evans, J.S., Goldberg, C.S., Robinson, S., Spear, S.F., Dezzani, R., Delmelle, E., Vierling, L. & Waits, L.P. (2007) Putting the "landscape" in landscape genetics. *Heredity* **98**, 128–42.

Storfer, A., Murphy, M.A., Spear, S.F., Holderegger R., & Waits, L.P. (2010) Landscape genetics: Where are we now? *Molecular Ecology* **19**, 3496–514.

ter Braak, C.J.F. (1986) Canonical correspondence analysis: a new eigenvector technique for multivariate direct gradient analysis. *Ecology* **67**, 1167–79.

Thomassen, H.A., Cheviron, Z.A., Freedman, A.H., Harrigan, R.J., Wayne, R.K., & Smith, T.B. (2010) Spatial modeling and landscape-level approaches for visualizing intra-specific variation. *Molecular Ecology* **17**, 3532–48.

Wahlund, S. (1928) Zusammensetzung von Population und Korrelationserscheinung vom Standpunkt der Vererbungslehre aus betrachtet. *Hereditas* **11**, 65–106.

Waits, L.P., Taberlet, P., Swenson, J., Sandegren, F., & Franzen, R. (2000) Nuclear DNA microsatellite analysis of genetic diversity and gene flow in the Scandinavian brown bear (*Ursus arctos*). *Molecular Ecology* **9**, 421–31.

Weir, B.S. & Cockerham, C.C. (1984) Estimating F-statistics for the analysis of population structure. *Evolution* **38**, 1358–70.

Wold, S. (1978) Cross-validatory estimation of the number of components in factor and principal components models. *Technometrics* **20**, 397–405.

Wright, S. (1943) Isolation-by-distance. *Genetics* **28**, 114–38.

Wright, S. (1951) The genetical structure of populations. *Annals of Eugenics* **15**, 323–54.

Yannic, G., Basset, P., & Hausser, J. (2008) Phylogeography and recolonization of the Swiss Alps by the Valais shrew (*Sorex antinorii*) inferred with autosomal and sex-specific markers. *Molecular Ecology* **17**, 4118–33.

# Chapter 8

# RESISTANCE SURFACE MODELING IN LANDSCAPE GENETICS

Stephen F. Spear,[1] Samuel A. Cushman,[2] and Brad H. McRae[3]

[1]The Orianne Society, USA
[2]Forest and Woodlands Ecosystems Program, Rocky Mountain Research Station, United States Forest Service, USA
[3]The Nature Conservancy, North America Region

## 8.1 INTRODUCTION

**Resistance surfaces** are an important component of the landscape genetics toolbox and can be defined as spatial layers that represent the extent to which the conditions at each grid cell constrain movement or gene flow. The basic framework for using resistance surfaces in landscape genetics involves the steps of variable selection, parameterizing values for resistance surfaces, correlating resistance with genetic data, and, in many cases, employing the best supported resistance surfaces in downstream analyses. Important considerations for variable selection include whether the variables are hypothesized to influence genetic connectivity, the appropriate spatial and thematic scale, and the accuracy of spatial data used to create resistance surfaces. Parameterizing resistance surfaces can be done through expert opinion or empirical methods. The ideal method for parameterizing resistance surfaces differs somewhat depending on the study system, although empirical methods are likely to provide the best outcome if they can be directly related to genetic connectivity. Such empirical methods can be based on tracking data, habitat suitability models, or optimization using genetic data. In addition to the method used to parameterize the surface, the translation of the resistance surface to a connectivity model is a critical step in most analyses, with the most common methods being least-cost paths and circuit theory. Each of these methods is based on different assumptions and thus the most appropriate method should be assessed for each study individually. Resistance surfaces that correlate strongly with genetic measures can then be used for applications such as corridor design, barrier detection, or predictions based on scenarios of future landscape change. We anticipate that resistance surfaces will continue to evolve as technology increases the accuracy and resolution of remotely sensed data, genomic advances present opportunities to investigate applicability of resistance surfaces to questions

*Landscape Genetics: Concepts, Methods, Applications*, First Edition. Edited by Niko Balkenhol, Samuel A. Cushman, Andrew T. Storfer, and Lisette P. Waits.
© 2016 John Wiley & Sons, Ltd. Published 2016 by John Wiley & Sons, Ltd.

of natural selection, and models of future landscape and climate scenarios become more refined. Overall, we expect that resistance surfaces will remain a key tool in landscape genetics and will be especially important for those whose interest in landscape genetics focuses on its potential as a predictive science for the various facets of global change that are occurring.

### 8.1.1   What is a resistance surface?

With respect to landscape genetics, a resistance surface is defined as a spatial layer whose values represent the extent to which the conditions at each grid cell constrain movement or gene flow (Spear et al. 2010). Thus, resistance values can represent the willingness of an organism to cross a particular environment, the physiological cost of moving through it, the reduction in survival for the organism, or an integration of all these factors (Zeller et al. 2012). Resistance surfaces can represent a single landscape variable, or several variables can be combined

into a single surface representing multiple variables. The creation of multivariate resistance surfaces, particularly when multiple surfaces need to be created for comparison or hypothesis testing, can be facilitated by the Gnarly Landscape Utilities toolkit (McRae et al. 2013).

### 8.1.2   Using resistance surfaces: a framework

The resistance surface has become a widely used tool in landscape genetics due to its applicability to multiple types of predictive models such as least-cost paths or circuit theory (see below). The underlying simplicity and ease of use of resistance surfaces belie the difficulty of developing, parameterizing (i.e., assigning cost values to landscape variables), and implementing a biologically justified resistance surface (Spear et al. 2010; Zeller et al. 2012). There is a general process involved in any landscape genetic study using resistance surfaces, although the methodology at each step certainly varies (Fig. 8.1). This process occurs in 4–5 steps,

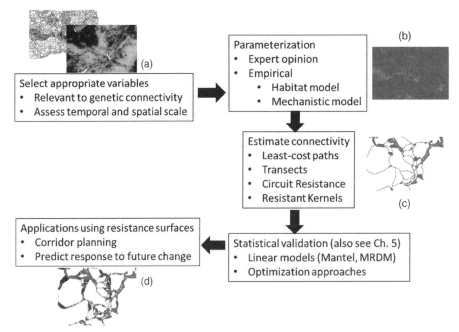

**Fig. 8.1** Framework for incorporating resistance surfaces into landscape genetic studies. The accompanying figures depict a scenario for each step on the same landscape. Inset (A) represents two spatial layers used as input for a resistance surface (in this case, roads and land cover). Inset (B) represents a resistance surface reflecting resistance due to roads and land cover, with lighter colors indicating higher resistance. Inset (C) shows a network of least-cost paths (connecting core areas in gray) for connectivity across the resistance surface. Finally, inset (D) is a network of connectivity corridors based on the least-cost paths, with corridor width and color representing the degree of connectivity expected along the corridor (greater resistance along narrow paths with lighter colors). (For a color version of this figure, please refer to the color plates section.)

beginning with the selection of appropriate landscape or environmental variables that will be used to derive the resistance surface. The second step is to use these variables to parameterize resistance surface(s). Third, the resistance surface is used to develop estimates or hypotheses of connectivity, and then a statistical model is used to test the relationship between the estimate of landscape connectivity and genetic connectivity. Optionally, researchers can then use the resistance surface with the greatest statistical support to conduct additional analyses, such as identifying conservation corridors or predicting genetic structure into the future. This chapter will focus on the various methods that have been used to achieve each of these steps and provide guidance on the benefits and drawbacks of each method depending on study objectives. We then conclude by discussing future directions for the use of resistance surfaces in landscape genetics.

### 8.1.3  Selecting variables for resistance surfaces: initial questions and assumptions

The most important first question to ask when developing a resistance surface is what factors are likely to create resistance for the species of interest (i.e., landscape definition; Chapter 2). In landscape genetics, researchers use genetic relationships as the dependent variable, so the resistance surface must directly relate to genetic connectivity to be meaningful in a landscape genetic analysis. Resistance surfaces can be applied to a number of different processes such as animal movement or habitat occupancy (Zeller et al. 2012), and these processes may vary in how directly they correlate with genetic structure. As described in Chapter 3, there are many processes and variables that can affect population genetic structure. Estimates of genetic connectivity are influenced by such factors as dispersal (either through active movement of animals or passive dispersal of seed or pollen by wind or animals), population history, population size and spatial distribution, mutation rate, and reproductive mode. A careful consideration of the most important factors influencing genetic structure is needed to inform the best approach to developing resistance surfaces, or whether a resistance surface is even appropriate for the study system. For instance, if dispersal among patches of good habitat is the primary driver for genetic connectivity, then a resistance surface should ideally represent the extent

and configuration of these habitat patches as well as properties of the matrix separating patches that influence movement among the patches. On the other hand, if the system of interest follows source-sink dynamics in which sinks have reduced reproduction relative to sources, then resistance surfaces depicting effects of the landscape structure on movement should be coupled with fitness surfaces that represent habitat quality and enable analysis to couple the separate processes that drive movement and differential mortality in the source-sink system (e.g., Landguth et al. 2012).

### 8.1.4  What factors dictate the utility of variables for resistance surfaces?

There are several factors that may influence whether the variables used to parameterize a resistance surface can be effectively used in a landscape genetic analysis. These include scale and resolution of landscape variables used, accuracy of remotely-sensed data, and temporal match of resistance surface and genetic data. Each of these should be considered when selecting variables for a landscape genetic study with resistance surfaces.

Scale is an essential consideration in any biological investigation and landscape genetics is no different (Anderson et al. 2010). Resistance surfaces are in many ways defined by the scale of the variables that comprise them (Chapter 2). This is most obvious with respect to the pixel size (the grain) of the resulting resistance surface. As most resistance surfaces are developed from remotely sensed layers, many resistance surfaces typically use the grain of the source GIS layer without transparent consideration of the biological relevance (Sawyer et al. 2011). The defense for this is that remotely sensed layers are often necessary to construct a resistance surface that has an extent large enough to measure differences in gene flow, and a researcher is restricted by the spatial resolution at which the data were collected. However, it is still useful to evaluate whether this is causing misleading results. In an attempt to address this question, Cushman and Landguth (2010a) performed a simulation study in which they examined how landscape genetic correlations (using Mantel tests) might change with different pixel sizes. Interestingly, they found that changing the pixel size did not have a large effect on the landscape genetic associations, at least in the case when the true scale was a 90 m pixel. Similarly, McRae et al. (2008) found high correlations between effective resistances calculated across a range of pixel sizes in circuit-based analyses. It is important to note that resampling to coarse pixel size through nearest neighbor interpolation can result in total loss of linear features, such as roads and rivers, which may have dominant influences on landscape connectivity. However, this can be addressed by using pixel resampling methods, such as bilinear interpolation, that retain the mean resistance value across the merged pixels and thus do not "lose" the signal of the linear features. An enhancement to this is to coarsen resistance layers based on satellite-derived land cover using average resistances and to coarsen layers based on linear features like roads and rivers using maximum resistances before combining into a multivariate resistance layer. This preserves the integrity of linear features by avoiding the introduction of erroneous weak spots into them that can result from aggregating using average resistances. Furthermore, while Cushman and Landguth (2010a) did not find dramatic differences, they did find that the smallest pixel size tested always had the strongest relationship. The exception to this would be if the data layer being used had a high error rate, in which it may be advisable to use a coarser pixel size than native resolution. In such situations, the Washington Wildlife Habitat Connectivity Working Group (WHCWG) (2012; www.waconnected.org) found that coarsening grid cell sizes can help to close spurious holes in barriers and remove single-cell patches with low resistance; such features often result from errors in classification of satellite data, and in any case would likely be too small to provide biologically viable movement pathways for many species.

Another scale-related factor for a resistance surface is the thematic resolution at which a pixel is defined (Chapter 2). In other words, is the landscape variable defined discretely, as a series of categorized classes, or as a continuous variable? This is especially relevant for variables whose native resolution is as a continuous variable (e.g., slope, elevation, percent forest cover, etc.) or a detailed category (e.g., lodgepole pine forest), but are represented as simpler categories (medium elevation or forest, respectively). In contrast to pixel size, Cushman and Landguth (2010a) found large influences of thematic resolution on the strength of apparent landscape–genetic relationships. As continuous variables were reclassified into categories, the support for the association dropped significantly. One

can imagine that the same result would occur when categorical variables such as land cover are simplified into fewer classes as well. Of course, there are also situations (such as generalist species) in which fine thematic resolution may not be appropriate. Ultimately, researchers need to consider both the most appropriate spatial and thematic scales relevant to the study system when preparing variables for incorporation into resistance surfaces.

A factor related to remotely sensed data that is rarely accounted for in landscape genetic studies is the accuracy of the map products derived from remote sensing analysis, such as classified land cover maps or continuous estimates of parameters such as canopy closure. This is a somewhat different issue than scale, as one may have a fine-scale, continuous landscape layer that is inaccurate and therefore not useful. Accuracy of data layers is often difficult to evaluate, and most landscape geneticists do not have training in interpreting and classifying remotely sensed data (Zeller et al. 2012). While it would be advantageous for landscape genetic researchers to receive training in these techniques, a reasonable and efficient first step is for studies to examine and report the accuracy assessments of the spatial data used and not use layers that do not meet the level of accuracy required by the researcher. All reliable data layers should have metadata that includes an accuracy assessment. While this information has rarely been reported in the landscape genetic literature, we strongly recommend that future studies include this information in published articles. While landscape genetic researchers often may not be able to fix such inaccuracies, they should be aware of errors in the spatial data they use to evaluate relationships between the genetic structure and landscape features. An important point to remember is that even if there is an accuracy estimate for a layer, different classes in that layer may have different error rates associated with them.

A fourth consideration related to both scale and accuracy is the temporal association of the spatial data with the genetic process being modeled. Most spatial data layers represent snapshots in time and so cannot account for factors of seasonality and environmental change over time unless multiple layers are used. In contrast, genetic data, although generally collected at a specific point in time, are influenced by processes occurring over many generations (see Chapter 3). This mismatch has been recognized by many scientists and several studies have addressed this issue by using spatial data layers across different time points (Anderson et al. 2010; Spear et al. 2010; see Chapter 12 for specific examples). The general result is that genetic data often have a time lag in which measures of genetic distance correlate more strongly with past landscapes, although it is also common for contemporary landscapes to have significant correlations with genetic data. One study that evaluated the temporal time lag issue in landscape genetics is Landguth et al. (2010), which used an individual-based, spatially explicit simulation model (Landguth & Cushman 2010) to quantify the number of generations for new landscape barrier signatures to become detectable and for old signatures to disappear after barrier removal. They found that the lag time for the signal of a new barrier to become established is short using Mantel's $r$ (1–15 generations), while $F_{ST}$ required approximately 200 generations to reach 50% of its equilibrium maximum. In strong contrast, the time scale for loss of signal following the removal of a barrier formerly dividing a population was highly dependent on dispersal distance and ranged from a few generations to several hundred generations. The degree to which a temporal mismatch is important will be most strongly tied to the process that is driving gene flow and how quickly the landscape has changed. For instance, if a species has seasonal reproduction and reproductive events are primarily influenced by vegetation, then spatial data collected during a different season may lead to misleading results, especially in temperate regions (Cushman & Lewis 2010). However, a more common concern is the fact that landscapes change more rapidly than patterns of genetic structure are likely to shift. This possibility can be tested by using multiple temporal resistance surfaces using a model selection approach (Vandergast et al. 2007; Spear & Storfer 2008; Epps et al. 2013), and, if possible, this can be an excellent way to ensure that resistance surfaces based on contemporary landscape data are suitable to use in landscape genetics context.

## 8.2 TECHNIQUES FOR PARAMETERIZING RESISTANCE SURFACES

A recent review of resistance surfaces (Zeller et al. 2012) considered that the process of parameterizing resistance surfaces could occur in one or two stages, and classified techniques into one of three categories.

These include a one-stage expert approach, a one-stage empirical approach, and a two-stage empirical approach. The one-stage expert approach is not relevant to landscape genetics, which by definition includes an empirical stage. The two-stage empirical approach may have either an expert or empirical first stage, but then includes a second empirical stage in which alternate models are statistically compared. Below, we discuss the use of both expert opinion and empirical approaches to develop resistance surfaces.

### Expert opinion

Expert parameterization of resistance values is a common method for developing a resistance surface and is often applied to systems with limited biological data or that cover broad spatial extents. Generally, it involves an expert or a group of experts estimating resistance values for the variables of interest. Clearly, this is the most subjective form of resistance surface development, although it does potentially take advantage of widespread expertise that is not represented in empirical data and may be more cost-efficient than gathering empirical data (Murray et al. 2009). The typical process is to develop several resistance surfaces parameterized through expert opinion and then use genetic data to select the best resistance surface based on statistical correlations (see statistical validation section below). This type of approach can involve model selection among resistance surfaces with different variable sets but only one set of resistance values for each variable (Spear et al. 2005, Vignieri 2005, and James et al. 2011 are some examples) or use some type of optimization that identifies which parameter values for each variable has the best correlation, either through a manual or iterative approach (Cushman et al. 2006; Epps et al. 2007; Wang et al. 2009; Shirk et al. 2010; Wasserman et al. 2010; Richardson 2012). Either method is generally used when researchers hypothesize that multiple landscape variables are affecting gene flow but lack non-genetic empirical data, which requires the use of expert opinion.

Studies that do not vary resistance values for each variable typically assume a linear relationship with continuous variables such as canopy cover or slope (Spear et al. 2005; James et al. 2011). Recent work has suggested that non-linear responses could be prevalent in landscape genetics, particularly asymptotic responses at the ends of the distributions (Fig. 8.2; Balkenhol 2009; Koen et al. 2012; Wasserman et al.

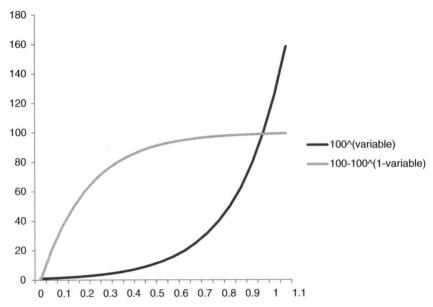

**Fig. 8.2** Example of non-linear responses that can be modeled using resistance surfaces, in this case using exponential functions. The dark gray line represents a scenario in which effective resistance remains low until variable values reach high levels and in which resistance becomes very high at the upper extreme of variable values. In contrast, the light gray line represents a scenario in which an organism responds strongly to small increases in variable cost, but that level of resistance plateaus at intermediate to high levels of that variable. These particular exponential transformations are taken from Balkenhol (2009) and Trumbo et al. (2013).

2010; Shirk et al. 2010; Trumbo et al. 2013). If the asymptotic tails intersect with a large portion of the variable range within the study landscape, then modeling resistance based on a linear relationship is unlikely to result in a significant correlation no matter the biological relevance of that variable. Therefore, even if thorough optimization approaches are not used, researchers should at least test some non-linear relationships to better understand how landscape variables will affect genetic structure.

Many variables of interest to landscape geneticists must be represented as categorical variables. Such variables include land cover classes and discrete landscape features (roads, rivers, etc.). The difficulty in using these variables in resistance surfaces is that resistances must be assigned to each variable, usually with much uncertainty as to the "true" value. Therefore, most studies that incorporate categorical variables have developed multiple resistance surfaces that represent different costs for the variable of interest and then use the correlation with genetic distance to select the best resistance value. The techniques to choose the

most appropriate resistance values vary widely. This is because there are an infinite number of cost values that could be given for any variable in a resistance surface. Researchers must therefore determine how extensively to test different variable costs. At the most basic end, a small number of values are chosen for each variable, with the cost value usually differing by a large amount. For example, Spear and Storfer (2010) tested two cost ratios (10:1 and 100:1) for a test of terrestrial versus stream movement in the Rocky Mountain tailed frog (*Ascaphus montanus*). Similarly, Richardson (2012) tested 5–8 different costs per variable that represented at least 1–2 orders of magnitude difference. The strategy behind this approach is to limit the number of alternative models (and possible false positives due to chance because of multiple comparisons) but be able to get a cost that approximates the true resistance of the variable (within an order of magnitude). A limitation of this approach is that the cost values are likely to be too inexact to be able to use the resistance surface for other purposes, such as using the resistance surface to map connectivity corridors or using it as the basis of future predictions (Rayfield et al. 2010). Instead, a small number of cost values is better suited for studies that wish only to identify variables that have a significant association with genetic structure amongst a large candidate list of potentially important variables.

### Empirical parameterization

Empirical approaches to the parameterization of resistance surfaces can include a variety of methods and are becoming increasingly popular in recent landscape genetic studies (Richards-Zawacki 2009; Shirk et al. 2010; Emaresi et al. 2011; Hagerty et al. 2011; Walpole et al. 2012). A strategy that attempts to use the genetic data to produce a final resistance surface is to measure landscape variables along straight-line transect buffers and use the correlation with genetic data to produce a resistance surface (Fig. 8.3; Braunisch et al. 2010; Murphy et al. 2010; Emaresi et al. 2011). The appeal of this method is that it does not assume any prior relationship with the landscape variables; the researchers must only select what variables are most relevant. It also facilitates the use of both continuous and categorical variables as categorical variables can be easily transformed into a continuous measure, such as proportion of cover type along a transect. Although the line transect analyses could be conducted without buffers, doing so would implicitly assume that individuals always moved in straight lines between sites, an assumption that would be routinely violated. Instead, calculating the landscape component within a buffer along the transect accounts for multiple movement paths between sites without needing to identify the exact movement routes (Emaresi et al. 2011). Furthermore, Murphy et al. (2010) demonstrated that using different buffer sizes could be informative for understanding the scale at which a landscape process is influencing gene flow. However, straight-line buffers may correlate very poorly with dispersal processes that tend to follow low-cost pathways – for example, a straight line might intersect an insurmountable barrier that could be sidestepped with only a slight detour through a low-cost habitat. Moreover, buffered transects do not account for the spatial configuration of land cover types within them; an impermeable cover type within a buffer may or may not bisect the buffer and disrupt movement. Once landscape variables within a transect are regressed against genetic distances, the results can then be used to build a resistance surface. For example, Braunisch et al. (2010) summed regression coefficients of significant landscape

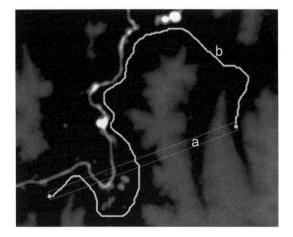

**Fig. 8.3** Illustration of the differences in assessing connectivity between two points on a resistance surface using transect (a) and least-cost path (b) methods. The resistance surface shown here is that produced by Cushman and Lewis (2010) predicting resistance to black bear movement based on movement pathway analysis. (a) This shows a belt transect between two points on the resistance surface. The transect method would measure the composition of the landscape contained within the belt transect and use that as the independent variable in predictions of genetic differences between the individuals or populations located at the two points. Conversely, (b) shows a least-cost path between the same two points on the resistance surface. The least-cost path method would calculate the lowest accumulative cost on the resistance surface between the two points and use this as the independent variable in analyses predicting genetic differences between the individuals or populations located at those two points. You can see that these methods would produce substantially different predictions of connectivity, with the belt transect indicating that individuals moving between the two points would cross two high ridges and twice cross a major highway, while the least-cost path approach would suggest that individuals would avoid these barriers and instead follow a route through low-cost habitat between the ridges and the highway.

variables to create a resistance surface that was then used to predict corridors for capercaillie (*Tetrao urogallus*). Murphy and Evans (2011) also used the results of the transect analysis in Murphy et al. (2010) to predict landscape connectivity for boreal toads (*Anaxyrus boreas*) in Yellowstone National Park.

There are a number of other empirical approaches for developing resistance surfaces in landscape genetics. An increasingly popular empirical approach is to use some form of a resource selection function (Boyce et al. 2002) or species distribution model (Phillips & Dudik 2008). Essentially, these models aim to describe habitat suitability based on the values of environmental variables. They can be built through a variety of data types. Species distribution models often use presence data only at broad scales to predict distribution (and, by assumption, habitat use), whereas individual radio telemetry observations can be used to build a fine-scale resource selection function. It is easy to see why models of habitat suitability are appealing for researchers interested in using resistance surfaces for landscape genetics: they can be easily represented as raster grids, they are based on objective empirical data, and they are created independently of genetic data (and therefore useful for testing genetic patterns). Some examples include using desert tortoise (*Gopherus agassizii*) observations to build a presence-only habitat model (Hagerty et al. 2011), trapping observations of spiny rats (*Niviventer coninga*) in a presence-only model (Wang et al. 2008), and resource selection functions based on radio telemetry points in mountain goats (*Oreamnos americanus*; Shafer et al. 2012). The disadvantage of using some form of a habitat suitability model is that there is a single best model output and so there are generally not comparable alternative models to test. This can be problematic because it can be difficult to assess whether a correlation is biologically significant without comparative models (Cushman & Landguth 2010b). One exception to this scenario is the investigation of mountain goat resource selection by Shafer et al. (2012). This study actually investigated 10 different resource selection functions (varying by combinations of sex and season) and tested the different models with genetic data (in addition to isolation-by-distance and isolation-by-barrier). This research found that summer habitat selection by both sexes had the highest correlation with genetic distance compared to winter habitat. It should be noted that one may often find poor correlation between habitat selection models and gene flow (e.g., Wasserman et al. 2010). This is because habitat selection and gene flow are governed by different behavioral

processes, typically different life stages, and often act at very different spatial scales. For example, habitat selection generally measures the features that promote occurrence or density of adults in fixed home ranges. Gene flow, on the other hand, is governed by dispersal, which is typically driven by juveniles. Juvenile dispersal behavior and adult territorial behavior are often very different, and individuals may be able to disperse through areas that represent poor habitat. Furthermore, regardless of life stage, landscape features that facilitate dispersal may be very different from features associated with core species habitat. As an example, Wasserman et al. (2010, 2012) showed that modeling connectivity using resistances based on an empirically optimized habitat quality map provided a very poor estimate of genetic resistance.

Experimental animal movement is a further empirical method to develop a resistance surface. This approach literally measures animal movement across different surfaces and uses this to estimate resistance. This method was used by Stevens et al. (2006a, 2006b) with the natterjack toad (*Bufo calamita*) in experimental arenas representing different land types such as roads, agricultural fields, and forests. In two separate studies, they measured both how easy it was to cross a given land type and how willing individuals were to enter a different land cover type. This research demonstrated that these two types of resistance were not always consistent. For example, forest cover was more difficult for toads to move through compared to other types, but was entered preferentially in the selection experiment. Interestingly, bare substrates such as concrete had both low resistance to movement and were selected by the toads. This type of research is quite useful because it provides a more mechanistic rationale behind a resistance surface, but because of the simplified experiment arenas, may not represent the true landscape reality. Add car traffic to roads, and permeability can dramatically decrease because of increased mortality risks. Furthermore, many species are simply too large or vagile to make use of even a simple experimental arena meaningful (but tracking translocated individuals may be occasionally possible and give similar information). As a result, this approach may be most fruitful for small invertebrates or extremely dispersal-limited organisms. Another example in which actual movement rates were incorporated into a resistance surface was conducted with *Daphnia* zooplankton by Michels et al. (2001). The researchers modeled zooplankton genetic connectivity among ponds using different resistance surfaces representing pond connections, flow rate among ponds, and measures of actual *Daphnia* dispersal. The functional measures of flow rate and dispersal had a stronger correlation with genetic distance than structural connectivity alone.

Another powerful method to empirically estimate resistance surfaces is to use movement pathway data to quantify association of movement path selection with environmental features. Movement pathway data are characterized by multiple sequential locations of the same individuals taken at sufficiently frequent intervals that enable one to treat each sequence as a movement pathway. Pathway data are much preferred over static detection data and relocation data when the focus is estimating resistance to movement of individuals through the landscape (Zeller et al. 2012), as they directly measure the behavioral choices an organism makes in selecting paths in relation to landscape features. There have been quite a few examples of using pathway data to estimate landscape resistance. Dickson and Beier (2007) identified the topographic composition of paths used by tracked cougars; such information could easily be translated into a resistance surface for the area. Cushman et al. (2010a) used pathway data, multivariate scaling, and randomization testing to develop resistance surfaces for elephant movement in Botswana. Similarly, Cushman et al. (2011) used matched case control logistic regression to quantify the effect of landscape fragmentation on landscape resistance to movement of American marten (*Martes americana*). Despite the clear advantages of pathway data, they have only been used in combination with landscape genetic data in two instances. Cushman and Lewis (2010) used matched case control logistic regression to test the resistance hypotheses evaluated using landscape genetics by Cushman et al. (2006) for black bears. They found that movement behavior explained patterns of genetic differentiation and that the same model of landscape resistance was supported both by genetic data and pathway data. In contrast, Reding et al. (2013), using similar methods, found that movement behavior and genetic differentiation of bobcats (*Lynx rufus*) in Iowa were related to different factors. Specifically, movement pathway selection was driven by the pattern of human activity and natural vegetation on the landscape, whereas genetic differentiation appeared to be governed by isolation-by-distance.

Pathway data can further be used to test some of the components of physical resistance that have previously been explored in the experimental arena discussed above. For instance, organisms may tend to avoid suboptimal habitats when a variety of types are available to move through, but at the same time be able to move quickly through the same suboptimal habitats if high-quality habitat types are not available (Dickson et al. 2005; Kuefler et al. 2010). For instance, agricultural areas might be avoided if adjacent to a natural grassland, but would be much more conducive to movement in a more developed matrix. Therefore, the resistance of such habitats may be highly variable depending on the overall landscape context. Movement rates across different landscapes inferred from pathway data could prove very useful in parameterizing resistance surfaces in such situations.

## 8.3   ESTIMATING CONNECTIVITY FROM RESISTANCE SURFACES

The choice of method to parameterize resistance surfaces is an extremely important consideration, as illustrated in the above sections. However, in landscape genetic studies, the resistance surface cannot be directly compared to genetic data. Genetic data are collected at the level of individuals or discrete sites and therefore genotypes do not form a continuous grid across the resistance surface. While it is possible to convert genetic data into a continuous grid that is directly comparable to a resistance surface (Murphy et al. 2008), it is more likely that both the resistance surface and the genetic data will need to be translated into a pairwise measure to represent genetic connectivity among individuals or sites. There typically have been two approaches to addressing this issue: using the resistance surface to draw least-cost paths (Adriaensen et al. 2003) or representing the resistance surface as nodes of an electrical circuit (McRae 2006). In both cases, the output is a pairwise measure that reflects the overall resistance to gene flow or movement among sites.

The least-cost path has been the most popular technique for studies applying resistance surfaces to landscape genetics (Storfer et al. 2007). The least-cost path approach is fairly intuitive: draw a line between two areas that minimizes the cumulative movement cost moving across the resistance surface (Figs. 8.3 and 8.4). Least-cost paths can be correlated with genetic data in a few different ways. The researcher can measure the Euclidean or topographical length of the least-cost path (Spear et al. 2005), calculate the total resistance cost along each path (Cushman et al. 2006;

Shafer et al. 2012), or simply use the least-cost path as a transect and calculate the value of landscape variables along the path, analogous to the straight-line transect approach discussed above (Spear et al. 2005; van Strien et al. 2012). In addition to the landscape genetic literature, least-cost paths have been commonly used in conservation corridor planning approaches (Singleton et al. 2002; Cushman et al. 2009; Beier et al. 2009; WHCWG 2012) to choose the optimal path to design linkages. However, there are a number of simplifying assumptions that are made with the use of least-cost paths. First, they assume that organisms have enough knowledge of the landscape to follow the ideal path or that the matrix outside the least-cost path is so hostile that it would force the organism to follow the ideal path. In other words, the modeled least-cost path determines movement and genetic structuring, whereas alternative pathways have little to no effect. The limitation of this assumption is addressed with the use of circuit theory (McRae 2006), discussed in the next section. This assumption is most important when using least-cost paths to design corridors, whereas it may be able to be relaxed if researchers are interested in relative differences between paths. However, beyond these assumptions, a common concern with least-cost path methodology is in regards to its sensitivity to differences in resistance surface parameterization and errors in base data. The sensitivity to different relative costs is especially important when the cumulative resistance is used as the independent variable or when the study is attempting to identify specific connectivity corridors that may have different spatial placements with different resistance costs (see examples in Chapter 2). For instance, Rayfield et al. (2010) demonstrated how spatial location of least-cost paths could vary with various relative costs, especially with increasing proportions of inhospitable matrix across the landscape (Chapter 2). Koen et al.

(2012) addressed a similar question of the effect of different relative cost values by examining the influence on both least-cost path length and cumulative least-cost path resistance. The authors found that the cumulative cost of the least-cost path increased linearly with increased relative resistance, but least-cost path length tended to quickly plateau with increasing resistance contrast. While this may provide an argument to use least-cost length as the least-cost measure when resistance parameterization is uncertain, it is probably far better to use an empirical model selection approach to choose the best parameterization scheme when using least-cost paths. In the latter case, cumulative resistance should be used as it will change linearly with resistance and lead to effective model selection (Koen et al. 2012). Least-cost methods can also be particularly sensitive to errors in base data (Beier et al. 2009; McRae et al. 2012).

The second common approach is the use of electrical circuit theory to model resistance across the landscape (McRae 2006). In most applications, landscapes are represented as grids of regularly-spaced nodes connected by resistors. The resistance surface determines the level of resistance among adjacent nodes. Pairwise resistances incorporate all possible pathways connecting patches or locations on the landscape (Fig. 8.4). Circuit theory analysis is typically used in the same fashion as least-cost paths for landscape genetics in that there is a summary pairwise resistance value that is correlated with some measure of genetic distance. In contrast to least-cost paths, circuit analyses incorporate many possible paths into the measure of resistance. This could be especially advantageous in landscapes in which there are several connections of similar total cost between sites such that a least-cost path would only represent a fraction of organismal connectivity. This fact has led many to suggest that circuit analyses are preferable for landscape genetic analyses as compared

Graph edges (thin lines)
Least-cost path (heavy line)

Resistance paths (heavier arrows indicate greater contribution)

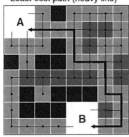 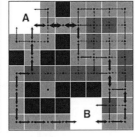

**Fig. 8.4** Resistance values, graph edges, and least-cost and circuit solutions for connectivity between two habitat patches, A and B. Per-cell resistance increases with darker colors. Both least-cost and circuit theory algorithms construct a graph that connects cells. Typically, graph edge weights are inversely proportional to average cost or resistance of cells being connected. Left-hand panel shows graph and least-cost path (this example shows only four-neighbour connections for simplicity). Right-hand panel shows pathways for effective resistance calculations based on circuit theory. Heavier arrows indicate higher contribution/importance of pathways.

to least-cost path analyses (McRae & Beier 2007; Lee-Yaw et al. 2009; Munshi-South 2012). Of course, circuit theory has its own set of assumptions. It does not assume complete knowledge of the landscape, but it does assume that organisms are capable of using the entire landscape (although completely unsuitable pixels can be masked out of the analysis). A modification to the circuit theory software to constrain movement to least-cost corridors has recently been implemented (McRae 2012), but this has yet to be evaluated for landscape genetic applications.

Several studies have compared least-cost paths and circuit theory using the same data set. The results from the different studies have not always been consistent, but largely fit within the expectations of each method. For instance, studies that have focused on wide-ranging species with portions of the matrix that are impermeable have often demonstrated a stronger correlation with circuit theory. McRae and Beier (2007) demonstrated that circuit theory modeled range-wide gene flow (i.e., range limits were the only barriers modeled) much better than least-cost paths. Similarly, but on a finer scale, boreal toad (*Anaxyrus boreas*) gene flow was primarily constrained by salt-water barriers and circuit theory described this pattern best (Moore et al. 2011). Furthermore, circuit theory best correlated with genetic distance in a generalist species (the white-footed mouse, *Peromyscus leucopus*) in an urban environment (Munshi-South 2012). On the other hand, gene flow of Cope's giant salamander (*Dicamptodon copei*), a stream-associated amphibian, provides an example of a study with consistently higher correlation with least-cost path models than circuit theory across three different regions (Trumbo et al. 2013). The tendency of this semi-aquatic species to use a linear stream network for dispersal and breeding likely explains this discrepancy. Other studies have found little difference between connectivity estimates using both least-cost paths and circuit resistances (Schwartz et al. 2009; Row et al. 2010). Therefore, while the best analytical method for estimating connectivity from resistance surfaces may be predicted based on the biological characteristics of the study system, more work is needed to demonstrate a consistent result. Ultimately, because different models that use resistance surfaces have different assumptions, no one model is likely to be best in all circumstances. For example, a seldom appreciated assumption of circuit methods that is violated in many landscape genetic applications is that each node (i.e., each grid cell in

raster analyses) is a population. This may be more realistic when predicting broad-scale patterns such as the effects of range size (e.g., McRae & Beier 2007). For finer-scale analyses, it may be that one model (e.g., Euclidean distance, least-cost path distance, or a hybrid of least-cost methods with circuit analyses) may be best at predicting movement between populations or individual locations and another model (e.g., circuit theory or simulation algorithms like those contained in CDPOP; Landguth & Cushman 2010; see also Chapter 6) may be needed to integrate the cumulative effects of gene flow across networks of populations or individuals. The multigenerational nature of genetic connectivity may also influence the appropriateness of least-cost paths or circuit theory approaches. Intuitively, circuit theory may represent multigenerational connectivity better due to the integration of multiple paths, and this has been borne out in simulations of networks of discrete populations with non-overlapping generations exchanging migrants with neighboring populations (McRae 2006). However, a rigorous study of the performance of the two approaches in systems with continuously distributed populations and more complex life history characteristics and dispersal behaviors has not been conducted, and is necessary to provide more concrete guidance on this issue.

## 8.4  STATISTICAL VALIDATION OF RESISTANCE SURFACES

The previous sections highlighted the development of resistance surfaces, but direct incorporation of genetic data is the necessary step when resistance surfaces are used for landscape genetic questions. This takes the form of some type of statistical correlation analysis (i.e., Mantel tests, multiple regression on distance matrices, etc.); the details on these techniques are discussed in Chapter 5 and so are not presented here. However, the process by which a researcher determines if a resistance surface is representative of patterns of genetic connectivity goes beyond the statistical correlative method chosen. The simplest type of study is to evaluate whether a specific variable has a significant correlation with gene flow – the researcher only needs to develop a resistance surface for the variable of interest, translate that resistance surface into some measure of connectivity, and then correlate that with a measure of genetic connectivity. For instance, Andrew et al. (2012)

evaluated how genetic divergence in prairie sunflowers (*Helianthis petiolaris*) was explained by a resistance surface representing either bare dune or vegetated dune habitat. The researchers found that a low resistance for bare dunes relative to vegetated dunes had the strongest correlation with genetic distance. In another example, timber rattlesnake (*Crotalus horridus*) gene flow was significantly correlated with basking habitat compared to Euclidean distance (Clark et al. 2008). It is important to note, however, the potentially severe pitfalls associated with seeking correlations between genetic structure and a single variable. Cushman and Landguth (2010b) and Cushman et al. (2013) showed that there are usually high spurious correlations between genetic distance and a very large number of incorrect resistance hypotheses due to the inherent high correlation of cost distances. *A priori* selection of a single landscape resistance model usually yields highly significant support, but would very likely lead to inferential error. Thus, evaluating support for a single landscape resistance model in isolation provides a very weak basis for inference. Reliable inference of the factors driving gene flow instead requires formal evaluation of support among a pool of realistic candidate models involving several variables and several functional responses for each variable (e.g., Cushman et al. 2006; Shirk et al. 2010; Wasserman et al. 2010).

The process to evaluate multiple resistance surfaces can be based on a model selection approach or through an iterative optimization algorithm. The first major attempt at a rigorous model selection framework in landscape genetics was conducted by Cushman et al. (2006). They tested 108 different multivariate resistance surfaces for black bear (*Ursus americanus*) gene flow across northern Idaho. The 108 surfaces represented every combination of four variables (land cover, roads, elevation, slope), each at 3–4 different relative costs. They then determined the combination of resistance values for each variable that had the strongest correlation with genetic differentiation based on Mantel and partial Mantel tests, through a framework described as causal modeling. This study is also instructive of the difficulty of optimizing resistance surfaces. The authors only tested a small number of variables at a small number of potential cost values and yet still had a large number of hypothetical resistance surfaces. Another example that more strongly illustrates this point is a study in which Wang et al. (2009) parameterized cost surfaces for three land cover types (chapparal, grassland, woodland) at every possible combination of values at

cost intervals of 0.1 ranging from 1 to 10 for California tiger salamander (*Ambystoma californiense*) breeding ponds. This created a total of 24,843 resistance surfaces. Interestingly, the authors found that a relatively narrow range of cost values was supported by using genetic estimates of contemporary migration. However, even this extensively optimized resistance surface would incompletely project California tiger salamander gene flow on other landscapes as it only includes three categories of natural land cover. A compromise to this optimization problem has been presented by Shirk et al. (2010). Using mountain goats (*Oreamnos americanus*) as a test case, they presented a framework in which expert opinion was used as a starting point, but then iteratively adjusted the resistance value and tested the correlation with genetic distance until a unimodal peak of correlation support was identified. Each variable was optimized separately and then run through a second set of optimization with all other variables held constant (i.e., a quasi-multivariate optimization). While still computationally intensive, it allows for a larger number of variables to be evaluated and takes advantage of prior knowledge that can be used with expert opinion approaches. However, there has been some recent concern that optimization approaches may not be producing resistance estimates that accurately reflect genetic connectivity. Graves et al. (2013) used a combination of a global optimization procedures based on Mantel correlations (the same metric used by Cushman et al. 2006 and Shirk et al. 2010) and landscape genetic simulations to demonstrate that optimized resistance surfaces rarely produced the true resistance estimates that parameterized the simulations. More recently Shirk et al. (2012) and Castillo et al. (2014) coupled simulation modeling with multivariate optimization using the reciprocal causal modeling method (Cushman et al. 2013) and showed that this method seems to perform quite well in identifying the correct drivers of gene flow and rejecting highly correlated alternative resistance hypotheses. Clearly, statistical optimization approaches may not always guarantee biologically relevant results and further development and testing of statistical analyses for landscape genetic work is of high priority.

### 8.4.1 Applications of resistance surfaces in landscape genetics

The identification of a resistance surface highly correlated with genetic connectivity is often used by researchers to then address a broader question. Here, we focus on two types of applications that resistance surfaces have great utility for: corridor identification or design and predictive modeling under future conditions. Both of these objectives are of increasing importance to understanding the conservation, ecology, and evolution of species in a changing environment characterized by increased fragmentation due to anthropogenic habitat alteration and global climate change.

Thus far, the resistance surface has served as the "foundation" for connectivity and corridor analyses, and resistance surfaces are by far the most commonly used tool in corridor planning (Clevenger et al. 2002; Epps et al. 2007; Beier et al. 2008, 2009, 2011; Cushman et al. 2009; Wasserman et al. 2012). Of course, the relevance of landscape genetics to this type of conservation planning is strongly dependent on what type of corridors or reserves are being considered. For instance, a migratory bird corridor would be unsuitable for a genetic study because the migrations do not represent a gene flow event. However, most connectivity corridor initiatives are interested in connecting populations through dispersal that does lead to gene flow. Therefore, a resistance surface that has been validated through genetic data is particularly useful, especially considering that most current connectivity efforts rely on expert opinion without

empirical validation (Beier et al. 2008). A genetically validated resistance surface can be used to map corridors based on least-cost paths or current flow, usually incorporating a cutoff value for maximum cost length or resistance value to identify possible corridors (Beier et al. 2011). An emerging methodology that has promise for using resistance surfaces for corridor planning (in addition to other applications) is the resistant kernel approach (Fig. 8.5; Compton et al. 2007; Cushman et al. 2010a, 2012; Cushman & Landguth 2012). This method is a modification of least-cost path principles that addresses some of the criticisms of least-cost paths. The idea behind a resistant kernel is to use a resistance surface to calculate the cumulative least-cost distance to reach each pixel from every source pixel, and then scale this so that it represents a probability of a disperser reaching each pixel. A probability distribution function is used to determine the maximum distance from the source that a pixel can be and still be included in the resistant kernel calculation. Essentially, the resistant kernel approach recognizes that, in addition to the habitat attributes of a pixel, the resistance of the surrounding pixels also affects the likelihood that the focal pixel will promote dispersal or gene flow. Resistant kernels have had extensive use in modeling dispersal synoptically across continuous landscapes (Compton et al. 2007) and mapping corridors and barriers to dispersal (Cushman et al. 2010a,

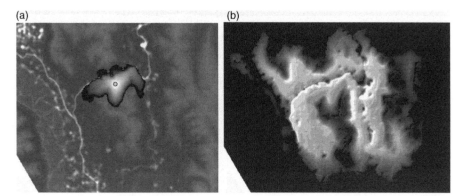

**Fig. 8.5** Illustration of resistant kernel modeling of connectivity across a resistance surface. The resistance surface is shown in pale gray-scale background in (a). A single-source location is shown in a point in (a). The dispersal kernel from that point is shown as a grey-scale kernel in (a) with white indicating areas with high probability of movement, black indicating areas with low probability of movement, and areas outside the kernel indicating areas that are not reachable from that source point given the dispersal ability of the species. (b) This shows the cumulative resistant kernel surface across the study area, with white areas indicating locations predicted to have high connectivity and high rates of movement of individuals through them and dark areas predicted to have low rates of movement. The structure of the resistant kernel surface can be used to identify corridors, barriers, core areas, and fracture zones of connectivity.

2012; Cushman & Landguth 2012). However, resistant kernels have yet to be used directly with genetic data in the same way as least-cost paths or circuit theory. This is because while it modifies the resistance surface, it still results in a continuous layer of pixel values, which is the same structure as any resistance surface. Researchers could apply a typical least-cost path approach to the resistant kernel surface or possibly even combine least cost and circuit approaches by running current through a resistant kernel surface. The use of resistant kernels has much promise and should correlate with gene flow, but we would recommend tests with combined empirical and simulation data sets to demonstrate their effectiveness in landscape genetic analyses.

Predicting the future genetic structure and connectivity of populations is a key question in understanding population responses to changing or novel landscape conditions. Resistance values for variables in their current state can be extrapolated to predicted future landscape conditions to develop a future resistance surface. Individual-based model (IBM) simulations (see Chapter 6) can then be implemented based on resistance surfaces based on future scenarios. IBMs based on resistance surfaces have produced ecological distances that correlate well with empirical genetic data (Vuilleumier & Fontanillas 2007; Shirk et al. 2012), which suggests that simulations on resistance surfaces can be representative of gene flow. For example, the software CDPOP (Landguth & Cushman 2010) has been used to simulate future scenarios for predicting reduced connectivity and genetic diversity under scenarios of climate change (Wasserman et al. 2012) and forest management (Spear et al. 2010). Other platforms for such simulations exist (see Chapter 6) and may provide even more flexibility for modeling a number of demographic and population parameters, as well as traits under selection, provided sufficient information is available for the species of interest (e.g., HexSim; Schumaker 2012).

### 8.4.2 Concise considerations for effective uses of resistance surfaces

Below we provide a short list of considerations that are important for researchers interested in using resistance surfaces for their research. Researchers should attempt to satisfactorily address all these points when designing a landscape genetic study with resistance surfaces, although they may not be easy to answer for many studies.

• What questions and hypotheses do I want to use a resistance surface to test?
• Are reliable data available for the variables to be used in the resistance surface?
• Is the thematic, spatial, and temporal scale of spatial data appropriate?
• Are appropriate empirical data available to guide choice of resistance values for each modeled variable?
• If empirical data are lacking, what is the strategy to ensure that the expert opinion process is as informative and objective as possible?
• Which model of effective connectivity based on resistance surfaces (least-cost, circuit resistance, etc.) is most suitable given my research questions and characteristics of my study system?

## 8.5 THE FUTURE OF THE RESISTANCE SURFACE IN LANDSCAPE GENETICS

The resistance surface has clearly been a foundation for landscape genetics to date, but, in a rapidly developing field such as landscape genetics, that alone does not guarantee that it will continue to be an important tool in the field. Below we discuss what we anticipate will be important or necessary directions for research using resistance surfaces in the future.

### 8.5.1 Advances in remote sensing

Resistance surfaces used in landscape genetics are currently largely driven by data derived from remote sensing technologies and this is unlikely to change. Remote sensing is using new technologies to substantially increase resolution and accuracy of spatial data layers, and the idea of a resistance surface at a submeter resolution is not outside the realm of possibility. Furthermore, tools such as LIDAR allow for three-dimensional representation and therefore allow modeling of factors such as vegetation height. This could potentially allow for breakthroughs in landscape genetic studies of species that have movement or gene flow over very fine scales, such as many invertebrates or small vertebrates. Of course, extremely fine resolution is of little use for systems in which the organisms exchange genes over broad scales. However, any technology that increases the accuracy of spatial data or allows for the reliable incorporation of more variables should be beneficial to any study. It should be noted that such advances would improve only the raw material for resistance surfaces – they may do little to improve parameterization and in some cases may lead the process to be more complicated.

### 8.5.2 Development of model selection and optimization methodologies

The incorporation of approaches to assess multiple alternative models is one of the most important directions that have improved landscape genetic studies (Cushman et al. 2006; Wang et al. 2009; Shirk et al. 2010). However, there is certainly much room for improvement in these model selection procedures. It is still very computationally intensive to test many possibilities for resistance surface parameters and exhaustive optimization can rapidly turn resistance surfaces into little more than a statistical model-fitting exercise divorced from biology. Furthermore, the non-independence inherent in both landscape and genetic data often violate assumptions of the most commonly used metrics of model selection employed in landscape genetics. The Mantel test, which has been commonly used in model selection with resistance surfaces (Cushman et al. 2006; Shirk et al. 2010) has been criticized on multiple fronts as inappropriate for landscape genetics (Legendre & Fortin 2010; Guillot & Rousset 2013), while others have found Mantel tests applied in a strict model comparison framework to have high power to correctly identify the driving process and reject highly correlated alternative models (e.g., Cushman & Landguth 2010b; Shirk et al. 2012; Cushman et al. 2013). Clearly, a widespread consensus has not been reached regarding this issue (see also Chapter 5). Breakthroughs in computational efficiency and statistical methodology will be invaluable in improving model selection. For instance, researchers are currently investigating possible approaches to directly translate genetic data into a parameterized resistance surface (Hand et al. 2014).

An intriguing method for parameterization that combines both mechanistic movement data with telemetry pathway data to create a resistance surface is the idea of an "energy landscape" (Shepard et al. 2013). An energy landscape directly translates the mechanistic cost of movement to the resistance surface and can be calculated by the energetic cost per unit of movement. Therefore, such a surface would be quite intensive to parameterize as it would require movement physiology studies for the focal species across a range of environments. Energy landscapes in landscape genetics would also assume that the physical ability to move through the landscape is the primary determinant of genetic connectivity, which of course may not be the case. However, if this assumption holds, a correctly parameterized energy landscape provides a highly empirical measure of resistance. Energy landscapes have not been used in a genetic study, but two recent studies have introduced a blueprint for how such a study might unfold. Wall et al. (2006) investigated African elephant (*Loxodonta africana*) movement in relation to slope. The study used estimates of the energy required to climb slopes accounting for body size and determined that climbing 100 vertical meters would require an extra half hour of foraging time, indicating that climbing slopes is costly to a large animal like an elephant. The researchers used radio telemetry to confirm that elephants avoided steep slopes in their movements. In an aquatic environment, McElroy et al. (2012) measured pallid sturgeon (*Scaphirhynchus albus*) movement at various depths and velocities in a large river. Using actual field measurements of velocity across the river and a radio-tracked sturgeon, the study demonstrated that the sturgeon used a least costly path for movement through the river. If key landscape factors affecting energetic relationships can be successfully modeled in a GIS database, then an energetic resistance surface should be a tool for developing a resistance surface for a landscape genetic analysis.

However, perhaps most important is a continued discussion of how parameterization should take place. For instance, what is the role of expert opinion in the advancement of resistance surfaces? It is clear that relying on expert opinion alone is not desirable and may lead to suboptimal models (Spear et al. 2010; Shirk et al. 2010; Zeller et al. 2012). However, a standardized expert opinion procedure may be useful to define the parameter space to be explored during an empirical model selection and thus reduce the chance of an overfit model that does not make biological sense. Regardless, the balance between expert opinion and empirical data will continue to be an important consideration for landscape genetic researchers using resistance surfaces.

### 8.5.3 Resistance surfaces in adaptive landscape genomics

One of the exciting directions in the field is the incorporation of questions regarding adaptive evolution, due in part to the advent of next-generation sequencing approaches (Manel et al. 2010a, 2010b; Parisod & Holderegger 2012; Chapter 9) and advanced spatially explicit simulation models (Landguth et al. 2012). This raises the possibility of developing adaptive resistance surfaces. Complex resistance surfaces have not been incorporated into adaptive landscape genetics, although gradients along variables such as elevation (Bonin et al. 2006) or precipitation and temperature (Manel et al. 2012; Yang et al. 2013) have shown an association with adaptive loci. Recently, Landguth et al. (2012), Schumaker (2012), and Rebaudo et al. (2013) have expanded individual-based, landscape genetic simulation modeling to explicitly incorporate selection gradients. Linking landscape effects to key evolutionary processes through individual organism movement and natural selection is essential to provide a foundation for evolutionary landscape genetics. Of particular importance is determining how spatially explicit, individual-based models differ from classic population genetics and evolutionary ecology models based on ideal panmictic populations in an allopatric setting in their predictions of population structure and frequency of fixation of adaptive alleles. The new models can incorporate all factors – mutation, gene flow, genetic drift, and selection – that affect the frequency of an allele in a population. The approach of explicitly coupling gene flow (governed by resistance surfaces) with natural selection (governed by selection surfaces) provides a powerful generalized framework for evaluating interactions between drift, migration, and selection in complex landscapes. The approach essentially formalizes Sewell Wright's adaptive landscape concept (Wright 1931). If selected genotypes can be mapped on an adaptive landscape, and those same genotypes can be associated with landscape variables, then the development of a resistance surface representing fitness might be possible. Of course, none of these prerequisites are particularly easy, as studying adaptive variation in wild populations is still in its relative infancy.

## 8.6 CONCLUSIONS

We have aimed to demonstrate both the process behind resistance surface development as well as to discuss the many advantages and disadvantages associated with the methodology. Resistance surfaces have been extremely important in the development landscape genetics, but there is much room for improvement in

the use of resistance surfaces. In particular, we echo past reviews as to the importance of using empirical data when possible and transparent expert opinion reviews. However, not all empirical data may be relevant to a resistance surface, and in some cases may result in misleading models. Ultimately, the choice of resistance surface parameterization needs to be driven by the larger question that the researcher is asking. We also strongly emphasize the need for effective and appropriate model selection and the comparison of several resistance surfaces and careful consideration of the best method to translate the resistance surface into a form that can be associated with genetic data. Overall, we predict that resistance surfaces will remain a key tool in landscape genetics and will be especially important for those whose interest in landscape genetics focuses on its potential as a predictive science for the various facets of global change that are occurring.

## REFERENCES

Adriaensen, F., Chardon, J.P., De Blust, G., Swinnen, E., Villalba, S., Gulinck, H., & Matthysen, E. (2003) The application of "least-cost" modeling as a functional landscape model. *Landscape and Urban Planning* **64**, 233–47.

Anderson, C.D., Epperson, B.K., Fortin, M.J., Holderegger, R., James, P., Rosenberg, M.S., & Spear, S. (2010) Considering spatial and temporal scale in landscape-genetic studies of gene flow. *Molecular Ecology* **19**, 3565–75.

Andrew, R.L., Ostevik, K.L., Ebert, D.P., & Rieseberg, L.H. (2012) Adaptation with gene flow across the landscape in a dune sunflower. *Molecular Ecology* **21**, 2078–91.

Balkenhol, N. (2009) Evaluating and improving analytical approaches in landscape genetics through simulations and wildlife case studies. Dissertation, University of Idaho.

Beier, P., Majka, D.R., & Spencer, W.D. (2008) Forks in the road: choices in procedures for designing wildland linkages. *Conservation Biology* **22**, 836–51.

Beier, P., Majka, D.R., & Newell, S.L. (2009) Uncertainty analysis of least-cost modeling for designing wildlife linkages. *Ecological Applications* **19**, 2067–77.

Beier, P., Spencer, W., Baldwin, R.F., & McRae, B.H. (2011) Toward best practices for developing regional connectivity maps. *Conservation Biology* **25**, 879–92.

Bonin, A., Taberlet, P., Miaud, C., & Pompanon, F. (2006) Explorative genome scan to detect candidate loci for adaptation along a gradient of altitude in the common frog (*Rana temporaria*). *Molecular Biology and Evolution* **23**, 773–83.

Boyce, M.S., Vernier, P.R., Nielsen, S.E., & Schmiegelow, F.K.A. (2002) Evaluating resource selection functions. *Ecological Modeling* **157**, 281–300.

Braunisch, V., Segelbacher, G., & Hirzel, A.H. (2010) Modeling functional landscape connectivity from genetic population structure: a new spatially explicit approach. *Molecular Ecology* **19**, 3664–78.

Castillo, J.A., Epps, C.W., Davis, A.R. & Cushman, S.A. (2014) Landscape effects on gene flow for a climate-sensitive montane species: the case of the American Pika. *Molecular Ecology*, **23**, 843–856.

Clark, R.W., Brown, W.S., Stechert, R., & Zamudio, K.R. (2008) Integrating individual behaviour and landscape genetics: the population structure of timber rattlesnake hibernacula. *Molecular Ecology* **17**, 719–30.

Clevenger, A.P., Wierzchowski, J., Chruszcz, B., & Gunson, K. (2002) GIS-generated, expert-based models for identifying wildlife habitat linkages and planning mitigation passages. *Conservation Biology* **16**, 503–14.

Compton, B.W., McGarigal, K., Cushman, S.A., & Gamble, L.R. (2007) A resistant-kernel model of connectivity for amphibians that breed in vernal pools. *Conservation Biology* **21**, 788–99.

Cushman, S.A. & Landguth, E.L. (2010a) Scale dependent inference in landscape genetics. *Landscape Ecology* **25**, 967–79.

Cushman, S.A. & Landguth, E.L. (2010b) Spurious correlations and inference in landscape genetics. *Molecular Ecology* **19**, 3592–602.

Cushman, S.A. & Landguth, E.L. (2012) Multi-taxa population connectivity in the Northern Rocky Mountains. *Ecological Modeling* **231**, 101–12.

Cushman, S.A. & Lewis, J.S. (2010) Movement behavior explains genetic differentiation in American black bears. *Landscape Ecology* **25**, 1613–25.

Cushman, S.A., McKelvey, K.S., Hayden, J., & Schwartz, M.K. (2006) Gene flow in complex landscapes: testing multiple hypotheses with causal modeling. *The American Naturalist* **168**, 486–99.

Cushman, S.A., McKelvey, K.S., & Schwartz, M.K. (2009) Use of empirically derived source-destination models to map regional conservation corridors. *Conservation Biology* **23**, 368–76.

Cushman, S.A., Chase, M., and Griffin, C. (2010a) Mapping landscape resistance to identify corridors and barriers for elephant movement in southern Africa. In: Cushman, S.A. & Huettmann, F. (eds.), *Spatial Complexity, Informatics and Wildlife Conservation*. Springer, Tokyo, Japan.

Cushman, S.A., Compton, B.W., & McGarigal, K. (2010b) Habitat fragmentation effects depend on complex interactions between population size and dispersal ability: modeling influences of roads, agriculture and residential development across a range of life-history characteristics. In: Cushman, S.A. and Huettmann, F. (eds.), *Spatial Complexity, Informatics and Wildlife Conservation*. Springer, Tokyo, Japan.

Cushman, S.A., Raphael, M.G., Ruggiero, L.F., Shirk, A.S., Wasserman, T.N., & O'Doherty, E.C. (2011) Limiting factors and landscape connectivity: the American marten in the Rocky Mountains. *Landscape Ecology* **26**, 1137–49.

Cushman, S.A., Landguth, E.L., & Flather, C.H. (2012) Evaluating the sufficiency of protected lands for maintaining wildlife population connectivity in the US northern Rocky Mountains. *Diversity and Distributions* **18**, 873–84.

Cushman, S.A., Wasserman, T.N., Landguth, E.L., & Shirk, A.J. (2013) Re-evaluating causal modeling with Mantel tests in landscape genetics. *Diversity* **5**, 51–72.

Dickson, B.G. & Beier, P. (2007) Quantifying the influence of topographic position on cougar (*Puma concolor*) movement in southern California, USA. *Journal of Zoology*, **271**, pp. 270–77.

Dickson, B.G., Jenness, J.S., & Beier, P. (2005) Influence of vegetation, topography and roads on cougar movement in southern California. *Journal of Wildlife Management* **69**, 264–76.

Emaresi, G., Pellet, J., Dubey, S., Hirzel, A.H., & Fumagalli, L. (2011) Landscape genetics of the Alpine newt (*Mesotriton alpestris*) inferred from a strip-based approach. *Conservation Genetics* **12**, 41–50.

Epps, C.W., Wehausen, J.D., Bleich, V.C., Torres, S.G., & Brashares, J.S. (2007) Optimizing dispersal and corridor models using landscape genetics. *Journal of Applied Ecology* **44**, 714–24.

Epps, C.W., Wasser, S.K., Keim, J.L., Mutayoba, B.M., & Brashares, J.S. (2013) Quantifying past and present connectivity illuminates a rapidly changing landscape for the African elephant. *Molecular Ecology* **22**, 1574–88.

Graves, T.A., Beier, P., & Royle, J.A. (2013) Current approaches using genetic distances produce poor estimates of landscape resistance to interindividual dispersal. *Molecular Ecology* **22**, 3888–903.

Guillot, G. & Rousset, F. (2013) Dismantling the Mantel tests. *Methods in Ecology and Evolution* **4**, 336–44.

Hagerty, B., Nussear, K., Esque, T., & Tracy, C. (2011) Making molehills out of mountains: landscape genetics of the Mojave Desert tortoise. *Landscape Ecology* **26**, 267–80.

Hand, B.K., Cushman, S.A., Landguth, E.L., & Lucotch, J. (2014) Assessing multi-taxa sensitivity to the human footprint, habitat fragmentation and loss by exploring alternative scenarios of dispersal ability and population size: a simulation approach. *Biodiversity and Conservation* **23**, 2761–79.

James, P.M.A., Coltman, D.W., Murray, B.W., Hamelin, R.C., & Sperling, F.A.H. (2011) Spatial genetic structure of a symbiotic beetle-fungal system: toward multi-taxa integrated landscape genetics. *PLoS One* **6**, e25359.

Koen, E.L., Bowman, J., & Walpole, A.A. (2012) The effect of cost surface parameterization on landscape resistance estimates. *Molecular Ecology Resources* **12**, 686–96.

Kuefler, D., Hudgens, B., Haddad, N.M., Morris, W.F., & Thurgate, N. (2010) The conflicting role of matrix habitats as conduits and barriers for dispersal. *Ecology* **91**, 944–50.

Landguth, E.L. & Cushman, S.A. (2010) cdpop: a spatially explicit cost distance population genetics program. *Molecular Ecology Resources* **10**, 156–61.

Landguth, E.L., Cushman, S.A., Schwartz, M.K., McKelvey, K.S., Murphy, M., & Luikart, G. (2010) Quantifying the lag time to detect barriers in landscape genetics. *Molecular Ecology* **19**, 4179–91.

Landguth, E.L., Cushman, S.A., & Johnson, N.A. (2012) Simulating natural selection in landscape genetics. *Molecular Ecology Resources* **12**, 363–8.

Lee-Yaw, J.A., Davidson, A., McRae, B.H., & Green, D.M. (2009) Do landscape processes predict phylogeographic patterns in the wood frog? *Molecular Ecology* **18**, 1863–74.

Legendre, P. & Fortin, M.-J. (2010) Comparison of the Mantel test and alternative approaches for detecting complex multivariate relationships in the spatial analysis of genetic data. *Molecular Ecology Resources* **10**, 831–44.

Manel, S., Joost, S., Epperson, B.K., Holderegger, R., Storfer, A., Rosenberg, M.S., Scribner, K.T., Bonin, A., & Fortin, M.-J. (2010a) Perspectives on the use of landscape genetics to detect genetic adaptive variation in the field. *Molecular Ecology* **19**, 3760–72.

Manel, S., Poncet, B.N., Legendre, P., Gugerli, F., & Holderegger, R. (2010b) Common factors drive adaptive genetic variation at different spatial scales in Arabis alpina. *Molecular Ecology* **19**, 3824–35.

Manel, S., Gugerli, F., Thuiller, W., Alvarez, N., Legendew, P., Holderegger, R., Geilly, L., & Taberlet, P. (2012) Broad-scale adaptive genetic variation in alpine plants is driven by temperature and precipitation. *Molecular Ecology* **21**, 3729–38.

McElroy, B., DeLonay, A., & Jacobson, R. (2012) Optimum swimming pathways of fish spawning migrations in rivers. *Ecology* **93**, 29–34.

McRae, B.H. (2006) Isolation by resistance. *Evolution* **60**, 1551–61.

McRae, B.H. (2012) Pinchpoint Mapper User Guide. [Online] Available at: http://www.circuitscape.org/linkagemapper.

McRae, B.H. & Beier, P. (2007) Circuit theory predicts gene flow in plant and animal populations, *Proceedings of the National Academy of Sciences of the United States of America*, vol. **104**, pp. 19885–90.

McRae, B.H., Dickson, B.G., Keitt, T.H., & Shah, V.B. (2008) Using circuit theory to model connectivity in ecology, evolution and conservation. *Ecology* **89**, 2712–24.

McRae, B.H., Hall, S.A., Beier, P., & Theobald, D.M. (2012) Where to restore ecological connectivity? Detecting barriers and quantifying restoration benefits. *PloS One* **7**, e52604.

McRae, B.H., A.J. Shirk, and J.T. Platt. 2013. Gnarly Landscape Utilities: Resistance and Habitat Calculator User Guide. The Nature Conservancy, Fort Collins, CO. Available at: http://www.circuitscape.org/gnarly-landscape-utilities.

Michels, E., Cottenie, K., Neys, L., De Gelas, K., Coppin, P., & De Meester, L. (2001) Geographical and genetic distances among zooplankton populations in a set of interconnected ponds: a plea for using GIS modeling of the effective geographical distance. *Molecular Ecology* **10**, 1929–38.

Moore, J.A., Tallmon, D.A., Nielsen, J., & Pyare, S. (2011) Effects of the landscape on boreal toad gene flow: does the pattern–process relationship hold true across distinct landscapes at the northern range margin? *Molecular Ecology* **20**, 4858–69.

Munshi-South, J. (2012) Urban landscape genetics: canopy cover predicts gene flow between white-footed mouse (*Peromyscus leucopus*) populations in New York City. *Molecular Ecology* **21**, 1360–78.

Murphy, M.A. & Evans, J.S. (2011) Boreal toad (Bufo boreas) population connectivity in Yellowstone National Park: quantifying matrix resistance and model uncertainty using landscape genetics. In: Drew, C.A., Wiersma, Y.F., & Huettmann, F. (eds.), *Predictive Species and Habitat Modeling in Landscape Ecology.* Springer.

Murphy, M.A., Evans, J.S., Cushman, S.A., & Storfer, A. (2008) Representing genetic variation as continuous surfaces: an approach for identifying spatial dependency in landscape genetic studies. *Ecography* **31**, 685–97.

Murphy, M.A., Evans, J.S., & Storfer, A. (2010) Quantifying Bufo boreas connectivity in Yellowstone National Park with landscape genetics. *Ecology* **91**, 252–61.

Murray, J.V., Goldizen, A.W., O'Leary, R.A., McAlpine, C.A., Possingham, H.P., & Choy, S.L. (2009) How useful is expert opinion for predicting the distribution of a species within and beyond the region of expertise? A case study using brush-tailed rock-wallabies Petrogale penicillata. *Journal of Applied Ecology* **46**, 842–51.

Parisod, C. & Holderegger, R. (2012) Adaptive landscape genetics: pitfalls and benefits. *Molecular Ecology* **21**, 3644–6.

Phillips, S.J. & Dudik, M. (2008) Modeling of species distributions with Maxent: new extensions and a comprehensive evaluation. *Ecography* **31**, 161–75.

Rayfield, B., Fall, A., & Fortin, M.-J. (2010) The sensitivity of least-cost habitat graphs to relative cost surface values. *Landscape Ecology* **25**, 519–32.

Rebaudo, F., Rouzic, A., Dupas, S., Silvain, J.-F., Harry, M., & Dangles, O. (2013) SimAdapt: an individual-based genetic model for simulating landscape management impacts on populations. *Methods in Ecology and Evolution* **4**, 595–600.

Reding, D.M., Cushman, S.A., Gosselink, T.E., & Clark, W.R. (2013) Linking movement behavior and fine-scale genetic structure to model landscape connectivity for bobcats (*Lynx rufus*). *Landscape Ecology* **28**, 471–86.

Richardson, J.L. (2012) Divergent landscape effects on population connectivity in two co-occurring amphibian species. *Molecular Ecology* **21**, 4437–51.

Richards-Zawacki, C.L. (2009) Effects of slope and riparian habitat connectivity on gene flow in an endangered Panamanian frog, *Atelopus varius. Diversity and Distributions* **15**, 796–806.

Row, J.R., Blouin-Demers, G., & Lougheed, S.C. (2010) Habitat distribution influences dispersal and fine-scale genetic population structure of eastern foxsnakes (*Mintonius gloydi*) across a fragmented landscape. *Molecular Ecology* **19**, 5157–71.

Sawyer, S.C., Epps, C.W., & Brashares, J.S. (2011) Placing linkages among fragmented habitats: do least-cost models reflect how animals use landscapes? *Journal of Applied Ecology* **48**, 668–78.

Schumaker, N.H. (2012) HexSim. United States Environmental Protection Agency, Environmental Research Laboratory, Cornvallis, OR. http://www.hexsim.net/index.html.

Schwartz, M.K., Copeland, J.P., Anderson, N.J., Squires, J.R., Inman, R.M., McKelvey, K.S., Pilgrim, K.L., Waits, L.P., & Cushman, S.A. (2009) Wolverine gene flow across a narrow climatic niche. *Ecology* **90**, 3222–32.

Shafer, A.B.A., Northrup, J.M., White, K.S., Boyce, M.S., Côté, S.D., & Coltman, D.W. (2012) Habitat selection in mountain goats. *Bulletin of the Ecological Society of America* **93**, 170–2.

Shepard, E.L.C., Wilson, R.P., Rees, W.G., Grundy, E., Lambertucci, S.A., & Vosper, S.B. (2013) Energy landscapes shape animal movement ecology. *The American Naturalist* **182**, 298–312.

Shirk, A.J., Wallin, D.O., Cushman, S.A., Rice, C.G., & Warheit, K.I. (2010) Inferring landscape effects on gene flow: a new model selection framework. *Molecular Ecology* **19**, 3603–19.

Shirk, A.J., Cushman, S.A., & Landguth, E.L. (2012) Simulating pattern–process relationships to validate landscape genetic models. *International Journal of Ecology* **2012**, 539109.

Singleton, P.H., Gaines, W.L., & Lehmkuhl, J.F. (2002) *Landscape Permeability for Large Carnivores in Washington: A Geographic Information System Weighted-Distance and Least-Cost Corridor Assessment.* United States Department of Agriculture, Forest Service, Pacific Northwest Research Station, Portland, OR.

Spear, S.F. & Storfer, A. (2008) Landscape genetic structure of coastal tailed frogs (*Ascaphus truei*) in protected vs. managed forests. *Molecular Ecology* **17**, 4642–56.

Spear, S.F. & Storfer, A. (2010) Anthropogenic and natural disturbance lead to differing patterns of gene flow in the Rocky Mountain tailed frog, *Ascaphus montanus*. *Biological Conservation*, **143**, 778–86.

Spear, S.F., Peterson, C.R., Matocq, M.D., & Storfer, A. (2005) Landscape genetics of the blotched tiger salamander (*Ambystoma tigrinum melanostictum*). *Molecular Ecology* **14**, 2553–64.

Spear, S.F., Balkenhol, N., Fortin, M.-J., McRae, B.H., & Scribner, K.T. (2010) Use of resistance surfaces for landscape genetic studies: considerations for parameterization and analysis. *Molecular Ecology* **19**, 3576–91.

Stevens, V.M., Leboulengé, E., Wesselingh, R.A., & Baguette, M. (2006a) Quantifying functional connectivity: experimental assessment of boundary permeability for the natterjack toad (*Bufo calamita*). *Oecologia* **150**, 161–71.

Stevens, V.M., Verkenne, C., Vandewoestijne, S., Wesselingh, R.A., & Baguette, M. (2006b) Gene flow and functional connectivity in the natterjack toad. *Molecular Ecology* **15**, 2333–44.

Storfer, A., Murphy, M.A., Evans, J.S., Goldberg, C.S., Robinson, S., Spear, S.F., Dezzani, R., Delmelle, E., Vierling, L., & Waits, L.P. (2007) Putting the "landscape" in landscape genetics. *Heredity* **98**, 128–42.

Trumbo, D.R., Spear, S.F., Baumsteiger, J., & Storfer, A. (2013) Rangewide landscape genetics of an endemic Pacific northwestern salamander. *Molecular Ecology* **22**, 1250–66.

Vandergast, A.G., Bohonak, A.J., Weissman, D.B., & Fisher, R.N. (2007) Understanding the genetic effects of recent habitat fragmentation in the context of evolutionary history: phylogeography and landscape genetics of a southern California endemic Jerusalem cricket (Orthoptera: Stenopelmatidae: Stenopelmatus). *Molecular Ecology* **16**, 977–92.

van Strien, M.J., Keller, D., & Holderegger, R. (2012) A new analytical approach to landscape genetic modeling: least-cost transect analysis and linear mixed models. *Molecular Ecology* **21**, 4010–23.

Vignieri, S.N. (2005) Streams over mountains: influence of Riparian connectivity on gene flow in the Pacific jumping mouse (*Zapus trinotatus*). *Molecular Ecology* **14**, 1925–37.

Vuilleumier, S. & Fontanillas, P. (2007) Landscape structure affects dispersal in the greater white-toothed shrew: inference between genetic and simulated ecological distances. *Ecological Modeling* **201**, 369–76.

Wall, J., Douglas-Hamilton, I., & Vollrath, F. (2006) Elephants avoid costly mountaineering. *Current Biology* **16**, R527–R529.

Walpole, A.A., Bowman, J., Murray, D.L., & Wilson, P.J. (2012) Functional connectivity of lynx at their southern range periphery in Ontario, Canada. *Landscape Ecology* **27**, 761–73.

Wang, Y.-H., Yang, K.C., Bridgman, C.L., & Lin, L.K. (2008) Habitat suitability modeling to correlate gene flow with landscape connectivity. *Landscape Ecology* **23**, 989–1000.

Wang, I.J., Savage, W.K., & Bradley Shaffer, H. (2009) Landscape genetics and least-cost path analysis reveal unexpected dispersal routes in the California tiger salamander (*Ambystoma californiense*). *Molecular Ecology* **18**, 1365–74.

Washington Wildlife Habitat Connectivty Working Group (2012) Washington Connected Landscapes Project: Analysis of the Columbia Plateau Ecoregion. [Online] Available at: http://waconnected.org [01 Oct 2014].

Wasserman, T.N., Cushman, S.A., Schwartz, M.K., & Wallin, D.O. (2010) Spatial scaling and multi-model inference in landscape genetics: *Martes americana* in northern Idaho. *Landscape Ecology* **25**, 1601–12.

Wasserman, T.N., Cushman, S.A., Shirk, A.S., Landguth, E.L., & Littell, J.S. (2012) Simulating the effects of climate change on population connectivity of American marten (*Martes americana*) in the northern Rocky Mountains, USA. *Landscape Ecology* **27**, 211–25.

Wright, S. (1931) Evolution in Mendelian populations. *Genetics* **16**, 97–159.

Yang, J., Cushman, S.A., Yang, J., Yang, M., & Bao. T. (2013) Effects of climatic gradients on genetic differentiation of Caragana on the Ordos Plateau, China. *Landscape Ecology* **28**, 1729–41.

Zeller, K.A., McGarigal, K., & Whiteley, A.R. (2012) Estimating landscape resistance to movement: a review. *Landscape Ecology* **27**, 777–97.

# Chapter 9

# GENOMIC APPROACHES IN LANDSCAPE GENETICS

*Andrew Storfer,[1] Michael F. Antolin,[2] Stéphanie Manel,[3] Bryan K. Epperson,[4] and Kim T. Scribner[5]*

[1]*School of Biological Sciences, Washington State University, USA*
[2]*Department of Biology, Colorado State University, USA*
[3]*Centre d'Ecologie Fonctionnelle et Evolutive (CEFE), France*
[4]*Department of Forestry, Michigan State University, USA*
[5]*Department of Fisheries and Wildlife & Department of Zoology, Michigan State University, USA*

## 9.1 INTRODUCTION

The recent development of next-generation sequencing platforms has helped to revolutionize landscape genetics by providing high statistical power for spatially-explicit descriptions of population structure and estimates of population genetic parameters such as gene flow (Coop et al. 2010; Safner et al. 2011). Availability of large numbers of genetic markers also led to the concept of population genomics, where estimates of spatial genetic differentiation (e.g., $F_{ST}$) of all loci was used to identify "outliers" – those loci with greater or lesser differentiation than expected purely from the balance between genetic drift and migration (Black et al. 2001; Luikart et al. 2003; Stinchcombe & Hoekstra 2008). Outlier loci are predicted to be linked to genomic regions that are under selection, with the possibility of being within the adaptive genes themselves, and can be the pathway to discovery of functionally adaptive genes within natural populations.

Four years after the term "landscape genetics" was introduced (Manel et al. 2003), Joost et al. (2007) introduced a spatial analysis method (SAM) based on logistic regression for testing the spatial correlation between large numbers of genetic markers and environmental variables, with the explicit goal of discovering genomic regions that include candidate genes for adaptations to those environmental conditions. The extension of landscape genetics to include large numbers of loci has been referred to as "landscape genomics" (e.g., Joost et al. 2007; Lowry 2010; Eckert et al. 2010; Manel et al. 2010a). A centerpiece of the landscape genomics approach is to survey hundreds to thousands of genetic loci scattered in the genomes of a large number of individuals in different habitats or with different phenotypes to discover genomic regions

*Landscape Genetics: Concepts, Methods, Applications*, First Edition. Edited by Niko Balkenhol, Samuel A. Cushman, Andrew T. Storfer, and Lisette P. Waits.
© 2016 John Wiley & Sons, Ltd. Published 2016 by John Wiley & Sons, Ltd.

under selection, either directly or more likely through physical linkage (Bonin et al. 2007; Stinchcombe & Hoekstra 2008). However, the inference of selection is often made without prior knowledge of which phenotypes are advantageous (Joost et al. 2013). Landscape genomics and the use of correlative approaches such as logistic regression add explicit spatial components and environmental variables for examining the agents of selection which can be the cause of local genetic differentiation and adaptation within species.

Although testing for selection was originally included in the landscape genetics definition (Manel et al. 2003), it was not until technological developments enabled collections of large numbers of loci that landscape genomics studies had the power to test for adaptation. As a result, landscape genetics and landscape genomics tend to focus on different aspects of how ecological variation shapes the spatial distribution of alleles among populations. Landscape genetics often focuses on ecological processes, such as landscape variables that restrict movement or gene flow (e.g., mountain ranges, ridges) and the influence of abiotic factors on population connectivity. In contrast, landscape genomics focuses on variation in abiotic factors, as well as how biotic factors such as the presence of different predators (e.g., Barrett et al. 2009) or competitor species in different habitats may influence functional adaptive genetic variation within populations. It is also important to note that the recent shift from genotyping relatively few markers (e.g., microsatellites) to thousands (e.g., SNPs) in landscape genomic studies has posed important challenges, chief among them the development of statistical/analytical frameworks better suited for large data sets and the many sources of uncertainty or error within such data sets. Further, the great statistical power to detect correlations between genetic and environmental variation brings with it the danger of finding spurious correlations (e.g. Atwell et al. 2010) and may erroneously inflate the number of potentially adaptive candidate loci that are identified (e.g. false positives).

The more recent development and cost-effectiveness of next-generation sequencing techniques like RAD-seq (restriction site associated DNA sequencing) have led to a proliferation of studies that include tens of thousands of single nucleotide polymorphisms (SNPs), including many that may be located within genes under selection (see Davey et al. 2011, 2013; Narum et al. 2013; Gagnaire et al. 2013). It is now feasible to study the spatial distribution of functional adaptive

genetic variation in virtually any organism without the extensive lead-in work required to identify candidates through traditional genetic studies (e.g., QTL studies) of laboratory or captive populations (Stinchcombe & Hoekstra 2008; Lowry 2010; Manel et al. 2010a; Schoville et al. 2012; Joost et al. 2013).

Landscape genomics also offers the possibility of incorporating a large number of environmental conditions or biotic factors in the analysis (e.g., Manel et al. 2010a, 2010b; Coop et al. 2010). These data can range from climatic factors to study the effect of climate on the spatial distribution of adaptive genetic variation (Hancock et al. 2010; Manel et al. 2010b) to other types of selection such as the effects of spatial variation in pathogen diversity or food availability on human evolution (Hancock et al. 2010; Fumagalli et al. 2011). For example, Manel et al. (2010b) investigated two AFLP (amplified fragment length polymorphism) data sets for the alpine rockcress (*Arabis alpina*) in Europe at two spatial scales: across the European Alps (large scale) and in three mountain massifs (local scale) of the French Alps. The study suggested that temperatures and precipitation were the major drivers of adaptive genetic variation at different spatial scales across Europe. However, when multiple environmental variables are considered, analyses should account for the various forms of co-linearity (e.g., high "global" correlations in topographic aspect and solar radiation) that can occur in regression analyses of environmental and genetic data (Wagner & Fortin 2005).

Our goals in this chapter are threefold. First, we review four general research approaches for linking adaptive genomic variation to spatial landscape factors (summarized in Table 9.1) and computational challenges and potential solutions for each. Second, we discuss some general challenges in landscape genomics. Third, we conclude with a consideration of current and future applications of landscape genomics studies.

## 9.2    CURRENT LANDSCAPE GENOMICS METHODS

Four general frameworks for linking adaptive variation to spatial variation in landscape characteristics have emerged. First, population genomics studies estimate genetic differences among populations using large numbers of molecular markers to identify outlier loci that are presumed to be under selection or are linked to

**Table 9.1** Characteristics of major approaches to landscape genomics.

| Method | Approach | Statistics | Inference | Limitations |
|---|---|---|---|---|
| Outlier (population genomics) | Estimate genetic distance (e.g., $F_{ST}$) among a large number of loci (e.g., AFLPs or SNPs) | Test for statistical $F_{ST}$ outliers by generating a frequency distribution of locus-specific $F_{ST}$s, Bayesian | Loci with significantly small $F_{ST}$ are under balancing selection and significantly large $F_{ST}$ under positive selection | Loci are anonymous, may be linked to actual loci; false positives due to high statistical power to detect outliers/population structure, unknown null distributions |
| Correlative | Test for correlations of allele frequencies of spatially diverged populations with environmental variables | Bayesian regression, latent factor mixed models, spatial generalized linear mixed models | Loci with allele frequency changes in geographic space that correlate with changes in environmental variables are under selection | Loci are anonymous, may be linked to actual loci; false positives due to high statistical power to detect outliers/population structure, unknown null distributions |
| QTL mapping | Development of large numbers of markers (e.g., microsatellites, AFLPs, SNPs), mapping them in the genome, crosses among individuals to detect genomic regions associated with phenotypic expression | ANOVA (marker regression), interval mapping, maximum likelihood | Identification of gene(s) associated with a phenotype by discovering genomic region(s) with a significant correlation with a phenotypic trait value | Genotype-by-environment interactions could produce false signal, labor intensive, generally reveal large genomic regions and not genes themselves, detects primarily genes of large phenotypic effect |
| GWAS | Even larger number of markers than QTLs (generally SNPs), genome scan comparison of individuals with different phenotypes (often disease-free and diseased individuals) | Search for linkage disequilibrium between alleles and phenotypes | Identification of particular SNP(s) associated (linked) with different phenotypes, generation of relative risk profiles (e.g., of developing a disease state) | False positives from population structure, false negatives from small sample sizes, reveal only candidate regions, linkage disequilibrium rates assumed constant or unknown |
| Candidate gene | Start with known loci (from other studies or species) and investigate whether allelic differences explain phenotypic differences among individuals | Genotyping of individuals with different phenotypes, correlation of genotype with phenotype | Allelic differences translate to phenotypic differences, may be associated with differences in environment | Gene(s) must be known a priori, often requires extensive upstream work, generally restricted to genes of large effect size |

(*continued*)

**Table 9.1**   *(Continued)*

| Method | Approach | Statistics | Inference | Limitations |
|---|---|---|---|---|
| Exomics, transcriptomics | High-throughput sequencing of cDNA or transcribed mRNA | Extensions of outlier approaches above, comparisons of mRNA copy number among different phenotypes or individuals under different conditions | Phenotypic differences are related to genotypic differences (in the exome) or gene expression differences (in the transcriptome) | Bias towards gene of large effect and against rare variants, potential for false positives when population structure is not accounted for, and collinearity of environmental factors that can make actual agents of selection difficult to identify |

adaptive genomic regions (Luikart et al. 2003). Much of what constitutes current landscape genomics projects involves studies of outlier loci and associated methods to detect correlations of outlier loci with environmental variables (Joost et al. 2007, 2013; Coop et al. 2010). Second, when phenotypes have also been measured, quantitative trait locus (QTL) mapping and/or genome-wide association studies (GWASs) can be used to identify the number and phenotypic effects of genomic regions that influence phenotypic and genetic variability in populations (Kruuk 2004; Stinchcombe & Hoekstra 2008). Depending upon availability of reference genomes, this approach may lead to identification of specific genes within genomes. Third, as a benefit of functional genetic studies of model organisms, candidate genes identified from comparative genomics among model organisms can be used to study phylogenetically related non-model organisms in which functionally adaptive phenotypes of individuals and populations have been identified. Fourth, with the advent of high-throughput sequencing, variation at the exome and transcriptome level can be identified by analyzing the types and abundances of RNAs that may reflect differences in gene regulation in different environments (e.g. Yi et al. 2010). Although the first approach is predominant in landscape genomics studies, technological advances have allowed increasing attention to the latter three. We discuss how each of these four approaches can be used in landscape genomic studies.

### 9.2.1   Population genomics

A predominant approach for assessing spatial adaptive genetic variation has been "population genomics" (Black et al. 2001; Luikart et al. 2003; Stinchcombe & Hoekstra 2008), whereby researchers have estimated between-population genetic differentiation for hundreds to thousands of loci such as amplified fragment length polymorphisms (AFLPs) and SNPs. Under a neutral island model with spatially uniform migration and gene flow, at migration-drift equilibrium it was shown how population differentiation of allele frequencies and related measures (e.g. $F_{ST}$) across a large number of loci can be used to infer the process of selection acting on a subset of loci (Lewontin & Krakauer 1973). Beaumont and Nichols (1996) implemented an algorithm called FDIST to identify outliers that exhibit strong differences from a null distribution of $F_{ST}$, using the null hypothesis (distribution of $F_{ST}$) of no differentiation (and implemented in the software LOSITAN, Antao et al. 2008, or ARLEQUIN v3.5.1.3, Excoffier & Lischer 2010). Statistical outlier loci, those with significantly greater or smaller $F_{ST}$ values than the background, can also be characterized by ln(RV) and ln(RH) statistics (natural log of the ratio of the variance and heterozygosity of alleles between two populations; Schlötterer 2002) and by the Ewens–Watterson test (Ewens 1972; Watterson 1977; Vigouroux et al. 2002). Outliers are presumed to be under selection or linked to loci under selection. Loci with values that are more diverged than expected from the distribution of genome-wide $F_{ST}$ values are presumed to be under diversifying or local selection and those less divergent than expected are inferred to be under stabilizing or purifying selection (Black et al. 2001; Luikart et al. 2003).

More recent methods for detecting statistical outliers have addressed long-standing problems in specifying appropriate null distributions for significance testing. That is, null distributions for $F_{ST}$ values among a suite of populations can be hard to estimate when population structure is high (Lowry 2010), thereby making it difficult to detect outliers. Beaumont and Balding (2004) developed a hierarchical Bayesian model (BayesFST), where correlation among loci and differences in locus-specific and population-specific parameters could be addressed through a multinomial Dirichlet likelihood function. Later, Foll and Gaggiotti (2008) introduced BayeScan, which estimates directly the probability that each locus is subject to selection using a Bayesian method. For example, an assessment of 392 AFLPs in the common frog (*Rana temporaria*) among populations at varying altitudes in the Alps revealed eight candidate loci with significantly higher $F_{ST}$ estimates than the remainder of loci, suggesting their potential involvement in adaptation to altitude (Bonin et al. 2006). Identification of outlier loci can also indicate the occurrence of selective sweeps (Schlötterer 2002), in which populations exposed to stressors, such as a highly virulent disease, exhibit significantly higher frequencies of certain alleles relative to populations without history of such a stressor. Genetic hitchhiking with the locally adaptive allele will result in reduced within-population diversity, whereas selection against maladaptive alleles will result in signatures of increased among-population divergence (Barton & Bengtsson 1986; Hilton et al. 1994; Charlesworth et al. 1997).

Much of landscape genomics now focuses on testing for correlations between particular alleles or allele frequencies with environmental variables (i.e., "correlative approaches"). Logistic regression has been useful for detecting selection using SNP data, since they have two alleles and are thus binary (Joost et al. 2007, 2013). Generalized estimating equations (GEEs) were also applied to deal with unknown covariance structures resulting from ***spatial autocorrelation*** in the data (Poncet et al. 2010). Both logistic regression in GEEs can be applied to individual or population-based samples. More widely used is Bayenv, a Bayesian regression method for detecting correlations of loci with environmental variables, which has been shown to be more powerful than $F_{ST}$ outlier analyses (Coop et al. 2010).

This method tests for large allele frequency differences across environmental gradients by comparing observed allele frequency differences to transformed normal distribution of underlying population frequencies. Covariance matrices are used to control for spatial autocorrelation (Coop et al. 2010). When considering population structure, however, these covariance matrices have to be based on different genetic data than those that are correlated with environmental variables. Newer correlative approaches, such as applications of latent factor mixed models can integrate population structure in analyses by limiting false positives (LFMM software; Frichot et al. 2013). Most recently, SGLMMs (spatial generalized linear mixed models) (Guillot et al. 2014) can be used to detect selection by testing for correlations between allele frequencies and environmental variables. Analyses using SGLMMs should prove faster than those using Markov chain Monte Carlo approaches because an INLA (integrated nested Laplace approximation) framework is instead used for parameter optimization in the model (Guillot et al. 2014).

A recent study compared several methods for detecting selection among populations. Using simulated data, De Mita et al. (2013) compared eight different methods for detecting selection that included outlier and correlative approaches. Some generalities emerged, including: (1) correlation methods are generally more powerful; (2) correlation methods should account for underlying correlation structure; (3) it is generally better to sample more populations than more individuals per population (De Mita et al. 2013).

Both outlier locus and correlative studies, however, have limitations, the primary one being that statistical outlier or correlation loci mark "candidate genomic regions" that may comprise several megabases of DNA, for which further study is needed. In humans, spatial studies of allele frequencies of hundreds of thousands of SNPs showed that remarkably few genes in the human genome had extreme allele frequency differences among populations (Hancock et al. 2008; Coop et al. 2009). In AFLP and SNP studies of non-model organisms, most of these putative loci are anonymous with unknown function. Without homologous sequences available in databases such as Genbank, candidate loci remain as those simply correlated with environmental variables at different sites. However, an increasing number of SNPs are being discovered within open reading frames that have a putative function or are homologous to those in databases such as Genbank or Swissprot. As an example, a recent study of redband trout

(*Oncorhynchus mykiss gairdneri*) found six candidate SNP loci associated with adaptation to dryer and lower altitude desert climates versus wetter montane climates at higher altitudes. The putative function of five of these genes was already known (Narum et al. 2010).

Outlier-detection methods are subject to high false-positive rates because of confounding effects such as hidden population structure or historical demographic change. Null distributions should be adjusted to reflect these effects (Narum et al. 2011). Excoffier et al. (2009) reduced the false-positive rate by explicitly modeling hierarchical population structure. To address the influence of demographic history, Bazin et al. (2010) developed an approximate Bayesian computational approach to estimate simultaneously demographic history and selection. However, if populations exhibit extremely complex population structure or demographic history (e.g., population expansions, admixture between species, or partial reproductive isolation), developing appropriate null distributions is likely to be challenging or controversial (Teshima et al. 2006; Hermisson 2009; Bierne et al. 2011).

It is especially important to note that outlier loci are not likely to be the candidate loci themselves, but are more likely to be in linkage disequilibrium with the genes under selection. As such, having genomic maps or reference genomes of study organisms becomes increasingly important to identify the actual gene(s) influencing variation in the phenotypic traits of interest. Even if genomic resources exist, such work can be extremely labor-intensive, depending on how tightly spaced the anonymous molecular markers are within a genomic map. Thus, regardless of the statistical method, these approaches yield candidate genes or genomic regions that often necessitate significant downstream work, including further exploration to identify the genes themselves, followed by functional genomic approaches to verify the ecological relevance of the candidate genes (Storz & Wheat 2010). Field or laboratory studies may further be needed to confirm fitness consequences of putatively adaptive allelic variation (Lowry 2010).

### 9.2.2 QTL to genome-wide association studies

Several decades ago, researchers began exploring the genomic basis for phenotypic variation identified as being adaptive (fitness-related) *a priori*, using quantitative trait locus (QTL) mapping. QTL mapping provided an early bridge between molecular genetic variation and quantitative genetic analysis of phenotypes in experimental populations or in natural population where genetic relationships among groups of individuals (pedigrees) could be ascertained (Kruuk 2004; Kruuk & Hill 2008). The approach required development of large numbers of marker loci, generation of a linkage map describing how the markers are arranged within the genome, and appropriate breeding designs with crosses among individuals from populations that vary in phenotypes of interest. Statistical association of those phenotypes with the location of molecular markers on chromosomes involved statistical analysis of co-segregation of phenotypes and markers within inbred lines and F1, backcrossed, or more advanced generations. A goal of QTL mapping was identification of the number of loci influencing traits including the distribution of allelic effects of those loci (Stinchcombe & Hoekstra 2008; Weiss 2008; Visscher et al. 2008). Map-based cloning of specific genes in QTL is possible – good early examples include Duchenne muscular dystrophy, cystic fibrosis, Huntington's disease in humans (Collins 1995), and the major sex-determining locus in honeybees (Beye et al. 2003). However, given that QTL may represent several megabases of DNA, positional cloning using this method is laborious and time-consuming, with much effort expended on fine-scaled mapping of markers in advanced crosses (e.g. F4, F5, and further intercrosses between inbred laboratory populations), followed by DNA sequencing (e.g. Atwell et al. 2010).

While an effective early tool in genomic analyses, QTL studies typically detect only those loci and alleles with large phenotypic effects and likely those at intermediate frequencies within populations (Bodmer & Bonilla 2008; Weiss 2008). Thus, QTL studies produce only a lower bound on the number of genes that determine phenotypic trait variation. In addition, it was discovered relatively quickly that genotype-by-environment interactions result in different QTLs being identified in different populations (e.g. Paterson et al. 1991). Finally, linkage maps of markers are usually species-specific and in some cases may be population-specific, so that linkage maps based on laboratory populations may be misleading in natural populations that do not share an immediate evolutionary history (Weiss 2008).

Many of these shortcomings with regard to the discovery of QTLs and their accuracy in accounting for genetic variation in phenotypic traits can be overcome by increasing the number of genetic markers (e.g. Atwell et al. 2010) and applying much finer-scaled genome-wide association studies (GWASs). GWASs tend to include many more molecular markers and allow finer-scale mapping than QTL studies, with higher power to detect statistical associations between genotype and phenotype. GWASs have primarily been applied to studies of human diseases, where genomic comparisons are between affected human cases and specifically matched control subjects (Wang et al. 2005; Balding 2006). In this methodology, patterns of linkage disequilibrium (LD) among loci are discerned among a large number of markers to test for statistical associations between molecular markers and phenotypic variation, with the stipulation that population structure and recombination rates are known (Mitchell-Olds & Schmitt 2006; Balding 2006; Stinchcombe & Hoekstra 2008).

Among the most important caveats for GWAS is the number of false positives that can arise because of problems such as undetected population structure (Novembre et al. 2008). Further, as is true for QTL studies, false negatives arise because of relatively small samples sizes (both numbers of individuals and numbers of markers), the tendency to detect only alleles of large effect (Balding 2006; Stinchcombe & Hoekstra 2008; Atwell et al. 2010), and the difficulty of detecting rare variants. Additionally, similar to population genomics and QTL studies in organisms without reference genomes, GWASs will yield candidate genes or regions within chromosomes, many of which are anonymous. Thus GWASs represent only the initial steps that may include genomic mapping, chromosomal sequencing, and functional genomic studies to determine whether such genes are truly important in an evolutionary context. For organisms without large-scale genomic resources, the good news is that fine-scaled linkage disequilibrium of the kind needed for GWASs appears to be common in natural populations (e.g. Flores et al. 2007; Hoffmann & Rieseberg 2008; Hohenlohe et al. 2012), and not only among laboratory inbred lines (e.g., Rockman & Krugylak 2009). Efforts to conduct GWAS of non-model organisms should thus succeed in natural populations that can be adequately sampled in different landscapes.

### 9.2.3 Candidate gene approaches

Perhaps the earliest of the candidate gene approaches involved the study of candidate loci themselves, such as investigation of the potential adaptive benefit of single locus protein polymorphisms identified in allozyme studies (Storz & Wheat 2010). Several well-known studies were able to link allelic polymorphism of allozymes to adaptive phenotypic variation, such as lactate dehydrogenase (LDH) in sow bugs (Mitton et al. 1997) and glucose-6-phosphate dehydrogenase (G-6-PDH) in *Drosophila* (Eanes 1984). As a specific example, variation among natural populations of *Colias eurytheme* butterflies in the phosphoglucose isomerase (PGI) allozyme translated into variation in the ability to mobilize ATP via glycolosis for flight (Watt 1977). Specific allelic combinations led to better flight performance, even at suboptimal temperatures, which appeared to result in fitness advantages for males who could mate many times more quickly and females who could lay more eggs (Carter & Watt 1988).

Another approach has involved the investigation of the direct ecological relevance of candidate genes, also termed the "gene first approach" (Lowry 2010). A well-known example of this approach resulted in the discovery of the Ectodysplasin-A (eda) gene among populations of threespine sticklebacks (Colosimo et al. 2005). The "low" allele has been associated with the parallel loss of external bony plates during multiple post-Pleistocene freshwater invasions from ancestral marine stickleback populations (Colosimo et al. 2004). Losses of bony plates were later shown to have a fitness advantage in freshwater habitats using a reciprocal transplant experiment (Barrett et al. 2009). Another well-known example is the melanocortin-1 receptor (Mc1r), a gene of large effect for coat color in beach mice that confers background matching in different habitats (Hoekstra et al. 2006). This gene is highly conserved among vertebrates and can be used to describe color variation in other species, such as coloration in three lizard species that live on different colored backgrounds in White Sands, New Mexico (Rosenblum et al. 2010).

While candidate genes have clearly been successful in mapping genotypes to functionally adaptive phenotypes in several cases, the approach has some important drawbacks. First, several years of work may be necessary to link genotype to phenotype after candidate loci are identified – for instance as DNA sequencing and allele identification followed by field-based experiments were necessary to confirm the importance of eda in sticklebacks. Second, candidate gene, or "gene first" approaches, tend to be biased toward discovery of genes of large effect that underlie discrete phenotypic traits

(Weiss 2008; Lowry et al. 2010). However, currently such examples are rare, as most phenotypes under selection are polygenic and it is unlikely that alleles of large effect inherited in a purely Mendelian fashion explain a large portion of functional phenotypic variation. Most phenotypic traits are continuous and the presumed distribution of genotype to phenotype relationships throughout the genome involves a few genes of large effect, with most genes having small effect and thus most traits being polygenic. In addition, despite the clear links between allelic variation and adaptive phenotypic differences among populations, it is important to note that adaptations do not arise solely as differences in structural proteins and that regulation of expression via variation in enzymes or RNA transcription factors is a major part of a complex metabolic network (Storz & Wheat 2010). Overall, as we continue to scan genomes for candidate loci involved in adaptation, empirical null distributions of gene expression are needed as backdrops for identifying candidate genes.

### 9.2.4 Exomes and transcriptomes

Next-generation sequencing approaches have fostered the development and utility of "genotyping-by-sequencing", which involves researchers obtaining large amounts of sequence information on individuals in population samples (Narum et al. 2013). As such, the ability to more directly assess functionally adaptive variation in populations will also become increasingly possible. We examine two approaches that each rely on analysis of "outliers" based on $F_{ST}$ estimates among populations.

First is sequencing the DNA from the known "exome", or the part of the genome that is transcribed and includes both structural and regulatory regions. In humans, scans of exomes were used in comparisons of low and high elevation populations to identify functional adaptive variation underlying adaptations to high elevation in Tibetans relative to the Han Chinese from which Tibetans diverged over the last 3000 years or so (Yi et al. 2010). Similar exome data from a Danish population were used as a comparison. Among the thousands of SNPs identified, many (>30) were found to occur in genes with significantly greater divergence than expected by chance alone. The most divergent gene is not structural, but rather a transcription factor in the hypoxia-induced family (EPAS1) that ultimately regulates the number of erythrocytes in the

blood. Further studies may determine the direction of gene regulation, but the SNP with the greatest divergence was in an intron, further illustrating that regulation of gene expression networks may be a major target of selection that may not have been otherwise detected in natural populations.

The second approach is an extension of the "outlier" approach to sequencing of the RNAs in organisms' transcriptomes (RNAseq). In this case, RNA is extracted and sequenced from individuals from several different environments or parts of a species range, and the previously described statistical outlier approach is used to identify sequence variation in RNAs that have either greater or less population-level differentiation than the background (e.g. de Wit & Palumbi 2013). In a study of abalone along the coast of California, potentially adaptive genes were identified by BLAST searches of gene sequences derived from RNAs of live organisms, with 34 that contained more than a single outlier SNP. Most of the identified genes are potentially adaptive to the environmental differences along the California coast, where the abalone cover several hundred kilometers (de Wit & Palumbi 2013).

Analysis of exomes and transcriptomes will be increasingly common in the future. These approaches use many of the same methodologies described above and represent the most direct advance in landscape genomic studies of functionally adaptive variation in natural populations. However, the same caveats listed above – bias towards genes of large effect and against rare variants, potential for false positives when population structure is not accounted for, and collinearity of environmental factors that can make the actual agents of selection difficult to identify – apply for this approach. However, this methodology will also elucidate novel data about the nature of adaptations themselves, in that environmental variation is also related to changes in gene transcription levels. Indeed, it is not out of the realm of possibility that transcription factors or promoter regions may be the most important factors, rather than allelic variants of structural genes. Even if transcriptome adaptation is only a minority factor, if substantial, it would not even begin to match the Fisherian view that the statistical association of allelic variants and particular environments is additive in nature, and the view that most phenotypes result from networks of genes for which transcription levels vary depending on both the gene network and the ecological context in which the network is expressed (Fisher 1930; Benfey & Mitchell-Olds 2008).

## 9.3 GENERAL CHALLENGES IN LANDSCAPE GENOMICS

Landscape genomics generally involves merging genotypic data over many loci with spatial environmental data. As such, there are inherent challenges in testing for patterns across these two, very different, data sources (e.g. Cushman 2014). Here, we discuss the importance of testing and accounting for, as necessary, spatial autocorrelation. In addition, we highlight the importance of understanding statistical associations in data sets that investigate local adaptation (or isolation-by-adaptation).

### 9.3.1 Spatial data collection

Because landscape genomics studies rely on high-resolution environmental data for analyses, it is important to test hypotheses regarding the spatial distribution of genetic variation for which landsat or other layers are available. Often, environmental data from existing GIS databases are readily available, whereas *de novo* collection and analysis of appropriate high-resolution GIS data are time consuming and challenging. Thus, whenever possible, researchers should capitalize on existing GIS data sets that typically include climatic vegetation and geological variables (e.g., The Global Map Project, http://www.iscgm.org/cgi-bin/fswiki/wiki.cgi, and WorldClim, http://www.worldclim.org) because recent increases in the availability of digital environmental data from remote sensors and weather stations have now made many global environmental data sets freely available. Previous reviews have cataloged useful GIS resources for landscape genomics (Manel et al. 2010a; Thomassen et al. 2010) and in the absence of local field-collected environmental data global environmental data may serve as valuable substitutes.

## 9.4 SPATIAL AUTOCORRELATION

Species distributional response to environmental conditions is a phenomenon that is often referred to as spatial dependence (Fortin et al. 2005; Legendre 1993). Spatial aggregation can occur due to biotic processes such as dispersal and species interactions. These spatial structures create spatially autocorrelated genetic data. As a result, statistical methods in

landscape genomics should consider both: (1) unaccounted spatially structured environmental variables resulting in a spatial structuring of allele frequency distribution (i.e., spatial dependence; Manel et al. 2010a) and (2) spatial autocorrelation in allele frequencies generated by biotic processes that are distance related.

To test for spatial autocorrelation in a data set on a global scale, Moran's I (Moran 1950) is often calculated to detect departure from spatial randomness. Moran's I values range from −1 (where samples close in space tend to be more unalike than those further away in space) to 1 (where nearby samples are much more similar than those far apart). Spatial randomness is determined when Moran's I values are not different from 0. Spatial autocorrelation can also occur at a local scale and local indicators of spatial association (LISA) can be used to test for such patterns (Anselin 1995).

In cases where spatial autocorrelation are found in a data set, spatial regression methods (e.g., conditional autoregressive models, or CAR and simultaneous autoregressive models, or SAR) are potentially appropriate for analyzing spatial dependence as landscape features can be explicitly built into the model structure (Diniz-Filho et al. 2009). One particularly promising approach uses Moran's eigenvector maps (MEM) (Borcard and Legendre 2002; Dray et al. 2006). MEM variables are the eigenvectors of a spatial weighting matrix calculated from the sampling location's geographic coordinates. MEM analysis produces uncorrelated spatial eigenfunctions used to dissect the correlations of allele frequencies with landscape variables into separate scales to be used as predictors in regressions. To detect loci potentially linked to genes under selection, Manel et al. (2010b) used multiple linear regressions to correlate single AFLP allele frequencies from a large genome scan of *Arabis alpina* with environmental variables. To consider the potential for unmeasured variables to create spatial structure in allelic distributions, they used only broad-scale principal coordinates of neighbor matrices (MEMs) as explanatory variables. When sample size is small, geographically weighted regression (GWR) (Brunsdon et al. 1996) is appropriate (Spear & Storfer 2008).

Regression tree methods such as RANDOMFOREST (Elith 2009) are an alternative when spatial autocorrelation and effects of environmental correlates are not constant across the study area (Murphy et al. 2010). Regression tree methods are based on an iterative procedure that splits the observations (samples) into a series of two groups in a hierarchical "tree" (dendrogram-like) structure where the values of dependent variables are similar within each group based on a specific value of one of the independent variables (quantitative or qualitative independent values). Usually the first deeper splits reflect mostly large spatial-scale processes while the later shallower splits in tree structure correspond to localized spatial effects.

While in principle spatial autocorrelation analysis may determine natural selection or other processes that govern allele frequency variation, experimental results may also be strongly affected by sampling scale and stochastic variation (Slatkin & Arter 1991) as well as confounded by population and family structure (Balding 2006; Excoffier et al. 2009). Bayesian geographical analysis approaches have been recently introduced to address this problem by testing for correlations between allele frequencies and environmental variables after correcting for population structure and differences in sample size by the inclusion of a covariance matrix of allele frequencies among populations (Yu et al. 2006; Hancock et al. 2008). Using this approach, Hancock et al. (2008) found evidence of a selective effect of climate on metabolism genes in humans from the analysis of the association between 973 SNPs and climatic variables. This approach requires that populations are known or defined in advance along with estimates of allele frequencies within those populations, which is not always possible depending on the species and the sampling design. For instance, Poncet et al. (2010) used GEEs with binomial error to test for correlations of allele frequencies to environmental variables. In this case, GEE was used to account for genetic relatedness of individuals at nearby locations using a quasi-likelihood extension of a general linear model that allowed for correlation in the response variable.

### 9.4.1 Isolation-by-adaptation

Spatial autocorrelation between genetic and landscape variation may arise if the movement of individuals is limited because of strong ecological selection. Alternatively, gene flow may act to reduce the ability of populations to respond genetically to changing conditions at the nodes (Slatkin 1985). Local adaptation will

occur in a form that could impede gene flow via selection against immigrants (Rundel et al. 2009). In this case, populations will diverge at both adaptive and neutral loci. For example, Lee and Mitchell-Olds (2011) estimated the relative contributions of environmental adaptation and isolation-by-distance (IBD) on genetic variation in *Boechera stricta*, a wild relative of *Arabidopsis*. By comparing patterns of neutral population differentiation to geographical and environmental distance, their study contrasted measures of isolation-by-distance and "isolation-by-adaptation" (Nosil et al. 2009). They identified specific environmental variables, including water availability, as the possible cause of differential local adaptation across geographical regions. More generally, identifying environmental adaptation from neutral genetic variation can be accomplished using a combination of ecological niche modeling with an analysis of the neutral genetic structure (Freedman et al. 2010).

Landscape heterogeneity exposes populations to varying environmental selection pressures, which, depending on the interplay of selection and gene flow, can result in adaptation to local environmental conditions (Lenormand 2002; Savolainen et al. 2007). Although there are some examples of ecological adaptation despite strong gene flow, such as heavy-metal tolerance in plants (Turner et al. 2010), host-race divergence in insects (Antolin et al. 2006) and diversification of fish ecotypes in lake ecosystems (Schluter 2000), both the likelihood of adaptation and the maintenance of adaptive variability are dramatically diminished when levels of gene flow exceed "migration selection equilibrium" (Bridle & Vines 2007). One of the main strengths of the landscape genomics approach is that it provides a set of tools for incorporating the spatial heterogeneity of the landscape and measures of gene flow with statistical models that elucidate patterns of adaptive variation.

Alternatively, non-adaptive factors may also affect gene flow between populations, including impediments to dispersal (habitat matrix or disperser behavior), differences in emigration and immigration rates due to density or local population dynamics, and reproductive barriers, confounding selective factors. This underscores the fact that a complete picture of functionally adaptive genetic variation ultimately depends on identifying the agents of selection – where environmental factors on the landscape cause phenotypic and genetic differentiation – rather than simply identifying those correlated with population structure.

## 9.5 APPLICATIONS OF LANDSCAPE GENOMICS TO CLIMATE CHANGE

One contemporary, yet challenging, question is whether adaptive evolution can keep pace with the rate and direction of environmental and climate changes induced by human activities (Lavergne et al. 2010). Several studies have shown that species have already shifted their geographic ranges in response to climate change (e.g., Walther et al. 2002; Frei et al. 2010), whereas the general potential of species to adapt to rapid environmental change is still being debated (Davis et al. 2005; Reusch & Wood 2007; Jay et al. 2012). The capacity of organisms to respond to changing climate is likely to vary widely as a consequence of variation among species in their degree of phenotypic plasticity and their adaptive evolutionary potential. Then, an important application of the study of adaptive variation on the landscape will be to predict how populations respond to a changing environment (Manel et al. 2010a; Manel & Holderegger 2013). One classical theoretical prediction for an adaptive response to climate change states that gene flow from populations in warmer climates will bring pre-adapted alleles to the leading edge and promote adaptation, whereas populations at the rear edge are likely to face local extinction (Davis et al. 2005).

Recent empirical observations in landscape genomics have utilized climate variables to determine putatively adaptive genes. For example, Hancock et al. (2011) conducted a genome-wide scan to identify climate-adaptive genetic loci in the plant *Arabidopsis thaliana*. They found that amino acid-changing variants were significantly enriched among the loci strongly correlated with climate, suggesting that their scan effectively detects adaptive alleles (Hancock et al. 2011). Their study provided a set of candidate loci for dissecting the molecular bases of climate adaptations. Manel et al. (2012) found correlations between AFLP markers and climatic variables for 13 alpine plant species consistently sampled across the entire European Alps. Temperature and precipitation were identified as the best environmental predictors of statistical associations. This was the first study to show that the same environmental variables are drivers of adaptation of multiple plant species at the scale of a whole biome (i.e., European Alps).

These empirical examples illustrate the importance of climate in the distribution of genes under selection potentially involved adaptation to climate change.

The next step is to integrate landscape genomics and predictions of species' geographic range shifts. Ecological niche modeling (or species distribution modeling) has been used extensively to predict future species' ranges based on climate shifts (Guisan & Thuiller 2005). Niche models frequently combine species' occurrence data with information about the climate and habitat at those locations to predict the geographic range of suitable habitat (Elith et al. 2006). Although niche models can predict the fundamental niche for a species, they fail to incorporate essential components in defining range boundaries, such as dispersal abilities and habitat barriers (Pearson & Dawson 2003; Angert et al. 2011). In addition, most niche-modeling approaches have the drawback of not accounting for gene migration, population differentiation, and niche variation within the species range (Aitken et al. 2008). Landscape genomics methods could be used to test the relative resistance of intervening habitat to species' dispersal between current ranges and expected future ranges based on niche modeling predictions. Although rarely, if ever, used together, the integration of landscape genetics and niche modeling is thus particularly powerful for assessing whether a species is capable of dispersing to the newly suitable habitat or if more active intervention, such as assisted migration, is needed to prevent declines or extinctions (McLachlan et al. 2007; Wiens et al. 2009). Furthermore, developing methods that include genetic diversity patterns, population structure (Mimura & Aitken 2007), and differences in adaptive variation between leading- and trailing-edge populations (Hampe & Petit 2005) could also yield different predictions of species responses to climate change. A study of the alpine herb, *Biscutella laevigta*, showed, for example, that gene flow was twice as restricted in trailing-edge populations relative to leading-edge populations (Parisod & Joost 2010). Genome scans revealed divergent selection in the trailing-edge population but no detectable signature of selection in the expanding population at the leading edge. Taken together, relatively low gene flow and an apparent lack of selection suggest limited ability of the leading-edge population to adapt to climate-induced range expansion (Parisod & Joost 2010).

With increases in availability of genomic data, it might become possible not only to predict species' distributions in response to climate change (Guisan & Thuiller 2005) but also to develop scenarios regarding the spatial distribution of adaptive genetic variation at the whole-genome level in response to changes in temperature and precipitation regimes, as proposed by Fournier-Level et al. (2011) and Hancock et al. (2011). Future steps will be to integrate the results on drivers of genetic adaptation into bioclimatic models and to test the evolutionary and functional relevance of temperature and precipitation under experimental conditions.

## REFERENCES

Aitken, S.N., Yeaman, S., Holliday J.A., Wang, T.L., & Curtis-McLane, S. (2008) Adaptation, migration or extirpation: climate change outcomes for tree populations. *Evolutionary Applications* **1**, 95–111.

Angert, A.L., Crozier, L.G., Gilman, S.E., Rissler, L.J., Tewksbury, J.J., & Chunco, J.J. (2011) Do species' traits predict recent shifts at expanding range edges? *Ecology Letters* **14**, 677–89.

Anselin, L. (1995) Local indicators of spatial association – LISA. *Geographical Analysis* **27**, pp. 93–115.

Antao, T., Lopes, A., Lopes, R.J., Beja-Pereira, A., & Luikart, G. (2008) LOSITAN: a workbench to detect molecular adaptation based on an $F_{st}$-outlier method. *BMC Bioinformatics* **9**, 323.

Antolin, M.F., Bjorksten, T.A., & Vaughn, T.T. (2006) Host-related fitness trade-offs in a presumed generalist parasitoid, Diaeretiella rapae (Hymenoptera: Aphidiidae). *Ecological Entomology* **31**, 242–54.

Atwell, S., Huang, Y.S., Vilhjalmsson, B.J., Willems, G., Horton, M., Li, Y., Meng, D., Platt, A., Tarone, A.M., Hu, T.T., Jiang, R., Muliyati, N.W., Zhang, X., Amer, M.A., Baxter, I., Brachi, B., Chory, J., Dean, C., Debieu, M., de Meaux, J., Ecker, J.R., Faure, N., Kniskern, J.M., Jones, J.D.G., Michael, T., Nemri, A., Roux, A., Salt, D.E., Tang, C., Todesco, M., Traw, M.B., Weigel, D., Marjoram, P., Borevitz, J.O., Bergelson, J., & Nordborg, M. (2010) Genome-wide association study of 107 phenotypes in *Arabidopsis thaliana* inbred lines. *Nature* **465**, 627–31.

Balding, D.J. (2006) 'A tutorial on statistical methods for population association studies', *Nature Reviews Genetics*, vol. **7**, pp. 781–791.

Barrett, R.D.H., Rogers, S.M., & Schluter, D. (2009) Environment specific pleiotropy facilitates divergence at the ectodysplasin locus in threespine stickleback. *Evolution* **63**, 2831–7.

Barton, N. & Bengtsson, B.O. (1986) The barrier to exchange between hybridizing populations. *Heredity* **57**, 357–76.

Bazin, E., Dawson, K.J., & Beaumont, M.A. (2010) Likelihood-free inference of population structure and local adaptation in a Bayesian hierarchical model. *Genetics* **185**, 587–602.

Beaumont, M.A. & Balding, D.J. (2004) Identifying adaptive genetic divergence among populations from genome scans. *Molecular Ecology* **13**, 969–80.

Beaumont, M.A. & Nichols, R.A. (1996) Evaluating loci for use in the genetic analysis of population structure. *Proceedings of the Royal Society of London, Series B* **1377**, 1619–26.

Benfey P.N. & Michell-Olds, T. (2008) From genotype to phenotype: systems biology meets natural variation. *Science* **320**, 495–7.

Beye, M., Hasselmann, M., Fondrk, M.K., Page, R.E. Jr. & Omholt, S.W. (2003) The gene CSD is the primary signal for sexual development in the honeybee and encodes an SR-type protein. *Cell* **114**, 419–29.

Bierne, N., Welch, J., Loire, E., Bonhomme, F., & David, P. (2011) The coupling hypothesis: why genome scans may fail to map local adaptation genes. *Molecular Ecology* **20**, 2044–72.

Black, W.C., Baer, C.F., Antolin, M.F., & DuTeau, N.M. (2001) Population genomics: genome-wide sampling of insect populations. *Annual Review of Entomology* **46**, 441–69.

Bodmer, W. & Bonilla, C. (2008) Common and rare variants in multifactorial susceptibility to common diseases. *Nature Genetics* **40**, 695–701.

Bonin, A., Miaud, C., Taberlet, P., & Pompanon, F. (2006) Explorative genome scan to detect candidate loci for adaptation along a gradient of altitude in the common frog (*Rana temporaria*). *Molecular Biology and Evolution* **23**, 773–83.

Bonin, A., Ehrich, D., & Manel, S. (2007) Statistical analysis of amplified fragment length polymorphism data: a toolbox for molecular ecologists and evolutionists. *Molecular Ecology* **16**, 3737–58.

Borcard, D. & Legendre, P. (2002) All-scale spatial analysis of ecological data by means of principal coordinates of neighbour matrices. *Ecological Modeling* **153**, 51–68.

Bridle, J.R. & Vines, T.H. (2007) Limits to evolution at range margins: when and why does adaptation fail? *Trends in Ecology and Evolution* **22**, 140–7.

Brunsdon, C., Fotheringham, A.S., & Charlton, M.E. (1996) Geographically weighted regression: a method for exploring spatial nonstationarity. *Geographical Analysis* **28**, 281–98.

Carter, P.A. & Watt, W.B. (1988) Adaptation at specific loci. V. Metabolically adjacent enzyme loci may have very distinct experiences of selective pressures. *Genetics* **119**, 913–24.

Charlesworth, B., Nordborg, M., & Charlesworth, D. (1997) The effects of local selection, balanced polymorphism and background selection on equilibrium patterns of genetic diversity in subdivided populations. *Genetic Resources* **70**, 155–74.

Collins, F.S. (1995) Positional cloning moves from the perditional to the traditional. *Nature Genetics* **9**, 347–50.

Colosimo, P.F., Peichel, C.L., Nereng, K., Blackman, B.K., Shapiro, M.D., Schluter, D., & Kingsley, D.M. (2004) The genetic architecture of parallel armor plate reduction in threespine sticklebacks. *PLoS Biology* **2**, 635–41.

Colosimo, P.F., Hosemann, K.E., Balabhadra, S., Villarrea, G. Jr. Dickson, M., Grimwood, J., Schmutz, J., Myers, R.M., Schluter, D., & Kingsley, D.M. (2005) Widespread parallel evolution in sticklebacks by repeated fixation of ectodysplasin alleles. *Science* **5717**, 1928–33.

Coop, G., Pickrell, J.K., Novembre, J., Kudaravalli, S., Li, J., Absher, D., Myers, R.M., Cavalli-Sforza, L.L., Feldman, M.W., & Pritchard, J.K. (2009) The role of geography in human adaptation. *PLoS Genetics* **5**, e1000500.

Coop, G., Witonsky, D., Di Rienzo, A., & Pritchard, J.K. (2010) Using environmental correlations to identify loci underlying local adaptation. *Genetics* **185**, 1411–23.

Cushman, S.A (2014) 'Grand challenges in evolutionary and population genetics: the importance of integrating epigenetics, genomics, modeling and experimentation.' *Frontiers in Genetics*, **5**, 197.

Davey, J.W., Hohenlohe, P.A., Etter, P.D., Boone, J.Q., Catchen, J.M., & Blaxter, M.L. (2011) Genome-wide genetic marker discovery and genotyping using next-generation sequencing. *Nature Reviews Genetics* **12**, 499–510.

Davey, J.W., Cezard, T., Fuentes-Utrilla, P., Eland, C., Gharbi, K., & Blaxter, M.L. (2013) Special features of RAD sequencing data: implications for genotyping. *Molecular Ecology* **22**, 3151–64.

Davis, M.B., Shaw, R.G., & Etterson, J.R. (2005) Evolutionary responses to changing climate. *Ecology* **86**, 1704–14.

De Mita, S., Thuillet, A.-C., Gay, L., Ahmadi, N., Manel, S., Ronfort, J., & Vigouroux, Y. (2013) Detecting selection along environmental gradients: analysis of eight methods and their effectiveness for outbreeding and selfing populations. *Molecular Ecology* **22**, 1383–99.

de Wit, P. & Palumbi, S.R. (2013) Transcriptome-wide polymorphisms of red abalone (*Haliotis rufescens*) reveal patterns of gene flow and local adaptation. *Molecular Ecology* **22**, 2884–97.

Diniz-Filho, J.A.F, Nabout, J.C., Telles, M.P.D., Soares, T.N., & Rangel, T. (2009) A review of techniques for spatial modeling in geographical, conservation and landscape genetics. *Genetics and Molecular Biology* **32**, 203–11.

Dray, S., Legendre, P., & Peres-Neto, P.R. (2006) Spatial modeling: a comprehensive framework for principal coordinate analysis of neighbour matrices (PCNM). *Ecological Modeling* **196**, 483–93.

Eanes, W.F. (1984) Viability interactions, *in vivo* activity and the G6PD polymorphism in *Drosophila melanogaster*. *Genetics* **106**, 95–107.

Eckert, A.J., Bower, A.D., Gonzalez-Martinez, S.C., Wegrzyn, J.L., Coop, G., & Neale, D.B. (2010) Back to nature: ecological genomics of loblolly pine (*Pinus taeda*, Pinaceae). *Molecular Ecology* **19**, 3789–805.

Elith, J. & Graham, C.H. (2009) Do they? How do they? Why do they differ? On finding reasons for differing performances of species distribution models. *Ecography* **32**, 66–77.

Elith, J., Graham, C.H., Anderson, R.P., Dudik, M., Ferrier, S., Guisan, A., Hijmans, R.J., Huettmann, F., Leathwick, J.R., Lehmann, A., Li, J., Lohmann, L.G., Loiselle, B.A., Manion, G., Moritz, C., Nakamura, M., Nakazawa, Y., Overton, J.M., Peterson, A.T., Phillips, S.J., Richardson, K., Scachetti-Pereira,

R., Schapire, R.E., Soberon, J., Williams, S. Wisz, M.S., & Zimmermann, N.E. (2006) Novel methods improve prediction of species' distributions from occurrence data. *Ecography* **29**, 129–51.

Ewens, W.J. (1972) The sampling theory of selectively neutral alleles. *Theoretical Population Biology* **3**, 87–112.

Excoffier, L. & Lischer, H.E.L. (2010) Arlequin suite ver 3.5: a new series of programs to perform population genetics analyses under Linux and Windows. *Molecular Ecology Resources* **10**, 564–7.

Excoffier, L., Hofer, T., & Foll, M. (2009) Detecting loci under selection in a hierarchically-structured population. *Heredity* **103**, 285–98.

Fisher, R.A. (1930) *The Genetical Theory of Natural Selection*. Oxford University Press, New York.

Flores, M., Morales, L., Gonzaga-Jauregui, C., Dominguez-Vidana, R., Zepeda, C., Yanez, O., Gutierrez, M., Lemus, T., Valle, D., Avila, C., Blanco, D., Medina-Ruiz, S., Meza, K., Ayala, E., Garcia, D., Bustos, P., Gonzalez, V., Girard, L., Tusie-Luna, T., Davila, G., and Palacios, R. (2007) Recurrent DNA inversion rearrangements in the human genome. *Proceedings of the National Academy of Sciences of the United States of America* **104**, 6099–106.

Foll, M. & Gaggiotti, O. (2008) A genome-scan method to identify selected loci for both dominant and codominant markers: a Bayesian perspective. *Genetics* **180**, 977–93.

Fortin, M.-J., Keitt, T.H., Maurer, B.A., Taper, M.L., Kaufman, D.M., & Blackburn, T.M. (2005) Species ranges and distributional limits: pattern analysis and statistical issues. *Oikos* **108**, 7–17.

Fournier-Level, A., Korte, A., Cooper, M., Nordborg, M., Schmitt, J., & Wilczek, A.M. (2011) A map of local adaptation in *Arabidopsis thaliana*. *Science* **334**, 86–9.

Freedman, A.H., Thomassen, H.A., Buermann, W., & Smith, T.B. (2010) Genomic signals of diversification along ecological gradients in a tropical lizard. *Molecular Ecology* **19**, 3773–88.

Frei, E., Bodin, J., & Walther, G.R. (2010) Plant species' range shifts in mountainous areas – all uphill from here? *Botanica Helvetica* **120**, 117–28.

Frichot, E., Schoville, S.D., Bouchard, G., & Francois, O. (2013) Testing for associations between loci and environmental gradients using latent factor mixed models. *Molecular Biology and Evolution* **30**, 1687–99.

Fumagalli, M., Sironi, M., Pozolli, U., Ferrer- Admettla, A., Pattini, L., & Nielsen, R. (2011) Signatures of environmental genetic adaptation pinpoint pathogens as the main selective pressure through human evolution. *PLoS Genetics* **7**, e1002355.

Gagnaire, P.A., Normandeau, E., Pavey, S.A., & Bernatchez, L. (2013) Mapping phenotypic, expression and transmission ratio distortion QTL using RAD markers in the Lake Whitefish (*Coregonus clupeaformis*). *Molecular Ecology* **22**, 3036–48.

Guillot, G., Vitalis, R., Rouzic, A.L., & Gautier, M. (2014) Detecting correlation between allele frequencies and environmental variables as a signature of selection. A fast computational approach for genome-wide studies. *Spatial Statistics* **8**, 145–55.

Guisan, A. & Thuiller, W. (2005) Predicting species distribution: offering more than simple habitat models. *Ecology Letters* **8** 993–1009.

Hampe, A. & Petit, R.J. (2005) Conserving biodiversity under climate change: the rear edge matters. *Ecology Letters* **8**, 461–7.

Hancock, A.M., Witonsky, D.B., Gordon, A.S., Eshel, G., Pritchard, J.K., Coop, G., & Di Rienzo, A. (2008) Adaptations to climate in candidate genes for common metabolic disorders. *PLoS Genetics* **4**, e32.

Hancock, A.M., Witonsky, D.B., Ehler, E., Alkorta-Aranburu, G., Beall, C., Gebremedhin, A., Sukernik, R., Utermann, G., Pritchard, J., Coop, G., & Di Rienzo, A. (2010) Human adaptations to diet, subsistence and ecoregion are due to subtle shifts in allele frequency. *Proceedings of the National Academy of Sciences of the United States of America* **107** (Suppl. 2), 8924–30.

Hancock, A.M., Brachi, B., Faure, M., Horton, M.W., Jarymowycz, L.B., Sperone, F.G., Toomajian, C., Roux, F., & Bergelson, J. (2011) Adaptation to climate across the *Arabidopsis thaliana* genome. *Science* **334**, 83–6.

Hermisson, J. (2009) 'Who believes in whole-genome scans for selection?' *Heredity*, vol. **103**, pp. 283–84.

Hilton, H., Kilman, R.M., & Hey, J. (1994) Using hitchhiking genes to study adaptation and divergence during speciation within the *Drosophila melanogaster* species complex. *Evolution* **48**, 1900–13.

Hoekstra, H.E., Hirschmann, R.J., Bundey, R.J., Insel, P., & Crossland, J.P. (2006) A single amino acid mutation contributes to adaptive color pattern in beach mice. *Science* **313**, 101–4.

Hoffmann, A.A and Rieseberg, L.H. (2008) Revisiting the impact of inversions in evolution: from population genetic markers to drivers of adaptive shifts and speciation? *Annual Review of Ecology, Evolution and Systematics* **39**, 21–42.

Hohenlohe, P.A., Bassham, S., Currey, M,. & Cresko, W.A. (2012) Extensive linkage disequilibrium and parallel adaptive divergence across threespine stickleback genomes. *Philosophical Transactions of the Royal Society B* **367**, 395–408.

Jay, F., Manel, S., Alvarez, N., Durand, E.Y., Thuiller, W., Holderegger, R., Taberlet, P., & Francois, O. (2012) Forecasting changes in population genetic structure of alpine plants in response to global warming. *Molecular Ecology* **21**, 2354–68.

Joost, S., Bonin, A., Bruford, M.W., Després, L., Conord, C., Erhardt, G., & Taberlet, P. (2007) A spatial analysis method (SAM) to detect candidate loci for selection: towards a landscape genomics approach to adaptation. *Molecular Ecology* **16**, 3955–69.

Joost, S., Vuilleumier, S., Jensen, J.D., Schoville, S., Leempoel, K., Stucki, S., Widmer, I., Melodelima, C., Rolland, J., & Manel, S.

(2013) Uncovering the genetic basis of adaptive change: on the intersection of landscape genomics and theoretical population genetics *Molecular Ecology* **22**, 3659–65.

Kruuk, L.E.B. (2004) Estimating genetic parameters in natural populations using the "animal model". *Philosophical Transactions of the Royal Society B* **359**, 873–90.

Kruuk, L.E.B and Hill, W.G. (2008) Evolutionary dynamics of wild populations: the use of long-term pedigree data. *Proceedings of the Royal Society B* **275**, 593–6.

Lavergne, S., Mouquet, N., Thuiller, W., & Ronce, O. (2010) Biodiversity and climate change: integrating evolutionary and ecological responses of species and communities. *Annual Review of Ecology, Evolution and Systematics* **41**, 321–50.

Lee, C.-R. & Mitchell-Olds, T. (2011) Quantifying effects of environmental and geographical factors on patterns of genetic differentiation. *Molecular Ecology* **20**, 4631–42.

Legendre, P. (1993) Spatial autocorrelation: trouble or new paradigm? *Ecology* **74**, 1659–73.

Lenormand, T. (2002) Gene flow and the limits to natural selection. *Trends in Ecology & Evolution* **17**, 183–9.

Lewontin, R.C. & Krakauer, J. (1973) 'Distribution of gene frequency as a test of the theory of the selective neutrality of polymorphisms', *Genetics*, vol. **74**, pp. 175–95.

Lowry, D.B. (2010) Landscape evolutionary genomics. *Biology Letters* **6**, 502–4.

Luikart, G., England, P.R., Tallmon, D., Jordan, S., & Taberlet, P. (2003) The power and promise of population genomics: from genotyping to genome typing. *Nature Reviews Genetics* **4**, 981–94.

Manel, S. & Holderegger, R. (2013) Ten years of landscape genetics. *Trends in Ecology and Evolution* **28**, 614–21.

Manel, S., Schwartz, M., Luikart, G., & Taberlet, P. (2003) Landscape genetics: combining landscape ecology and population genetics. *Trends in Ecology and Evolution* **18**, 189–97.

Manel, S., Joost, S., Epperson, B.K., Holderegger, R., Storfer, A., Rosenberg, M.S., Scribner, K.T., Bonin, A., & Fortin, M.E. (2010a) Perspectives on the use of landscape genetics to detect genetic adaptive variation in the field. *Molecular Ecology* **19**, 3760–72.

Manel, S., Poncet, B.N., Legendre, P., Gugerli, F., & Holderegger, R. (2010b) Common factors drive adaptive genetic variation at different scale in *Arabis alpina*. *Molecular Ecology* **19**, 2896–907.

Manel, S., Gugerli, F., Thuiller, W., Alvarez, N., Legendew, P., Holderegger, R., Geilly, L., & Taberlet, P. (2012) Broad-scale adaptive genetic variation in alpine plants is driven by temperature and precipitation. *Molecular Ecology* **21**, 3729–38.

McLachlan, J.S., Hellman, J.J., & Schwartz, M.W. (2007) A framework for debate of assisted migration in an era of climate change. *Conservation Biology* **21**, 297–302.

Mimura, M. & Aitken, S.N. (2007) Adaptive gradients and isolation-by-distance with postglacial migration in *Picea sitchensis*. *Heredity* **99**, 224–32.

Mitchell-Olds, T. & Schmitt, J. (2006) Genetic mechanisms and evolutionary significance of natural variation in *Arabidopsis*. *Nature* **441**, 947–52.

Mitton, J.B., Carter, P.A., & Digiacomo, A. (1997) Resting oxygen consumption varies among lactate dehydrogenase genotypes in the sow bug, Porcellio scaber. *Proceedings of the Royal Academy of Sciences B* **1387**, 1543–6.

Moran, P.A.P. (1950) Notes on continuous stochastic phenomena. *Biometrika* **37**(1–2), 17–23.

Murphy, M.A., Evans, J.S., & Storfer, A. (2010) Quantifying *Bufo boreas* connectivity in Yellowstone National Park with landscape genetics. *Ecology* **91**, 252–261.

Narum, S.R., Campbell, N.R., Kozfkay, C.C., & Meyer, K.A. (2010) Adaptation of redband trout in desert and montane environments. *Molecular Ecology* **19**, 4622–37.

Narum, S.R., Buerkle, C.A., Davey, J.W., Miller, M.R., & Hohenlohe, P.A. (2013) Genotyping-by-sequencing in ecological and conservation genomics. *Molecular Ecology* **22**, 2841–7.

Nosil, P., Funk, D.J., & Ortiz-Barrientos, D. (2009) Divergent selection and heterogeneous genomic divergence. *Molecular Ecology* **18**, 375–402.

Novembre, J., Johnson, T., Bryc, K., Kutalik, Z., Boyko, A.R., Auton, A., Indap, A., King, K.S., Bergmann, S., Nelson M.R., Stephens, M., & Bustamante, C.D. (2008) Genes mirror geography within Europe. *Nature* **456**, 98–101.

Parisod, C. & Joost, S. (2010) Divergent selection in trailing-versus leading-edge populations of *Biscutella laevigata*. *Annals of Botany* **105**, 655–60.

Paterson, A.H., Damon, S., Hewitt, J.D., Zamir, D., Rabinowitch, H.D., Lincoln, S.E., Lander, E.S., & Tanksley, S.D. (1991) Mendelian factors underlying quantitative traits in tomato: comparison across species, generations and environments. *Genetics* **127**, 181–97.

Pearson, R.G. & Dawson, T.P. (2003) Predicting the impacts of climate change on the distribution of species: are bioclimate envelope models useful? *Global Ecology and Biogeography* **12**, 361–71.

Poncet, B., Herrmann, D., Gugerli, F., Taberlet, P., Holderegger, R., Gielly, L., Rioux, D., Thuiller, W., Aubert, S., & Manel, S. (2010) Tracking genes of ecological relevance using a genome scan: application to *Arabis alpina*. *Molecular Ecology* **19**, 2896–907.

Reusch, T.B.H. & Wood, T.E. (2007) Molecular ecology of global change. *Molecular Ecology* **16**, 3973–92.

Rockman, M.V. & Kruglyak, L. (2009) Recombinational landscape and population genomics of *Caenorhabditis elegans*. *PLoS Genetics* **5**, e1000419.

Rosenblum, E.B., Römpler, H., Schöneberg, T., & Hoekstra, H.E. (2010) Molecular and functional basis of phenotypic convergence in white lizards at White Sands. *Proceedings of the National Academy of Sciences* **107**, 2113–17.

Rundel, P.W., Graham, E.A., Allen, M.F., Fisher, J.C., & Harmon, T.C. (2009) Environmental sensor networks in ecological research. *New Phytologist* **182**, 589–607.

Safner, T., Miller, M.P., McRae, B.H., Fortin, M., & Manel, S. (2011) Comparison of Bayesian clustering and edge detection methods for inferring boundaries in landscape genetics. *International Journal of Molecular Sciences* **12**, 865–89.

Savolainen, O., Pyhäjärvi, T., & Knürr, T. (2007) Gene flow and local adaptation in trees. *Annual Review of Ecology Evolution and Systematics* **38**, 595–619.

Schlötterer, C. (2002) A microsatellite-based multilocus screen for the identification of local selective sweeps. *Genetics* **160**, 753–63.

Schluter, D. (2000) *The Ecology of Adaptive Radiation*. Oxford University Press, Oxford.

Schoville, S.D., Bonin, A., Francois, O., Lobreaux, S., Melode-lima, C., & Manel, S. (2012) Adaptive genetic variation on the landscape: methods and cases. *Annual Review of Ecology, Evolution and Systematics* **43**, 23–43.

Slatkin, M. (1985) Gene flow in natural populations. *Annual Review of Ecology and Systematics* **16**, 393–430.

Slatkin, M. & Arter, H.E. (1991) Spatial autocorrelation methods in population genetics. *American Naturalist* **138**, 499–517.

Spear, S.F. & Storfer, A. (2008) Landscape genetic structure of coastal tailed frogs (*Ascaphus truei*) in protected vs. managed forests. *Molecular Ecology* **17**, 4642–56.

Stinchcombe, J.R. & Hoekstra, H.E. (2008) Combining population genomics and quantitative genetics: finding the genes underlying ecologically important traits. *Heredity* **100**, 158–70.

Storz, J.F. & Wheat, C.W. (2010) Integrating evolutionary and functional approaches to infer adaptation at specific loci. *Evolution* **64**, 2489–509.

Teshima, K.M., Coop, G., & Przeworski, M. (2006) How reliable are empirical genomic scans for selective sweeps? *Genome Research* **16**, 702–12.

Thomassen, H.A., Cheviron, Z.A., Freedman, A.H., Harrigan, R.J., Wayne, R.K., & Smith, T.B. (2010) Spatial modeling and landscape-level approaches for visualizing intra-specific variation. *Molecular Ecology* **19**, 3532–48.

Turner, T.L., Bourne, E.C., von Wettberg, E.J., Hu, T.T., and Nuzhdin, S.V. (2010) Population resequencing reveals local adaptation of *Arabidopsis lyrata* to serpentine soils. *Nature Genetics* **42**, 260–3.

Vigouroux, Y., McMullen, M., Hittinger, C.T., Houchins, K., Schulz, L., Kresovich, S., Matsuoka, Y., & Doebley, J. (2002) Identifying genes of agronomic importance in maize by screening microsatellites for evidence of selection during domestication. *Proceedings of the National Academy of Sciences of the United States of America* **99**, 9650–5.

Visscher, P.M., Hill, W.G., & Wray, N.R. (2008) Heritability in the genomics era – concepts and misconceptions. *Nature Reviews Genetics* **9**, 255–66.

Wagner, H.H. & Fortin, M.-J. 2005. Spatial analysis of landscapes: concepts and statistics. *Ecology* **86**, 1975–87.

Walther, G.-R., Post, E., Convey, P., Menzel, A., Parmesan, C., Beebee, T.J.C., Fromentin, J.M., Hoegh-Guldberg, O., & Bairlein, F. (2002) Ecological responses to recent climate change. *Nature* **416**, 389–95.

Wang, W.Y.S., Barratt, B.J., Clayton, D.G., & Todd, J.A. (2005) Genome-wide association studies: theoretical and practical concerns. *Nature Reviews Genetics* **6**, 109–18.

Watt, W.B. (1997) Adaptation at specific loci I. Natural selection on phosphor glucose isomerase ec- 5. 3. 1. 9 of colias butterflies: biochemical and population aspects. *Genetics* **87**, 177–94.

Watterson, G.A. (1977) Heterosis or neutrality? *Genetics* **85**, 789–814.

Weiss, K.M. (2008) Tilting at quixotic trait loci (QTL): an evolutionary perspective on genetic causation. *Genetics* **179**, 1741–56.

Wiens, J.A., Stralberg, D., Jongsomijt, D., Howell, C.A., & Snyder, M.A. (2009) Niches, models and climate change: assessing the assumptions and uncertainties. *Proceedings of the National Academy of Sciences* **106**, 19729–36.

Yi, X., Liang, Y., Huerta-Sanchez, E., Jin, X., Cuo, P., Pool, J.E., Xu, X., Jiang, H., Vinckenbosch, N., Korneliussen, T.S., Zheng, H., Liu, T., He, W., Li, K., Luo, R., Nie, X., Wu, H., Zhao, M., Cao, H., Zou, J., Shan, Y., Li, S., Yang, O., Asan, Ni, P., Tian, G., Xu, J., Liu, X., Jiang, T., Wu, R., Zhou, G., Tang, M., Qin, J., Wang, T., Feng, S., Li, G., Huasang, Luosang, J., Wang, W., Chen, F., Wang, Y., Zheng, X., Li, Z., Bianba, Z., Yang, G., Wang, X., Tang, S., Gao, G., Chen, Y., Luo, Z., Gusang, L., Cao, Z., Zhang, Q., Ouyang, W., Ren, X., Liang, H., Zheng, H., Huang, Y., Li, J., Bolund, L., Kristiansen, K., Li, Y., Zhang, Y., Zhang, X., Li, R., Li, S.H., Yang, S., Nielsen, R., Wang, J., & Wang, J. (2010) Sequencing of 50 human exomes reveal adaptation to high altitude. *Science* **329**, 75–8.

Yu, J.M., Pressoir, G., Briggs, W.H., Bi, I.V., Yamasaki, M., Doebley, J.F., McMullen, M.D., Gaut, B.S., Nielsen, D.M., Holland, J.B., Kresovich, S., & Buckler, E.S. (2006) A unified mixed-model method for association mapping that accounts for multiple levels of relatedness. *Nature Genetics* **38**, 203–8.

*Chapter 10*

# GRAPH THEORY AND NETWORK MODELS IN LANDSCAPE GENETICS

*Melanie Murphy,[1] Rodney Dyer,[2] and Samuel A. Cushman[3]*

[1]*Department of Ecosystem Science and Management, Program in Ecology, University of Wyoming, USA*
[2]*Department of Biology, Virginia Commonwealth University, USA*
[3]*Forest and Woodlands Ecosystems Program, Rocky Mountain Research Station, United States Forest Service, USA*

## 10.1 INTRODUCTION

Landscape genetic data are by nature **graph**-like in structure; graph approaches are extremely powerful for a wide range of landscape genetic applications (Dyer et al. 2010; Garroway et al. 2008; Murphy et al. 2010a, 2010b; Neel 2008; Dyer & Nason 2004). Graph theory, a branch of combinatorial mathematics that describes how discrete objects are connected, was first introduced in 1741 by Leonhard Euler who provided a mathematical proof that it is impossible to traverse the seven bridges of Königsberg only once in a continuous walk (Biggs et al. 1986). Euler's geometrical interpretation translated the problem into a discrete set of **nodes** (islands and shorelines) and their connections as **edges** (bridges), laying the foundation for graph theory. Over the last 175 years, graph theory has been continually applied to our understanding of the underlying mechanisms in biological systems. Even Darwin's original notebook contains a bifurcating graph as a hypothesis of species evolution (1837), a precursor to now common phylogenetic trees (e.g., Penny & Hendy 1985).

Graphs are a mathematical approach that allow inference based on overall shape (topology, see Box 10.1), a unique contribution when compared to analytical methods applied in ecology that focus on parameter estimation. While graph theory is arguably underutilized in ecology (Dale & Fortin 2010), it has been applied to address a wide range of ecological questions including describing nutrient transport (Bebber et al. 2007), defining metacommunity relationships (Economo & Keitt 2008), describing habitat mosaics (Urban et al. 2009), and quantifying habitat connectivity as a function of dispersal ability (Keitt et al. 1997).

Graph applications have great potential to address landscape genetics questions in evolution, ecology, and

*Landscape Genetics: Concepts, Methods, Applications*, First Edition. Edited by Niko Balkenhol, Samuel A. Cushman, Andrew T. Storfer, and Lisette P. Waits.

## Box 10.1  What is graph topology?

The processes governing interactions among nodes within a graph result in an overall *graph topology* (Albert & Barabasi 2002). Graph theoretic analyses allow inference based upon this overall topology, a unique contribution of graph approaches when compared to analytical methods focused on parameter estimation. For example, if a graph represents functional connectivity amongst a set of demes within a metapopulation (Hanski 1998), then the underlying biological processes is migration, which will dictate the location and qualities of the edges (Bode et al. 2008). Graph topology can provide biologically meaningful inference at two major levels: shape of the entire graph and structure within a localized region (for more extensive review of graph metrics, see Rayfield et al. 2011).

At the level of the entire graph, there are two common measures used in the comparison of the topological structure: *diameter* and *degree* distribution. Diameter is an overall description of graph width quantified as the length of the path separating the two most distant nodes (Wasserman & Faust 1994). Path length can be quantified as either a count of the number of edges separating nodes or as based on a quantitative measure such as habitat difference, genetic distance, or geographic distance. The degree distribution is one of the most commonly reported graph-level metrics. A degree is the number of connections a node has to other nodes. Degree distribution evaluates the statistical distribution of node degrees for a given graph (Fig. 10.2). Random graphs (Erdös & Rényi 1960) are those where nodes are randomly selected for connection (with probability *p*), yielding a degree distribution conforming to a Poisson distribution (when the number of nodes is large (Erdös & Rényi 1960). The Internet (Barabasi & Albert 1999), citation networks (De Solla 1976), protein–protein interaction networks (Nacher et al. 1999)), pollination networks (Dyer et al. 2012), and some social networks (e.g., "six degrees of Kevin Bacon"; Collins & Chow 1998) follow a power law degree distribution. Of particular interest in landscape genetics, the central parameter of the degree distribution ($P(k)$) is highly correlated with Wright's $F_{ST}$ under some conditions (Dyer 2007).

The goal of local metrics is to quantify the relative contribution of nodes, collections of nodes, or edges in the overall structure of the graph topology (Rayfield et al. 2011; Dale and Fortin 2010). The interpretations of either localized metrics of those describing the total graph depend upon what the graph is intended to represent. In analyses of connectivity, researchers may be interested in how connected an individual node or edge may be within the larger graph (Dyer et al. 2010), a property called *centrality* (Wasserman & Faust 1994). There are multiple metrics assessing aspects of centrality (see Glossary) including: degree centrality, *betweenness* centrality, and closeness centrality (Wasserman & Faust 1994). Nodes of high degree centrality are those that have a high degree of overall connectivity (e.g., Node 4 in Graph A), which allows them to have a potentially larger relative influence than those that are more sparsely connected (Graph B).

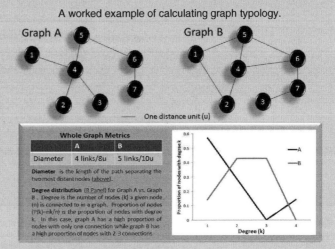

A worked example of calculating graph typology.

Graph A     Graph B

—— One distance unit (u)

**Whole Graph Metrics**

|          | A         | B          |
|----------|-----------|------------|
| Diameter | 4 links/8u | 5 links/10u |

**Diameter** is the length of the path separating the two most distant nodes (above).

**Degree distribution** (R Panel) for Graph A vs. Graph B. Degree is the number of nodes (k) a given node (n) is connected to in a graph. Proportion of nodes (P(k)=nk/n) is the proportion of nodes with degree k. In this case, graph A has a high proportion of nodes with only one connection while graph B has a high proportion of nodes with 2-3 connections

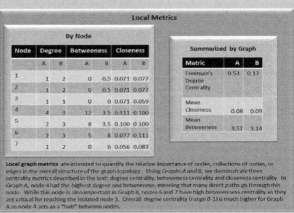

**Local Metrics**

| Node | Degree A | Degree B | Betweeness A | Betweeness B | Closeness A | Closeness B |
|------|----------|----------|--------------|--------------|-------------|-------------|
| 1 | 1 | 2 | 0 | 0.5 | 0.071 | 0.077 |
| 2 | 1 | 2 | 0 | 0.5 | 0.071 | 0.077 |
| 3 | 1 | 1 | 0 | 0 | 0.071 | 0.059 |
| 4 | 4 | 3 | 12 | 3.5 | 0.111 | 0.100 |
| 5 | 2 | 3 | 8 | 3.5 | 0.100 | 0.100 |
| 6 | 2 | 3 | 5 | 8 | 0.077 | 0.111 |
| 7 | 1 | 2 | 0 | 6 | 0.056 | 0.083 |

**Summarized by Graph**

| Metric | A | B |
|--------|------|------|
| Freeman's Degree Centrality | 0.53 | 0.17 |
| Mean Closeness | 0.08 | 0.09 |
| Mean Betweeness | 3.57 | 3.14 |

**Local graph metrics** are intended to quantify the relative importance of nodes, collections of nodes, or edges in the overall structure of the graph topology. Using *Graphs A and B*, we demonstrate three centrality metrics described in the text: degree centrality, betweenness centrality and closeness centrality. In Graph A, node 4 had the highest degree and betweenness, meaning that many direct paths go through this node. While this node is also important in Graph B, nodes 6 and 7 have high betweenness centrality as they are critical for reaching the isolated node 3. Overall degree centrality (range 0-1) is much higher for Graph A as node 4 acts as a "hub" between nodes.

Within the larger context of the graph, it is possible to have highly connected subcomponents that occupy peripheral locations on the topology. Despite having a high degree they may have low betweenness centrality (Freeman 1977) as the vast majority of connections, or shortest paths through the graph, do not traverse those nodes (or edges). Betweenness and the next measure, closeness centrality, can be conceived within a transport network context such as a road system in a city. Intersections (nodes) or streets (edges) with high betweenness would be those that are the most heavily travelled and are where urban planners want to put essential services that everyone tries to get to, such as the post office or a grocery store. Closeness centrality is the converse; it measures how far a node or edge has to travel to reach the rest of the topology and in an urban context may be where a fire station would be placed. Within this urban movement analogy, it is not necessary for graph elements to have similar levels of degree, betweenness, and closeness centrality as the topology of the entire graph determines each property (see the worked example this box).

Measures of centrality for nodes within a graph depicting ecological connectivity (e.g., effective distances; see the examples in this chapter and Chapter 8) are highly informative in quantifying organismal movement within and among patches. Network measures are broadly informative in development of multispecies connectivity models (Estrada et al. 2008). Dyer and Nason (2004) showed that despite some populations of the Sonoran endemic cactus *Lophocereus schottii* having low population genetic diversity (a common measure of relative conservation genetic importance), their centrality on a graph defined by genetic covariance made them integral to overall genetic connectivity. The ubiquity of graphical interpretations of data and processes across broad reaches of science provides a wealth of existing approaches that can be repurposed for use in landscape genetic analyses.

conservation. Graph theory has been applied frequently to assess both structural and functional connectivity of landscapes (e.g., O'Brien et al. 2006; Theobald 2006; Urban & Keitt 2001). Connectivity affects species' movement (Chetkiewicz et al. 2006; Crooks & Sanjayan 2006; Cushman et al. 2010; Gaston & Fuller 2008) and dispersal ability (McRae et al. 2008), components well addressed using a landscape genetics approach (see

Chapter 8). Graphs can represent *a priori* hypotheses of landscape influence on genetic structure, describe the genetic structure itself, or be used in an inferential modeling framework to estimate the effect of independent variables on functional connectivity (Table 10.1).

## 10.2  BACKGROUND ON GRAPH THEORY

### 10.2.1  What is a graph?

Graphs are a collection of nodes connected by edges (Fig. 10.1). Nodes are sample units that may have spatial location, size, shape, and characteristics (Dale & Fortin 2010). Edges are connections among nodes, which may be aspatial theoretical connections or represent physical structures on the landscape connecting nodes (e.g., roads (edges) connecting settlements (nodes), rivers (edges) connecting lakes (nodes), dispersal corridors (edges) connecting habitat patches) that have explicit spatial location, size, shape, and/or characteristics (Dale & Fortin 2010). While edges may be complex (Fig. 10.1), the space surrounding the graph (the landscape matrix, see Chapter 2) is not directly included in the graph description of connectivity among nodes (Cantwell & Forman 1993). In landscape genetic applications, genetic data may be associated with the nodes or the edges (Table 10.2). In addition, a multitude of genetic distance measures can be associated with edges (Table 10.3).

**Table 10.1 Landscape genetic objectives suited to graph analysis.** A wide range of objectives can be addressed in landscape genetics using graph approaches. Table 10.1 presents four major objectives we will address in this chapter and the general type of graph assessment that will be addressed in more detail throughout this chapter.

| Objective | Graph application |
| --- | --- |
| Hypothesis of connections | Topological comparison |
| Describe population structure | Topological metrics (see Box 10.1) |
| Relate functional connectivity (pattern) to landscape (process) | Network models |
| Local adaptation | Network models |

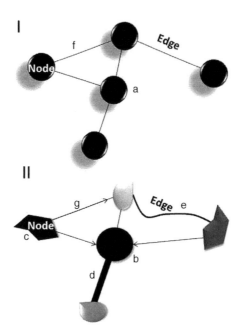

**Nodes** are a point location. In addition to location (a), nodes can have size (area, b), shape (polygon, c), and additional characteristics such as habitat type or quality (II, represented by color variation).

**Edges** are connections between nodes. They can be represented as a-spatial connections (familial relationships among individuals), straight-line connections (I), or more spatially complex connections (II) that have size (d) and/or shape (e). In addition, edges can be bidirectional (f) or directional (g).

**Fig. 10.1** Graphs are a collection of nodes and edges. Graphs range in level of spatially-explicit information for both nodes and edges. This figure presents an example of a relatively simple graph (I) and a graph (II) with more spatial information associated with both nodes and edges.

## 10.2.2  What are the assumptions of graph theoretic approaches?

While application of graph theory to ecological and evolutionary problems using genetic data is compelling, graph analyses have a series of assumptions and considerations. Nodes need to be a clearly defined unit such as habitat patches (e.g., meadows; Mennechez et al. 2003) or ponds (Murphy et al. 2010a), individuals (Rayfield et al. 2010), plots (McIntire & Fortin 2006), or specific features that are connected by edges (Vasas et al. 2009). Some graph metrics are

**Table 10.2 What do nodes and edges represent in a landscape genetic analysis?** Landscape genetic data are a natural fit to the graph framework. Genetic and landscape data can both be associated with nodes or edges in a graph framework (Fig. 10.1) in landscape genetic analyses. Genetic data at nodes can include measures of genetic diversity (including presence–absence of an allele), while associated edges are generally measures of genetic exchange. "Landscape data" (e.g., environmental variables, metrics of habitat composition/configuration; see Chapter 5) at nodes would include metrics describing the patch, while edges would describe covariates for that edge.

|  | Genetic | Landscape |
| --- | --- | --- |
| Node | Heterozygosity | Percent suitable habitat |
|  | Allelic diversity | Habitat quality |
| Edge | Genetic distance | Geographic distance |
|  | Gene flow | Resistance/conductance |
|  | Number of migrants | Environmental limits |

**Table 10.3** Measures of genetic dissimilarity that can be associated with graph edges. Measures of genetic distance can be based on individuals as nodes (A) or groups as nodes (B). Each genetic distance has associated considerations that may govern when it is most efficiently and effectively applied in graph applications of landscape genetics.

**(A) Nodes represent individuals**

| Metric | Considerations |
|---|---|
| Co-ancestry | Observed similarities are actually due to identity by descent rather than identity by state (Loiselle et al. 1995) |
| Bray-Curtis[1] | A simple similarity measure. Developed in ecology and does not make any genetic assumptions (Bray & Curtis 1957) |
| AMOVA | A statistical distance between multilocus genotypes with no genetic assumptions (Smouse & Peakall 1999) |
| Ladder distance | Distances among individuals due to incremental mutational processes (e.g., microsatellites; Bowcock et al. 1994) |
| Proportion of shared alleles[1] | Generally given as 1-Dps to convert into a dissimilarity index (Bowcock et al. 1994) |
| Relatedness | Large samples sizes are needed for parentages analysis (Lynch & Ritland 1999) |

**(B) Nodes represent groups**

| Metric | Considerations |
|---|---|
| Cavalli-Sforza distance | "Genetic chord distance". Shown to work well under some migration models, particularly for microsatellite-like loci. (Cavelli-Sforza & Edwards 1967) |
| Conditional genetic distance | Not based upon population genetic assumptions but requires sufficient within stratum genetic variability to get a good variance estimate (Dyer & Nason 2004) |
| Nei's | Distances are thought to arise because of drift/migration processes (Nei 1973) |
| Pairwise structure ($F_{ST}$, $G_{ST}$, $\Phi_{ST}$, etc). | These measures have equilibrium assumptions and are sensitive to small sample sizes (Hedrick 2005; Wright 1951; Excoffier et al. 1992; Nei 1972) |

[1]Measure of genetic distance that can be applied to both individuals and groups.

intended to be calculated on a complete sample of nodes, so as in all landscape genetic approaches, sample design is critically important (see Chapter 4). Edges must represent a reasonable estimate of some processes connecting nodes. A saturated (e.g. complete) graph has an edge connecting every node to every other node (see Fig. 10.2a), and thus may contain edges that do not represent actual ecological or biological processes.

### 10.2.3   What edges are relevant?

To address the issue of unnecessary edges, saturated graphs (Fig. 10.2A) can be reduced ("pruned") via a wide range of approaches to include only relevant edges, avoid long edges, or not allow overlapping

edges (Dale & Fortin 2010). If, and how, to prune a saturated graph depends on the research question and sample distribution. Rule-based structures include: a biologically relevant limit (e.g., some function of maximum dispersal distance; Murphy et al. 2010a), rule-based graph type (e.g., minimum spanning tree, Delaunay triangulation; Goldberg & Waits 2010; Naujokaitis-Lewis et al. 2009), a hypothesis of what nodes are ecologically connected (Treml et al. 2008), or a threshold connectivity value (e.g., maximum genetic distance). Rule-based pruning imposes a given network structure on the data and the selected graph type may have a large impact on results of subsequent analysis (Naujokaitis-Lewis et al. 2013). Alternatively, graphs may be pruned based upon a statistical model to define which set of edges are retained in the graph (Dyer & Nason 2004).

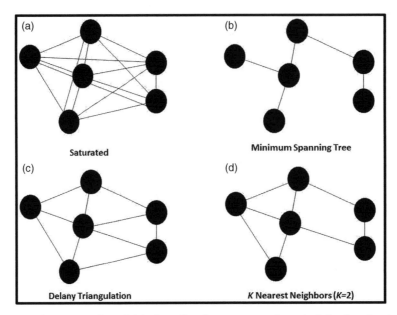

**Fig. 10.2** Pruning a graph. A saturated graph (a) where all nodes are connected may include edges that do not represent a biological or ecological process. Therefore, graphs may be pruned via a variety of methods. Here we illustrate pruning with example graphs formed from the same set of nodes using just a few of the rule-based pruning methods described in the literature (for a review, see Dale & Fortin 2010). (b) Minimum spanning tree that includes the minimum edges to connect all nodes, (c) Delany triangulation, and (d) K nearest neighbor graph ($K = 2$), where each node is connected to the nearest K neighbors. Graphs could be pruned based on geographic distance, genetic distance (Table 10.3), or an ecological distance.

## 10.3 LANDSCAPE GENETIC APPLICATIONS

In this section of the graph theory chapter, we will discuss applying graph approaches in three major applications: (1) using graph methods to describe population genetic structure, (2) testing hypotheses of gene flow among nodes, and (3) assessing functional connectivity using a graph structure (Table 10.1).

### 10.3.1 Describing population structure

**Theoretical background**

The pattern of connectivity in genetically defined graph topologies is based upon some level of statistical genetic distance (or covariance) among either sets of individuals (social networks or pedigrees) or demes (Dyer & Nason 2004). Topological analyses of genetic structure (*sensu lato*) can be informative for several kinds of landscape genetic hypotheses (see Table 10.1). While

perhaps not generally acknowledged, the underlying quantitative tools, analysis, and algebra of graphs defined upon genetic similarities has been in use since the very beginning of population genetics and is used in the vast majority of landscape and population genetic studies today. Analyses of isolation-by-distance (IBD) based upon pair-wise distance matrices (genetic, spatial, and more recently ecological), which are *de facto* adjacency matrices for saturated graphs (Slatkin 1987; Wright 1943). One major strength of the graph-theoretic approach to landscape genetics for assessing genetic structure is the ability to move away from testing a hypothesis such as $H_O$: $F_{ST} = 0$ (e.g., no structure) towards one asking how variation is spatially distributed, thus providing a level of inference that was previously unattainable (Dyer & Nason 2004).

**Conceptual framework**

Depiction of distance-based analyses as implicit graph topologies have recently become more prevalent, due in large part to methodologies for pruning edges from a saturated graph (e.g., see Dyer & Nason 2004). Under-lying these newer approaches is the acknowledgement that across increasingly larger spatial and ecological scales, sets of individuals or demes may not directly interact but may do so in an indirect fashion. While this move mirrors how the stepping stone model (Kimura & Weiss 1964) extended the original island model (Wright 1943), the distinction here is that the topological shape of the genetic graph may be based upon either data external to the genetics of the organism (rule based, Fig. 10.2) or may actually be estimated using the genetic data.

Rule-based approaches seek to identify the subset of edges in the saturated graph for subsequent analyses with spatial and/or ecological data (Fig. 10.2). Rule-based genetic graphs date back to the earliest uses of "trees" in population genetics. Algorithms such as neighbor joining (Saitou & Nei 1987), minimum span-ning trees (Guiller et al. 1998), splits-trees (Huson 1992), and others take estimates of genetic dissimilarity and define a particular type of graph. Extensions of these more classical approaches do not generally require the resulting graph to be tree-like (e.g., one with no *cycles*) but instead impose restrictions on which subset of edges in the saturated graph are important and which are not. These restrictions can be based upon data either external to the graph or based upon the graph topological struc-ture directly. Murphy et al. (2010a) pruned their satu-rated graph based upon a function of maximum dispersal distance in mountain frog populations, thereby remov-ing potential connections between ponds that are beyond their estimate of directional migration (see below). In that study, the saturated graph performed as well as the spatially pruned topology when analyzing ecological/genetic correlations, suggesting that dispersal limitations are not a governing overall genetic structure. Rule-based approaches continue to be applicable in landscape genetic analyses though they suffer the same potential problems as other "expert opinion" meth-ods (Chapter 8).

Graphs can also be pruned by the internal structure of the topology itself. The percolation threshold, for exam-ple, is the point at which removal of edges within a graph topology increases the number of disconnected components (Rozenfield et al. 2008). In a study of the threatened seagrass *Posidonia oceanica* (Posidoniaceae), edges were removed based upon the decreasing mag-nitude of genetic distances between populations (Rozenfield et al. 2008). Points at which the topology broke into two and more disconnected subgraphs were

used as critical points, which they interpreted as being indicative of underlying disconnected metapopulation structure.

Alternatively, graphs can be pruned based upon genetic data as in the case of *population graphs* (Dyer & Nason 2004), which uses conditional genetic independence to define the overall topology. This statistical approach is similar to that used in multiple regression where one determines whether the addi-tion of an additional variable (e.g., the edge in the graph) aids significantly to the statistical performance of the model (the overall genetic covariance among strata). Dyer (2007) further showed that several aspects of a population graph topology are correlated with more traditional population genetic parameters (e.g., degree is proportional to $F_{ST}$ and diameter is correlated with IBD). Moreover, because conditional genetic covariance is determined by the underlying migration model, the overall shape of a population graph itself is informative. Populations experiencing one-dimensional stepping stone-like patterns of migration result in topologies that are topologically elongated, whereas two-dimensional models are more sheet-like.

### Data requirements

Data requirements for graph-based estimates of popu-lation structure estimations are primarily based upon the stability of the distance metric used (Table 10.3) and the kind of genetic marker. Behavior of these parameters is in line with normal parametric statistical parameters with respect to sample sizes and intensity. For a given sample size, slowly evolving genetic loci yield estimates of genetic distance with higher coeffi-cients of variation than more rapidly evolving loci (e.g., genetic equilibrium takes longer for low mutation rates; Kalinowski 2005). Moreover, the less structure you have in the system, the more samples you will need to get a good estimate (Kalinowski 2005). For population graphs, there is a critical need for within-population genetic variation as the topology is based upon condi-tional genetic covariance. All estimates of covariation require within-group variance to prevent a divide by a zero mathematical error. As for measures of genetic distance, the stability of within-site variance will be asymptotic in the number of individuals sampled as well as the rate of mutation for the loci. However, it should be noted that with respect to within-site genetic variance, it is measured within a multilocus context rather than at a single locus or as averages of single locus estimates. Therefore it is somewhat robust to a mix of loci of various levels of polymorphism.

## Software

Currently, there are a host of software applications that can be used to create graphs based upon genetic structure (e.g., gstudio; Dyer 2014), particularly since many of the types of genetic graphs are based simply upon common measures of genetic distance (or covariance). Software for graph visualization is more constrained. Standalone applications like Pajek (de Nooy et al. 2012) and Gephi (Heyman et al. 2003) can be used to visualize network topologies relatively easily. As an analytical toolbox, R (R development core team) has a large number of packages aimed at estimating genetic parameters (ade4; Dray & Dufour 2007) and visualizing graphs (igraph; Csardi and Nepusz 2006), or both (gstudio; Dyer 2014).

## Case study

There are various situations in which genetic-based graph topologies may provide supportive evidence of biological processes (Garrick et al. 2009; Garroway et al. 2008; Koen et al. 2010; Sork et al. 2010) and the definition of "graph theoretic" in population genetics can be quite broad. We present a case study from Dyer et al. (2010) who used population graphs to identify intervening landscape features that influence interpopulation gene flow in the Sonoran Desert endemic *Euphorbia lomelii* (Euphorbaceae). Given limited dispersal of seeds, overall connectivity in this system is primarily dictated by the movement of hummingbird pollinators. Bioclimatic factors that mediate flowing synchrony, the hypothesized regulating factor of gene flow, were used to create cost surfaces that would be highly correlated with interpopulation genetic covariance (see Chapter 8).

The decomposition of genetic structure in this broad-scale study system required a two-step approach. A previous phylogeographic analysis of the system (Garrick et al. 2009) showed at least two separate Pleistocene refugia for this plant on the peninsula of Baja California, one in the north and another in the Cape region (see Fig. 1 in Dyer et al. 2010). Consequently, across the entire species range, a potentially large fraction of the observed spatial genetic variation is due to Pleistocene phylogeography and subsequent range expansion processes masking contemporary patterns of genetic connectivity. The spatial patterns identified from the phylogeographic analysis were regressed out of among population genetic covariance from population graphs. These "phylogeographically corrected" genetic covariance and bioclimatic data were analyzed under a multiple regression on distance matrices (MRM) model (Legendre et al. 1994), with AIC used for model evaluation (see Chapter 5). The results of these analyses showed that when working at large spatial scales, failure to remove phylogeographic patterns can have significant impacts on landscape genetic inferences. Models using genetic covariance uncorrected for phylogeography included not only different subsets of landscape features but also changed their relative importance. Once corrected for broad-scale patterns, precipitation and temperature coinciding with peak flowering times were identified as significant contributions to describing conditional genetic covariance.

### 10.3.2 Hypothesis of connectivity

#### Theoretical background

The goal for a "hypothesis of connectivity" objective is to compare hypothesized graph topology to observed graph topology (based on genetic data) to determine if graph structure diverges from the hypothesis (for potential metrics, see Box ). These graph(s) represent an *a priori* hypothesis of connections through the landscape based on: barrier, directional flow, or some hypothesis of landscape factors limiting connections (Fig. 10.3).

#### Implementation

To implement this modeling framework, a reasonable *a priori* graph (or set of graphs) must be identified that represents the hypothesized processes. In order to develop an *a priori* graph based on the landscape, some limited landscape data (topography, potential barriers, stream channels, etc.) may be required. In addition, an "observed" graph topology must be constructed based on genetic data (Arnaud et al. 1999) using a threshold in genetic distance value, treating the genetic distance value as a probability of connection or potentially implimenting population graphs as the observed graph (see the previous section). Graph

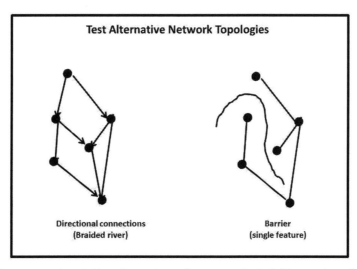

**Fig. 10.3** Example of two potential *a priori* hypotheses of a graph structure. On the left is a graph topology representing connectivity among nodes along a braided river where there are multiple paths between nodes, but flow is directional (downstream only). On the right is a graph topology representing a barrier to connectivity where direct gene flow has been cut off between some nodes.

comparison can be conducted in Pajek (de Nooy et al. 2005), Gelphi (Heyman et al. 2003), and R (e.g., igraph package; Csardi and Nepusz 2006).

### Case studies

Arnaud et al. (1999) hypothesized that land snails were connected through hedgerows versus an isolation-by-distance pattern. They compared a graph of land snail connectivity restricted to hedgerows to IBD and found that hedgerows were a better fit to measures of gene flow based on neutral markers (Arnaud et al. 1999). Stream networks are of particular interest for fish – especially if there may be barriers to movement through some portion(s) of the stream network (Meeuwig et al. 2010). In Glacier National Park (USA), genetic distance among groups of bull trout more closely followed barrier graph topology (one- or two-way barriers) than the stream network alone.

This type of graph comparison is not common in landscape genetics studies, likely due in part to the complexity of testing for differences among graphs that are not bifurcating tree topologies. However, comparison of the observed pattern of genetic distance with *a priori* graphs of functional connectivity in a landscape has huge potential applications on

landscape planning and conservation. Researchers could test alternative graph topologies against observed genetic distance to evaluate if specific effects of interest (roads, novel hot areas due to climate change, etc.) fit the observed data. However, there is a risk of "affirming the consequent" without the addition of a robust simulation framework to create a null distribution (Cushman & Landguth 2010). Without this component, analyses are more confirmatory in nature.

### 10.3.3   Functional connectivity

#### Theoretical background

***Gravity models*** (Fotheringham & O'Kelly 1989), or spatial interaction models, are one example of functional connectivy analyses using networks. This approach allows for inferential modeling of functional connectivity that incorporates landscape data both at site (nodes) and between sites (edges). The gravity form (see equation 10.1) predicts flow through a graph based on "mass" – a measure of flow potential from a given node (Anderson 1979). While gravity models have a long history in transportation and economics (Voorhees 1956; Willig & Bailey 1979), applications in ecology are relatively new (Xia et al. 2004; Bossen-broek et al. 2001, 2007; Ferrari et al. 2006). In the gravity form, graph flow ($Ti_j$ – measure of genetic differentiation for landscape genetics) is a function of three elements: distance between sites ($w$ spatial weights matrix, see Chapter 5), production/attraction ($v$, i.e., nodes characteristics), and resistance ($c$, i.e., edge characteristics; see equation 10.1). Parameter estimates are $\mu$, $\alpha$, and $\beta$, respectively (in equation 10.1, $\mu$ and $\beta$ may represent a vector of variables).

$$T'_{ij} = k'v_\mu[w_{ja} \cdot c_{ij}(-\beta)]$$

(Anderson 1979; Fotheringham & O'Kelly 1989)

(10.1)

### Modeling framework

Gravity models (equation 10.1) can be estimated as (i) unconstrained, (ii) singly constrained, or (iii) doubly constrained (Fotheringham and O'Kelly 1989; Table 10.4). (i) Unconstrained models fit a single model to all of the data simultaneously (Fotheringham and O'Kelly 1989) often by taking the natural log of equation (10.1) and solving with linear regression (Murphy et al. 2010a). While signally constrained gravity models use all of the data in a single estimate, they do not take full advantage of the information contained in the graph topology. In addition, linear regression does not account for the non-independent nature of data from a graphs structure (Table 10.4). (ii) To incorporate localized effects and make more spatially distributed estimates of flow, the gravity equations can be singly constrained ("grouped") by nodes as either production (landscape processes influencing flow potential from a node) or attraction (landscape processes influencing potential to attract flow from nodes) effects. With a single constraint, the natural log of the gravity form (equation 10.1) results in a linear equation with a unique $k$ (intercept) for each node, which can be estimated using a mixed effects models (Murphy et al. 2010a). As each node has an independent intercept ($k$), the singly constrained gravity model accounts for the non-independence of matrix data. (iii) The doubly constrained form fits estimates to all combinations of pairs (each origin and destination) and is generally solved using maximum likelihood or Bayesian methods (Fotheringham et al. 2002).

### Data requirements and assumptions

*Edge data.* Gravity models, as applied in landscape genetics thus far, assume the flow of neutral (as opposed to adaptive) genetic variation through the graph. Sensitivity of gravity models to the choice of genetic distance metric (Table 10.3) is currently unknown but the proportion of shared alleles (Bowcock et al. 1994) was used in the original rational method employed (Murphy et al. 2010a). Graph topology must be defined (see "What edges are relevant?", Fig. 10.2). Gravity models assume autocorrelation (see Chapter 5) – there must be some form of spatial relationship among nodes (Voorhees 1956; Wilson 1967; Murphy et al. 2010a). This distance relationship may be described with any spatial weights matrix from simple (distance between sites) to complex (model of distance die-off, see Chapter 5).

*Node data.* Gravity models may be sensitive with two missing nodes from the network. Sensitivity to missing nodes can be tested by iteratively omitting nodes and assessing effects on resulting models (Murphy et al. 2010a).

---

**Table 10.4** Trade-offs in gravity model constraints. Gravity models can be unconstrained, singly constrained, or doubly constrained (Fotheringham & O'Kelly 1989). Constraints to the gravity model result in trade-offs between model quality and quantity of data used.

|  |  | Data quantity | |
| --- | --- | --- | --- |
|  |  | **Large** | **Small** |
| **Model quality** | **High** | Singly constrained | Doubly constrained |
|  | **Low** | Unconstrained |  |

*Statistical assumptions.* As both unconstrained and singly constrained gravity models are commonly solved using forms of linear regression (although this is not a requirement), all standard regression assumptions apply (Zar 1999). Independent variables (both at nodes and along edges) must be continuous measures or categorical data approaching a continuous distribution (Zar 1999). Residual variation is assumed to be independently, identically distributed with mean of zero and constant variance (Zar 1999).

### Software implementation

Once data are transformed (i.e., natural log of dependent and independent variables), gravity models can be estimated using any statistical software capable of linear regression. The specific framework for gravity model analyses with genetic data is in the process of being implemented in an R package – GenNetIt – which includes the ability to extract and calculate statistical moments from associated landscape data represented by rasters (Murphy & Evans, in preparation).

### Case study

Gravity models were first applied in landscape genetics in a study of Columbia spotted frogs (*Rana luteiventris*) in high mountain lakes of central Idaho, USA (Murphy et al. 2010a). In this system, explaining functional connectivity among occupied breeding ponds is crucial for understanding metapopulation dynamics and species' ecology (Chapter 2). Murphy et al. (2010a) tested hypotheses of functional connectivity including both node and edge landscape characteristics (8 microsatellite loci, 37 nodes). The study system was a network of nutrient-poor high mountain lakes where a short growing season and complex topography was thought to constrain *R. luteiventris* functional connectivity. Production of offspring at breeding sites was thought to be limited by introduction of predatory fish and site productivity (warm open water for algae and tadpole growth).

*Rana luteiventris* connectivity was negatively correlated with distance between nodes, presence of predatory fish, and major topography between sites. Node productivity (as measured by heat load index; McCune & Keon 2002) and growing season (measured by a mean frost-free period along the edges) were positively correlated with functional connectivity. The negative effect of predation and positive effect of site productivity at nodes supported the presence of source-sink metapopulation dynamics.

## 10.4 RECOMMENDATIONS FOR USING GRAPH APPROACHES IN LANDSCAPE GENETICS

### 10.4.1 Recommendation 1: Clearly identify research questions

While graph applications in landscape genetics have vast potential, there are several important considerations for researchers and managers wishing to apply graphs in a landscape genetics context. First, graphs can be applied in multiple fashions (e.g., evaluating genetic structure, hypotheses of graph topology, or as inferential models of network flow). When evaluating genetic structure, connections (i.e., edges) arise from the data. In hypotheses of graph topology, one compares observed connections among nodes to some *a priori* hypothesis of connections. Finally, in inferential models of network flow researchers want to (a) describe the observed "weight" (e.g., measure of genetic distance) across all connections in relation to covariates or (b) evaluate only edges that represent contributing connections among nodes. Each of these approaches makes slightly different assumptions and is prompted by disparate questions. Thus, it is critical to clearly identify the motivating research or management objectives and then match them with the appropriate analytical framework.

### 10.4.2 Recommendation 2: Choosing an adequate study design

The more effectively a sample design captures the spatial distribution of the study species, the more accurate is the spatial prediction (Chapter 4). Considerations for robust study design of graph approaches in landscape genetics include: identification of the appropriate unit of analysis (nodes), sample size (nodes), and if/how to prune edges. Because graph nodes can represent individuals, groups, or populations, researchers and managers should carefully consider what unit of observation is ecologically and statistically justified when designing a study. Multiple individuals collected from the same location (e.g., frogs from a pond) cannot be represented as unique nodes. However, individual trees within a habitat matrix may function as nodes. In determining sample size, there are two considerations (1) the number of spatial locations (nodes) and (2) the number of samples taken at each sample location (node). Increasing the number of nodes improves the ability to estimate the spatial process. In addition, graphs pruned via a rule-based approach may be sensitive to unsampled nodes (Naujokaitis-Lewis et al. 2013).

Deciding if and how to prune edges of a graph is not straightforward (see "What edges are relevant"). We suggest that if researchers choose to prune, that they compare multiple pruning methods to test the sensitivity of the results to pruning method (e.g., Murphy et al. 2010a).

### 10.4.3 Recommendation 3: Testing underlying assumptions

Graphs assume that edges are a reasonable estimate of connections among nodes, that graph extent is not arbitrary (Jelinski & Wu 1996), and thus may be sensitive to missing nodes (Dale & Fortin 2010). It is rare to have all nodes within an extent sampled, especially for continuously distributed abundant species (e.g., tree species within a study area) or elusive species. Missing nodes have a larger impact on rule-based graphs than unpruned graphs as network topology is inherently related to the distribution of nodes (Naujokaitis-Lewis et al. 2013). In graphs based upon statistical models, like population graphs, the statistical power of the analysis is based upon the sampling (Koen et al. 2013). If sampling is sparse across a dynamic landscape, the ability to pick up trends is reduced (Koen et al. 2013), just as it would be for any other statistical model.

The effect of missing nodes for a given application can be assessed using a sensitivity analysis (Murphy et al. 2010a) where the effect of $1 - k$ missing nodes can be assessed. While this approach does not determine the effect of the full network versus a subsample, it does indicate the overall effect of additional subsampling. When networks are applied in inferential modeling, graphs pruned using rule-based methods appear to introduce bias when compared to saturated graphs. Therefore, it may be advisable to use unpruned graphs when nodes represent a statistical sample ($n$) and not a population ($N$; Naujokaitis-Lewis et al. 2013).

### 10.5   CURRENT RESEARCH NEEDS

The resolution of several research needs will facilitate the increased application of graph theory in landscape genetics. Additional work is needed to determine: optimal spatial sampling strategies (Murphy et al. 2008), the effects of missing nodes (Naujokaitis-Lewis et al. 2013), the minimum number of samples needed per node, which methods are appropriate for individual-based analysis (see Table 10.3 for genetic metrics), and the use of weighted edges in landscape genetics. Genetic differentiation is a proxy for gene flow – other processes, such as small population sizes and breeding structure, can influence pair-wise genetic differentiation (Wright 1951). Identification of real-time migrants identified from assignment tests (see Chapter 7) could be used as flow rates between nodes in a graph analysis in cases where power is sufficient (Faubet et al. 2007; Pritchard et al. 2000).

Some systems are "natural" graphs where nodes are clearly identifiable, such as frog ponds. For other species, nodes may be much less clearly defined (e.g., Garroway et al. 2008). More research is needed on how to identify nodes in graph approaches when the node is unclear (continuous habitat, overlapping home range, etc.). In addition, some systems may have clearly defined patches with individuals that move freely among multiple patches in a given season.

The influence of a pruned graph type also requires additional study (see Fig. 10.2).

### 10.6   CONCLUSION – POTENTIAL FOR APPLICATION OF GRAPHS FOR CONSERVATION

The potential of graphs to aid in conservation and landscape planning decisions is immense (Minor & Urban 2007). In a graph analysis, it is possible to identify which nodes are most important to graph structure, what nodes are needed to maintain connectivity, and how the landscape impedes or promotes connectivity through the network (Murphy & Evans 2011; Minor & Urban 2008). Graph analysis could be applied to identify which areas are critical for conservation or restoration activities using software such as Conefor (Saura & Torne 2009). For example, those areas that have high betweeness are likely to be important for maintaining the current connectivity (Minor & Urban 2008). Graphs could also be used to evaluate the effectiveness of alternative habitat acquisitions on overall reserve networks. Finally, estimates based on a graph framework have been used to predict resistance to the landscape, resulting in model-based "cost" surfaces (Murphy & Evans 2011), an alternative approach for resistance surfaces in landscape genetics (see Chapter 8).

# REFERENCES

Albert, R. & Barabasi, A. (2002) Statistical mechaniscs of complex networks. *Reviews of Modern Physics.* **74**, pp. 47–97.

Anderson, J.E. (1979) 'A theoretical foundation for the gravity equation', *American Economic Review.* **69**, pp. 106–16.

Arnaud, J.-F., Madec, L., Beludo, A., & Guiller, A. (1999) Microspatial genetic structure in the land snail *Helix aspersa* (Gastropoda: Helicidae). *Heredity* **83**, 110–19.

Barabasi, A. & Albert, R. (1999) Emergence of scaling in random networks. *Science* **286**, 509–12.

Bebber, D., Hynes, J., Darrah, P., Boddy, L., & Fricker, M. (2007) Biological solutions to transportation network design. *Proceedings of the Royal Society B* **274**, 2307–15.

Biggs, N., Lloyd, E., and Wilson, R. (1986) *Graph Theory.* Oxford University Press, Oxford.

Bode, M., Burrage, K., & Possingham, H.P. (2008) Using complex network metrics to predict the persistence of metapopulations with asymmetric connectivity patterns. *Ecological Modeling* **214**, 201–9.

Bossenbroek, J., Kraft, C.E., & Nekola, J.C. (2001) Prediction of long-distance dispersal using gravity models: zebra mussel invasion of inland lakes. *Ecological Applications* **11**, 1778–88.

Bossenbroek, J., Johnson, L.E., Peters, B., & Lodge, D.M. (2007) Forecasting the expansion of zebra mussels in the United States. *Conservation Biology* **21**, 800–10.

Bowcock, A.M., Ruiz-Linares, A., Tomfohrde, J., Minch. E., Kidd, J.R., & Cavalli-Sforza, L.L. (1994) High resolution of human evolutionary trees with polymorphic microsatellites. *Nature* **368**, 455–7.

Bray, J.R. & Curtis, J.T. (1957) An orgination of upland forest communities of southern Wisconsin. *Ecological Monographs* **27**, 325–49.

Cantwell, M.D. & Forman R.T.T. (1993) Landscape graphs: ecological modeling with graph theory to detect configurations common to diverse landscapes. *Landscape Ecology* **8**, 239–55.

Cavalli-Sforza, L.L. & Edwards, A.W.F. (1967) Phylogenetic analysis: models and estimation procedures. *American Journal of Human Genetics* **19**, 233–57.

Chetkiewicz, C.-L.B., St. Clair, C.C., & Boyce, M.S. (2006) Corridors for conservation: integrating pattern and process. *Annual Review of Ecology and Systematics* **37**, 317–42.

Collins, J. & Chow, C. (1998) 'It's a small world', *Nature* **393**, 409–10.

Crooks, K.R. and Sanjayan, M. (eds.) (2006) *Connectivity Conservation.* Cambridge University Press, Cambridge.

Csardi, G. & Nepusz, T. (2006) The igraph software package for complex network research. *InterJournal, Complex Systems* **1695**.

Cushman, S.A. & Landguth, E.L. (2010) Spurious correlations and inference in landscape genetics. *Molecular Ecology* **19**, 3592–602.

Cushman, S.A., Compton, B.W., and McGarigal, K. (2010) Habitat fragmentation effects depend on complex interactions between population size and dispersal ability: modeling influences of roads, agriculture and residential development across a range of life-history characteristic. In: Cushman, S.A. and Huettmann, F. (eds.), *Spatial Complexity, Informatics and Wildlife Conservation.* Springer, Tokyo, Japan.

Dale, M.R.T. & Fortin, M.J. (2010) From graphs to spatial graphs. *Annual Review of Ecology and Systematics* **41**, 21–38.

de Nooy, W., Mrvar, A. and Batagelj, V. (2005) *Exploratory social network analysis* with Pajek, A., Bateagelj, V., 1st edition. Cambridge University Press, Cambridge.

de Nooy, W., Mrvar, A., and Batagelj, V. (2012) *Exploratory Social Network Analysis with* Pajek, A., Bateagelj, V., 2nd edition. Cambridge University Press, Cambridge.

De Solla, P.D. (1976) A general theory of bibliometrics and other cumulative advantage processes. *Journal of the American Society for Information Science* **27**, 292–306.

Dray S. & Dufour, A.B. (2007) The ade-4 package: implementing the duality diagram for ecologists. *Journal of Statistical Software* **22**, 1–20.

Dyer, R.J. (2007) The evolution of genetic topologies. *Theoretical Population Biology* **71**, 71–9.

Dyer, R.J. (2014) gstudio: analyses and functions related to the spatial analysis of genetic marker data. R package, version 1.3.

Dyer, R.J. & Nason, J.D. (2004) Population graphs: the graph theoretic shape of genetic structure. *Molecular Ecology* **13**, 1713–27.

Dyer, R.J., Nason J.D., & Garrick, R.C. (2010) Landscape modeling of gene flow: improved power using conditional genetic distance derived from the topology of population networks. *Molecular Ecology* **19**, 3746–59.

Dyer, R.J., Chan, D.M., Gardiakos, V.A., & Meadows, C.A. (2012) Pollination graphs: quantifying pollen pool covariance networks and the influence of intervenins landscape on genetic connectivity in the North American understory tree, *Cornus florida* L. *Landscape Ecology* **27**, 239–51.

Economo, E. & Keitt, T.H. (2008) Species diversity in neutral metacommunities: a network approach. *Ecology Letters* **11**, 52–62.

Erdös, P. & Rényi, A. (1960) On the evolution of random graphs. *Publications of the Mathematical Institute of the Hungarian Academy of Sciences* **5**, 17–61.

Estrada, A., Real, R., & Vargas, J.M. (2008) Using crisp and fuzzy modeling to identify favourability hotspots useful to perform gap analysis. *Biodiversity and Conservation* **17**, 857–71.

Excoffier, L., Smouse, P.E., & Quattro, J.M. (1992) Analysis of molecular variance inferred from metric distance among DNA haplotypes: application to human mitochondrial DNA restriction data. *Genetics* **131**, 479–91.

Faubet, P., Waples, R.S., & Gaggiotti, O.E. (2007) Evaluating the performance of a multilocus Bayesian method for the estimation of migration rates. *Molecular Ecology* **16**, 1149–66.

Ferrari, M.J., Bjornstad, O.N., Partain, J.L., & Antonovics, J. (2006) A gravity model for the spread of a pollinator-borne plant pathogen. *American Naturalist* **168**, 294–303.

Fotheringham, A.S. & O'Kelly, M.E. (1989) *Spatial Interaction Models: Formulation and Applications*. Kluwer Academic, Dordrecht.

Fotheringham, A.S., Brunsdon, C., & Martin C. (2002) *Geographically Weighted Regression: The Analysis of Spatially Varying Relationships*. John Wiley & Sons, Ltd, Chichester.

Freeman, L. (1977) A set of measures of centrality based on betweenness. *Sociometry* **40**, 35–41.

Garrick, R.C., Nason, J.D., Meadows, C.A., & Dyer R.J. (2009) Not just vicariance: phylogeography of a Sonaroan desert euphorb indicates a major role of range expansion along the Baja penisula. *Molecular Ecology* **18**, 1673–6.

Garroway, C.J., Bowman, J., Carr, D., & Wilson, P.J. (2008) Applications of graph theory to landscape genetics. *Evolutionary Applications* **1**, 620–30.

Gaston, K.J. & Fuller, R.A. (2008) Commonness, population depletion and conservation biology. *Trends in Ecology and Evolution* **23**, 14–19.

Goldberg, C. & Waits, L.P. (2010) Comparative landscape genetics of two pond-breeding amphibian species in a highly modified agricultural landscape. *Molecular Ecology* **19**, 3650–63.

Guiller, A., Bellido, A., & Madec, L. (1998) Genetic distances and orginations: the land snail *Helix aspera* in North Africa as a test case. *Systematic Biology* **47**, 208–27.

Hanski, I. (1998) Metapopulation dynamics. *Nature* **396**, 41–9.

Hedrick, P. (2005) A standardized genetic differentiation measure. *Evolution* **59**, 1633–8.

Heyman, O., Gaston, G., Kimerling, A.J., & Campbell, J.T. (2003) A per-segment approach to improving aspen mapping from high-resolution remote sensing imagery. *Journal of Forestry* **101**, 29–33.

Huson, D.H. (1992) Splitstree: analyzing and visualzing evolutionary data. *Bioinformatics* **14**, 68–73.

Jelinski, D.E. & Wu, J. (1996) The modifiable areal unit problem and implications for landscape ecology. *Landscape Ecology* **11**, 129–40.

Kalinowski, S.T. (2005) Do polymorphic loci require large samples sizes to estimate genetic distance? *Heredity* **94**, 33–6.

Keitt, T.H., Urban, D.L., & Milne, B.T. (1997) Detecting critical scales in fragmented landcapes. *Ecology and Society* **1**, 1–19.

Kimura, M. & Weiss, G. (1964) The stepping stone model of population structure and the decrease of genetic correlation with distance. *Genetics* **49**, 561–76.

Koen, E.L., Garroway, C.J., Wilson, P.J., & Bowman, J. (2010) The effect of map boundary on estimates of landscape resistance to animal movement. *PLoS One* **5**, e11785.

Koen, E.L., Bowman, J., Garroway, C.J., & Wilson, P.J. (2013) The sensitivity of genetic connectivity measures to unsamples and under-sampled sites. *PLoS One* **8**, e56204.

Legendre, P., Lapointe, F.J., & Casgrain, P. (1994) Modeling brain evolution from behaviour: a permutation regression approach. *Evolution* **48** 1487–99.

Loiselle, B.A., Sork, V.L., Nason, J.D., & Graham, C.H. (1995) Spatial genetic structure of a tropical understory shrub, *Psychotria officinalis* (Rubiaceae). *American Journal of Botany* **82**, 1420–5.

Lynch, M. & Ritland, K. (1999) Estimation of pairwise relatedness with molecular markers. *Genetics* **152**, 1753–66.

McCune, B. & Keon, D. (2002) Equations for potential annual direct incident radiation and heat load. *Journal of Vegetation Science* **13**, 603–6.

McIntire, E.J.B. & Fortin, M.-J. (2006) Structure and function of wildfire and mountain pine beetle forest boundaries. *Ecography* **29**, 309–18.

McRae, B.H., Dickson, B.G., Keitt, T.H., & Shah, V.B. (2008) Using circuit theory to model connectivity in ecology, evolution and conservaiton. *Ecology* **89**, 2712–24.

Meeuwig, M.H., Guy, C.S., Kalinowski, S.T., & Fredenberg, W.A. (2010) Landscape influences on genetic differentiation among bull trout populations in a stream-lake network. *Molecular Ecology* **19**, 3620–33.

Mennechez, G., Schtickzelle, N., & Baguette, M. (2003) Metapopulation dynamics of the bog fritillary butterfly: comparison of demographic parameters and dispersal between a continuous and a highly fragmented landscape. *Landscape Ecology* **18**, 279–91.

Minor, E.S. & Urban, D.L. (2007) Graph theory as a proxy for spatially explicit population models in conservation planning. *Ecological Applications* **17**, 1771–82.

Minor, E.S. & Urban, D.L. (2008) A graph-theory framework for evaluating landscape connectivity and conservation planning. *Conservation Biology* **22**, 297–307.

Murphy, M.A. and Evans, J.S. (2011) Genetic connectivity as a function of landscape connectivity: application of landscape genetics for predictive modeling. In: Drew, C.A., Huettmann, F., and Wiersma, Y.F. (eds.), *Predictive Species and Habitat Modeling in Landscape Ecology*. Springer.

Murphy, M.A. & Evans, J.S. (in preparation) GenNetIT: gravity analysis in R for landscape genetics. *Molecular Ecology Resources*.

Murphy, M.A., Evans, J.S., Cushman, S., & Storfer, A. (2008) Representing genetic variation as continuous surfaces: an approach for identifying spatial dependency in landscape genetic studies. *Ecography* **31**, 685–97.

Murphy, M.A., Dezanni, R.J., Pilliod, D., & Storfer, A. (2010a) Landscape genetics of high mountain frog metapopulations. *Molecular Ecology* **19**, 3634–49.

Murphy, M.A., Evans, J.S., & Storfer, A. (2010b) Quantifying *Bufo boreas* connectivity in Yellowstone National Park with landscape genetics. *Ecology* **91**, 252–61.

Nacher, J., Hayashida, M., & Akutsu, T. (1999) Emergence of scale-free distribution in protein–protein interaction networks based upon random seleciton of interacting domain pairs. *Biosystems* **95**, 155–9.

Naujokaitis-Lewis, I.R., Curtis, J.M.R., Arcese, P., & Rosenfeld, J. (2009) Sensitivity analyses of spatial population viability analysis models for species at risk and habitat conservation planning. *Conservation Biology* **23**, 225–9.

Naujokaitis-Lewis, I.R., Rico, Y., Lovell, J.L., Fortin, M.-J., & Murphy, M. (2013) Implications of incomplete networks on estimation of landscape genetic connectivity. *Conservation Genetics* **14**, 287–98.

Neel, M.C. (2008) Patch connectivity and genetic diversity conservation in the federally endangered and narrowly endemic plant species *Astragalus albens* (Fabaceae). *Biological Conservation* **141**, 938–55.

Nei, M. (1972) Genetic distance between populations. *The American Naturalist* **106**, 283–92.

Nei, M. (1973) Analysis of gene diversity in subdivided populations. *Proceedings of the National Academy of Science* **70**, 3321–3.

O'Brien, D., Manseau, M., Fall, A., & Fortin, M.-J. (2006) Testing the importance of spatial configuration of winter habitat for woodland caribou: an application of graph theory. *Biological Conservation* **130**, 70–83.

Penny, D. & Hendy, M.D. (1985) The use of tree comparison metrics. *Systematic Zoology* **34**, 75–82.

Pritchard, J.K., Stephens, M., & Donnelly, P. (2000) Inference of population structure using multilocus genotype data. *Genetics* **155**, 945–59.

Rayfield, B., Fortin, M.-J., & Fall, A. (2010) The sensitivity of least-cost habitat graphs to relative cost surface values. *Landscape Ecology* **25**, 519–32.

Rayfield, B., Fortin, M.-J., & Fall, A. (2011) Connectivity for conservation: a framework to classify network measures. *Ecology* **92**, 847–58.

Rozenfield, A., Arnaud-Haond, S., Hernández-Garcia, E., Eguiluz, V., Serrão, E., & Duarte, C. (2008) Network analysis identifies weak and strong links in a metapopualtion system. *Proceedings of the National Acadamy of Science* **105**, 18824–9.

Saitou, N. & Nei, M. (1987) The neighbor-joining method: a new method for reconstructing phylogenetic trees. *Molecular Biology and Evolution* **4**, 406–25.

Saura, S. & Torne, J. (2009) Conefor Sensinode 2.2: a software package for quantifying the importance of habitat patches for landscape connectivity. *Environmental Modeling and Software* **24**, 135–9.

Slatkin, M. (1987) Gene flow and the geographic structure of the natural populations. *Science* **236**, 787–92.

Smouse, P.E. & Peakall, R. (1999) Spatial autocorrelation analysis of individual multiallele and multilocus genetic structure. *Heredity* **82**, 561–73.

Sork, V.L., Davis, F.W., Westfall, R., Flint, A., Ikegami, M., Wang, H.F., & Grivet, D. (2010) Gene movement and genetic association with regional climate gradients in California valley oak (*Quercus lobata* Née) in the face of climate change. *Molecular Ecology* **19**, 3806–23.

Theobald, D.M. (2006) Exploring functional connectivity of landscapes using landscape networks. In: Crooks, K.R. and Sanjayan, M. (eds.), *Connectivity Conservation*. Cambridge University Press, Cambridge.

Treml, E., Halpin, P., Urban, D., & Pratson, L. (2008) Modeling population connectivity by ocean currents, a graph-theoretic approach for marine conservation. *Landscape Ecology*, **23**(Suppl. 1), 19–36.

Urban, D.L. & Keitt, T.H. (2001) Landscape connectivity: a graph-theoretic perspective. *Ecology* **82**, 1205–18.

Urban, D.L., Minor, E.S., Treml, E.A., & Schick, R.S. (2009) Graph models of habitat mosaics. *Ecology Letters* **12**, 260–73.

Vasas, V., Magura, T., Jordán, F., & Tóthmérész, B. (2009) Graph theory in action: evaluating planned highway tracks based on connectivity measures. *Landscape Ecology* **24**, 581–6.

Voorhees, A.M. (1956) *A Genetial Theory of Traffic Movement*. Institute of Traffic Engineering, New Haven, Connecticut.

Wasserman, S. and Faust, K. (1994) *Social Network Analysis: Methods and Applications*. Cambridge University Press, Cambridge.

Willig, R.D. & Bailey, E.E. (1979) The economic gravity model. *American Economic Review* **69**, 96–101.

Wilson, A.G. (1967) Statistical theory of spatial trip distribution models. *Transportation Research* **1**, 253–69.

Wright, S. (1943) Isolation-by-distance. *Genetics* **28**, 114–38.

Wright, S. (1951) The genetical structure of populations. *Annals of Eugenics* **15**, 323–54.

Xia, Y., Ottar, N.B., & Grenfell, B.T. (2004) Measles metapopulation dynamics: a gravity model for epidemiological coupling and dynamics. *American Naturalist* **164**, 267–81.

Zar, J.H. (1999) *Biostatistical Analysis*. Simon & Schuster, Upper Saddle River, New Jersey.

# PART 3

# APPLICATIONS

# Chapter 11

# LANDSCAPES AND PLANT POPULATION GENETICS

*Rodney J. Dyer*

*Department of Biology, Virginia Commonwealth University, USA*

## 11.1 INTRODUCTION

Landscape genetics has become a nearly ubiquitous term, applied to a broad range of studies which investigate the microevolutionary processes of gene flow and drift within a spatially and ecologically explicit characterization of landscape features. Coined by Manel et al. (2003), "landscape genetic" studies were originally considered to be focused on identifying structural features in nature that either impede or at least modify gene flow, resulting in statistically identifiable spatial genetic discontinuities and/or obvious hierarchical nesting. However, predating this moniker was an existing emphasis in many plant population genetic laboratories to look at microevolutionary processes at landscape spatial scales (e.g., Sork et al. 1999) with the intent to understand how ecological context (e.g., where gene flow occurs) influenced overall genetic variation. In the intervening years it became increasingly clear that classical approaches to population genetics, often those that relied upon equilibrium assumptions and idealized populations (e.g., Slatkin 1985a, 1985b; Neigel 1997), were not sensitive to contemporary alterations of landscape features, how these

changes may influence ongoing genetic exchange, and the large-scale consequences for population genetic structure. While these "indirect approaches" (see Neigel 1997) were most beneficial in describing deep time evolutionary change, the power of their generality was bought at the expense of their ability to capture the intricacies of contemporary microevolutionary processes such as asymmetry in dispersal (Endler 1973; Papa & Gepts 2003), heterospecific mitigation of gene flow (e.g., Dyer & Sork 2001), and realized consequences of population fragmentation (Fore et al. 1992; Young et al. 1996; Nason & Hamrick 1997). Moreover, these indirect approaches generally ignored metapopulation dynamics (Hanski & Simberloff 1997; Chapter 2), a paradigm applied widely to plant studies at the time.

Direct methods for characterizing gene flow, such as sampling of individual pollen grains (Castaing & Vergeron 1974; Campbell 1991), marker aided paternity analysis (Meagher 1986; Meagher & Thompson 1987; Adams & Birkes 1991; Streiff et al. 1999), and mating system analysis (Mitton 1992), focused on a more limited number of mating events and when applied in an experimental context, rather than as a descriptive measure, often showed that spatial and ecological

*Landscape Genetics: Concepts, Methods, Applications*, First Edition. Edited by Niko Balkenhol, Samuel A. Cushman, Andrew T. Storfer, and Lisette P. Waits.
© 2016 John Wiley & Sons, Ltd. Published 2016 by John Wiley & Sons, Ltd.

heterogeneity can have significant influences on future population genetic structure. This was not a surprising finding, as population geneticists have long suspected that the qualities of the intervening habitat influence connectivity and, by extension, genetic structure, across a wide variety of species (see, for example, Sokal et al. 1986). However, there was mounting concern that the ubiquity of indirect approaches that provide robust generalities may actually mask underlying specifics of how ecological features and evolutionary processes interact. These concerns intensified as research was increasingly directed at contemporary processes, such as fragmentation and anthropogenic landscape conversion, where demographic and genetic equilibrium conditions were not readily assumable.

While direct approaches to the analysis of gene flow continued to be applied at small spatial scales, their application at the landscape level was fraught with logistical challenges from two sources. First, direct methods are specific because they often employ a thorough sampling of mating individuals interacting over one or a few mating events; once the genotypes of all potential pollen donors are determined, what remains is a routine analysis of seed paternity. However, a drawback to having this level of precision is that the logistical requirements associated with scaling this to a landscape level are prohibitory (though see Schuster & Mitton 2000 for an extreme example) because the sheer number of individuals needing to be both identified and genotyped rises with the square of the increase in radius. Applications of parentage approaches also require a high level of genetic resolution to differentiate among potential pollen donors, who often were both related and spatially proximate. Second, it was becoming common to show that ecological context covaries with measures of gene flow, making the circumscription of limiting ecological features and their temporal variation exceedingly difficult (Sorensen 1994; Aldrich et al. 1998; Burczyk 1998). The resulting generality of process and interaction in different ecological contexts – perhaps with different sets of species – gained from concise yet spatially and ecologically restricted studies is largely speculative. Motivated in part by the realization that analyses gain both realism and applicability if they incorporate "habitat in which an individual population is located and the qualities of the intervening landscape" (Sork et al. 1999), research since the turn of the century has seen a significant shift towards context-specific plant population genetic approaches.

Statistically, there have been several advances that have aided in expanding population genetic studies to truly landscape scales that have provided new levels of biological inferences. Some of these methods are novel in population genetics (e.g., circuit theory, McRae & Beier 2007; population graphs, Dyer & Nason 2004; principal components on neighbor matrices, Borcard & Legendre 2002), some are extensions and refinements of currently used approaches (e.g., multiple regression on distance matrices, Lichstein 2007; extending Manly 1986, the Mantel test restrictions, Legendre & Fortin 2010), and the rest have been borrowed and repurposed from other disciplines like landscape ecology (e.g., gravity models, Murphy et al. 2010; causal modeling, Cushman et al. 2006). Underlying all of these advances is the notion that more sophisticated, context-specific treatments of ecological and spatial settings are required if we hope to address context-dependent biological hypotheses. Somewhat ironically, the vast majority of statistical advances salient to landscape genetic studies in plants and animals both are found on the ecological side of the equation rather than on the genetic side and as such population genetics has perhaps received a disproportionate benefit from its coupling with landscape ecology.

Another key distinction is the increasingly important role that modeling microevolutionary processes plays in the development and refinement of these approaches. Classic population genetic models based upon simplifying equilibrium conditions made the algebra tractable. However, contemporary landscape-level analyses often do not have the luxury to assume large population sizes or huge numbers of generations necessary to satisfy these equilibrium assumptions. The approaches being used in contemporary landscape genetic studies are increasingly taking stochasticity and integrating it into the analysis rather than iterating it out of the equation. Given the broad range of potential questions, study species, life history traits, and habitats, a robust approach to modeling landscape genetic processes cannot be confined to a single analytical framework or computational program. This has functionally resulted in the development of software frameworks rather than individual software solutions and underlies the increasing need for practitioners of landscape genetics to be fluent in both modern analytical tools such as R and programming grammars such as Perl or Python. In Chapter 6, Landguth et al. provided a nice overview of how simulation modeling can aid in the development of testable hypotheses.

### 11.1.1 Plant systems as models for landscape genetics

Understanding the extent to which landscape context can influence population genetic structure in plants comes from specific features of their life history. Plants are generally immobile (though see *Stenocereus eruca*, Cactaceae), which affixes them within a specific and potentially definable spatial and ecological context. Stationarity allows mating processes, and by extension potential influences on population genetic structure, to be locally influenced (e.g., Thomson 1978; Watkins & Levin 1990; Dyer & Sork 2001) and can homogenize selection pressures (e.g., Bazzaz 1991; Anderson et al. 2011). More importantly, though, reproduction in plants is a two-step process, allowing the analysis and understanding of population-level consequences of mating to be partitioned into effects due to either pollen or seed dispersal. The pollen vector in most plant species is the most pervasive vector in that it has the ability to disperse gametes the farthest and as such broadly dictates the level of genetic isolation among populations, whereas the seed dispersal vector is often more spatially restricted. Functionally, these vectors also have different consequences for standing population genetic structure. Independent of the extent that pollen is dispersed (from extremely limited to panmictic), limitations in seed dispersal are the only mechanism that will create fine-scale genetic structure measured as spatial autocorrelation (e.g., Smouse & Peakall 1999). The net effects of these two vectors may, however, be in opposition. Broad dispersal of pollen will homogenize spatial population genetic structure. However, with spatial limitations to seed dispersal, fine-scale structure will characteristically arise since individuals will be germinated in spatial proximity to both their maternal individuals as well as potential siblings. If both pollen and seed dispersal are limited, adult autocorrelation and local pollen donors tend towards an increase in inbreeding either directly or through biparental inbreeding (Dyer et al. 2010), resulting in a more restricted spatial granularity in genetic structure and potential fitness consequences.

A large fraction of landscape genetic studies have been applied to characterizing how habitat influences animal behavior in terms of movement (Cushman et al. 2006; Cushman & Lewis 2010; Row et al. 2012; Coster & Kovach 2012). While it is unclear the extent to which behavioral associations directly lead to changes in population genetic structure (see Yannic et al. 2014), the approaches being used on animal systems are directly applicable to understanding biotically assisted pollination and seed dispersal processes. However, a pollinator is not a "bear" and there are specific reasons why generalizations made from studies of individual movement in large mammal studies may not be appropriate for studying plant connectivity and subsequent population genetic structure. Dispersal in plants is typically described in terms of the distribution of dispersal events rather than characterization of individual movements. An individual anther in a flower produces as much as 700–1200 pollen grains per millimeter (Agnihotri & Singh 1975). A single individual plant may produce tremendous volumes of pollen each reproductive cycle (e.g., *Fraxinus angustifolia* $1.12–16.17 \times 10^{10}$ grains, *Pinus pinaster* $2.09–3.23 \times 10^{10}$ grains, *Shorea robusta* $2.0–3.3 \times 10^{14}$ grains; Bera 1990; Molina et al. 1996).

Tracking individual pollinators from flower to flower does not necessarily translate into fertilizations and subsequent fruit set either. Pollen carryover, the process by which pollen from a single anther may be spread across a sequence of subsequent stigma visited by a pollinator (e.g., Price & Waser 1982) is an all too common occurrence. Perhaps more problematic than carryover is the fact that most animal pollinated plant species utilize a broad range of pollinators, each of which may be interacting with local ecological heterogeneity in its own fashion.

Functionally, analyses of pollen-mediated gene flow in plants has been performed by either looking at collections of pre-dispersed seeds (Smouse et al. 2001) or recently established seedlings (Dyer and Sork 2001), treating these as the unit of interest in population analyses. Consequently, the focus here is on the collection of individuals who are potentially the "next generation" and, as such, we are looking directly at genetic consequences from a population perspective. In contrast, landscape genetic studies on animals are often focused on the genotypes of individuals sampled on a landscape rather than populations. Of particular distinction are species that are largely mobile across the landscape, such as fishers (Garroway et al. 2011), lynx (Row et al. 2012), land tortoises (Latch et al. 2011), etc. In these studies, inferences regarding landscape heterogeneity are largely concerning behavioral choices that influence individual movement. However, the result of this behavior is only indirectly salient to the subsequent population genetic structure as the genetic makeup of the next generation is not actually being sampled (e.g., Cushman & Lewis 2010). In plant studies, which often sample the next generation directly, the interaction between landscape and population genetic structure is somewhat more direct.

A final distinction that should be appreciated when studying plant systems is the way in which relevant landscape features are treated, particularly in a temporal context. For example, elevation as a feature of the landscape has been found to be a significant factor contributing to movement of a variety of animal species ranging from potentially low dispersing species, such as grasshoppers (Ortego et al. 2012) and blotched tiger salamanders (Spear et al. 2005), to species that have the ability to move considerable distances relative to the spatial scale of elevation heterogeneity, such as mink (Zalewski et al. 2009), marten (Wasserman et al. 2012), black bears (Coster & Kovach 2012), and bighorn sheep (e.g., Epps et al. 2007). Often, it is not elevation that directly influences movement but rather its indirect association with changes in vegetation and structural features of the landscape and as such it is a static feature of the landscape. For plants, a feature like elevation is not static. Gene flow is curtailed, temporally, through mating phenology. Plants that inhabit lower elevations flower earlier than those at higher elevations and can potentially be reproductively isolated even if they share the same set of pollinators. As a consequence, when considering the development of resistance models, a common approach in modern landscape genetics, features such as elevation may not be treated as static features as they may be for animal movement studies. Rather, for any population, it may be the similarity in elevational contours that correspond to synchrony in phenology (species x environment interactions) rather than the absolute value of elevation that best capture features influencing connectivity (e.g. Yang et al. 2014).

This chapter examines landscape genetic studies in plant population genetics. The main focus here is in the discussion of approaches and examples showing how population genetic structure can be influenced by spatial and ecologically heterogeneity. The case studies presented will be directed towards the biological inferences regarding population genetic structure that are gained from the inclusion of spatial and ecological contexts. This chapter will not focus on specific methodological aspects; rather it is partitioned into sections associated with the temporal depth of the analysis – shallow time processes versus historical processes. Despite being a relatively new field, landscape genetics already has a disproportionate number of reviews in the literature (see, for example, Holderegger & Wagner 2008; Holderegger et al. 2010; Storfer et al. 2010), of which many contain methodological recommendations (e.g., Balkenhol et al. 2009; Dyer et al. 2010; Manel & Holderegger 2013), so adding yet another such review to that mix would be unwarranted. This chapter also takes a broader, and perhaps more comprehensive, view of what would be considered *landscape genetics* than is commonly used. Given the depth of the plant population genetic literature – and requests from the editors – I have limited this chapter largely to studies of neutral genetic variance and the inferences gained from its analysis. In the sections that follow, an arbitrary partition has been imposed based upon the spatial and temporal extent of the processes being examined, either contemporary or historical, straddling both population genetic and phylogeographic domains. I then close with a short discussion on future research and potentially understudied aspects of plant landscape genetic analyses.

## 11.2 CONTEMPORARY POPULATION GENETIC PROCESSES

For the purposes of this chapter, I will define a contemporary landscape genetic process as one that operates at the temporal scale of one or a few mating events. The spatial extent of these mating events is solely determined by the dispersal vectors of the species being studied. At this limited temporal scale, microevolutionary processes do not have sufficient time to trend toward equilibrium conditions. Processes such as selection and inbreeding, while potentially important locally, generally do not have significant impacts on broad-scale restructuring of genetic variance. At this limited scale, it is the effects of the intervening landscape on both pollen and seed movement that have the dominant effect.

### 11.2.1   Pollen pools

Human-mediated changes in land-use often result in population fragmentation and reductions in local population sizes, both of which can significantly impact population genetic structure and subsequent species persistence (e.g., Ledig 1992). Functionally, these changes are mediated through detrimental effects such as reductions in the genetic effective population size, increases in inbreeding (either direct or biparental), loss of both neutral and adaptive genetic variation, and reduced reproductive output. Quantitatively, these changes will first be identified by changes in both genetic diversity and structure of the pollen donor population and can readily be measured by genetic assay of spatially separated pollen pools.

In this context, maternal individuals act as spatially fixed sampling locations of the larger pollen pool. Under the condition that pollen dispersal is not panmictic throughout the landscape (or at least the spatial extent to which individuals are being sampled), maternal individuals in close proximity will sample pollen pools comprised of a more similar set of pollen donor individuals than maternal individuals situated more distantly. In fact, the degree to which multilocus genotypes differ spatially is inversely related to both pollen donor density and variance in dispersal distance $(\Phi_{FT} = (8\pi\sigma^2 d)^{-1}$; Smouse et al. 2001; Austerlitz & Smouse 2001). Analysis of pollen structure in this manner directly links the process of contemporary gene flow to both genetic effective population size $(N_{ep})$ and effective pollination neighborhood size, $A_{ep}$, both of which are parameters central for management and conservation planning.

Of particular importance is the role these parameters they play in predicting future population genetic structure as a function of potential landscape change. When applied to declining populations of California Valley oak (*Quercus lobata* Née, Fagaceae), Sork et al. (2002) showed that even for this wind-dispersed species, pollination was very localized. Derived from pollen pool genetic structure, they estimated a remarkably short average contemporary dispersal distance of 64.8 m. Given the specific spatial arrangement of pollen donor individuals on the landscape, this population had a genetic effective population size of $N_{ep} = 3.68$ individuals. Using aerial photography to reconstruct donor densities from 50 years prior, they estimated that the current effective population size is 19.5% smaller than a half century ago $(N_{ep;1944} = 4.57)$. As a result, even for a wind-dispersed species that has the potential for long-distance dispersal, pollen movement can be very localized and, more importantly, local conditions can have significant impacts on evolutionary potential through modification in genetic effective population sizes. Despite the prevailing notion that the most proximate pollen sources may be best adapted to localized conditions, remediation and conservation programs should take into account relatively small genetic neighborhoods

when implementing reforestation programs such that population genetic variance, and effective population size is increased rather than diminished.

While aggregate differentiation in pollen pool genetic structure does provide an estimate of genetic effective population size $(N_e)$, its foundation is firmly set in Wright's genetic effective neighborhood paradigm. This model requires that the dispersal function itself be both continuous and isotropic across the landscape (e.g., Austerlitz & Smouse, 2001). Algebraically, relying upon a dispersal function with these characteristics makes the derivations more tractable from a theoretical basis (see, for example, Okubo & Levin 1989), though empirical support has been questioned. For example, Dyer and Sork (2001) showed that the density of heterospecific canopy tree density (*Quercus sp.*, *Carya sp.*, etc.) is inversely proportional to the genetic diversity of *Pinus echinata* pollen reaching a stand. Similarly, Young et al. (1993) showed that forest fragmentation (e.g., removing the intervening canopy structure) actually increased average dispersal distance for *Acer saccharum*. Given the inherent heterogeneity in natural landscapes with respect to spatial placement of heterospecifics and canopy openings, continuity and isotropy are not common features of the natural landscape, which has led to recent focus on anisotropic methods to estimate genetic neighborhood size in complex landscapes (e.g. Shirk & Cushman 2014). The ability of pollen to be dispersed across the landscape is differentially influenced by intervening forest features, none of which are homogeneous. Using the understory, insect-pollinated tree *Cornus florida* (Cornaceae), Dyer et al. (2012) constructed a covariance network based upon pollen pool genetic structure among maternal trees across a 60 ha site in Virginia, USA. Pollinator preference/avoidance of particular forest structure features was estimated by spatial association of pollen pool covariance and intervening forest canopy and understory features (e.g., conifer canopy, mixed hardwood canopy, canopy gaps, roads, *Cornus* canopy, and open fields). Genetic covariance was assumed to be the aggregate result of pollinator movement and decreased among sites separated by open primary and secondary canopy, whereas it was increased in routes moving in a trap-line fashion across the landscape. The relative importance and direction at which different intervening habitat features influenced pollen pool genetic covariance provided insights into how a suite of generalist pollinators are interacting with features of forest understory and canopy. However, they also point towards a context-specific association between dispersal and genetic effective neighborhood size. Not all habitat features influence genetic connectivity in exactly the same way. While it may be easier to generalize using a continuous and isotropic dispersal function, it may not necessarily represent the intricacies of how pollinators interact, *en masse*, with their plants and the intervening landscape. Pollen pool analysis will continue to be a powerful approach, allowing the direct analysis of how dispersal vectors (biotic or abiotic) interact with intervening landscape features and will play an increasingly important role in the development and implementation of evolutionarily sensitive conservation and management strategies.

### 11.2.2   Seed pools

While pollen-mediated dispersal can generally maintain genetic connectivity over larger distances than connectivity by seeds, seed dispersal is the only mechanism available for colonization and potential rescue of declining populations. Moreover, it is the only process that determines the spatial genetic structuring among adult plants, the patterns of which influence potential mating and inbreeding patterns throughout the population. Landscape-level analyses of seed dispersal are less developed than those for pollen movement, primarily because they are confronted with several logistical challenges that are still being addressed. Once a seed is dispersed away from the maternal individual, correctly identifying both parents can be difficult if the potential population of adults is large or the distance is far. Moreover, once the set of putative parents is identified, it can be difficult to determine which role, pollen donor, or ovule donor, each individual played in the parentage of the seed (though see Provan et al. 2001).

In the animal dispersed understory tree *Prunus mahaleb*, Godoy and Jordano (2001) used SSR markers and genotyped the maternally inherited endocarp of individual seeds defecated or regurgitated into an array of seed traps across a landscape in southern Spain. The leptokurtic frequency distribution of seed dispersal distances showed that 60% of the seeds were deposited within 20 m of the maternal tree. Despite the preponderance of locally dispersed seeds, seeds dispersed greater than 100 m were associated with either conifer trees or forest edges presumably due to perching preferences of dispersing animals. In a subsequent analysis, Jordano et al. (2007) partitioned the overall dispersal distribution for *P. mahaleb* seeds by agent, combining mammals (foxes, martens, and badgers), medium-sized birds (thrushes and crows), and small frugivorous birds (warblers, redstarts, and robins) into homogeneous dispersal groups, all of which had different landscape preferences and consequent net effects. Mammals had the largest weighted contribution to the dispersal distribution and along with crows preferentially deposited into open habitats. Deposition by thrushes was primarily in pines and small frugivorous birds preferred high shrub habitat for seed deposition. Heterogeneity in seed dispersal vectors confounds landscape preferences for either movement or deposition. As a consequence, projection of seed-mediated gene flow from seed trap data alone will only provide accurate characterizations of dispersal potential if all sites have the exact same set of dispersing species *and* they occur in the same relative frequencies – a notion that is doubtful at best.

Even if all dispersers respond similarly, local heterogeneity in other landscape features may result in differential survival and subsequent genetic structure. In the case of the understory tree *Cornus florida*, Kwit et al. (2004) showed that seed dispersers preferentially dropped seeds under the canopy of the co-occurring tree *Ilex opaca*, a largely unsuitable habitat. As a consequence, density-dependent mortality was not the sole result of conspecific density but rather the community within which dispersal was occurring.

Given the wide range of potential habitats that dispersing vectors may deposit seeds, habitat-mediated survival determines who will contribute to the next generation. A study of recruitment of *Carapa guianensis* (Meliaceae) seedlings in dry and wet forests (Martins et al. 2012) showed that realized dispersal distances in wet stands was more restricted than dry forest stands despite no significant differences in pollen dispersal distance. Heterogeneity in genetic effective population size ($N_e$) driven by differential site-specific factors, biotic or abiotic, can lead towards broad-scale granularity in population genetic structure and subsequently divergent evolutionary potential (e.g., Rogers et al. 1999; Jansen et al. 2004; Forget and Cuiljpers 2008).

### 11.2.3   Species interactions

Studies that integrate landscape features into the decomposition of population genetic structure tend to focus on a single species in isolation despite the fact that dispersal generally does not occur in the absence of co-occurring species. In fact, both pollen and seed dispersal studies have shown that context-specific features of the landscape can have a significant impact on how and where genes are dispersed and as such can have a long-term impact on standing genetic variance. These interspecific interactions can be either indirect, mediated through some other feature or features of the landscape, or direct.

A direct heterospecific example was shown by Dyer and Sork (2001) who examined pollen pool genetic diversity in *Pinus echinata* Mill in southern Missouri, USA. Genetic diversity of *P. echinata* offspring in second growth forests was shown to have an inverse relationship with heterospecific (*Quercus* spp., *Carya* spp.) stand density. Recruitment within dense stands of co-occurring canopy tree species reduced within stand genetic variation in the pine, presumably through decreased penetration of wind-dispersed pollen into these dense stands, a finding consistent with the pollen turbulence models of Okubo and Levin (1989). Gram and Sork (2001) surveyed a broader range of forest structure, soil, and aspect variables in the same site, looking for site-specific impacts on population genetic structure in *Quercus alba* (Fagaceae), *Carya tomentosa* (Juglandaceae), and *Sassafras albidum* (Lauraceae). They found significant environmentally associated patterns in genetic variance in each species, though each was aligned along different canonical axes representing different components of forest structure.

Perhaps the most spectacular example of indirect species interactions that have the potential to influence the genetic structure of populations is the experimental results from Knight et al. (2005) in which they showed that the presence of fish in the aquatic habitat negatively controlled dragonfly larvae populations across a set of ponds in an experimental landscape in Florida, USA. Reductions in the local abundance of predatory dragonflies significantly increased pollination success for the perennial shrub *Hypericum fasciculatum* (Clusiaceae) by allowing local pollinator populations to increase in size. As a consequence, the trophic cascade of interactions from fish to dragonfly, dragonfly to pollinator, and finally pollinator to plant resulted in significant differences in plant reproductive output and presumably subsequent population genetic structure.

While local interactions may influence recruitment and the ability of genes to reach a particular location within the landscape, simultaneous analysis of genetic covariance among co-occurring taxa provides insights into the homogeneity of dispersal processes. Similarity in the spatial distribution of genetic variance would be indicative of commonality in factors influencing gene flow across taxa and support the hypothesis that multispecies management strategies could be developed and broadly applied rather than creating recommendations on a species-by-species basis. Fortuna et al. (2009), who analyzed four common woody Mediterranean species to look at spatial similarity in genetic covariance, found modularity within genetic covariance networks (aka population graphs, Dyer & Nason 2004; Murphy et al., Chapter 10) in three of the taxa and interpreted these patterns as representing evolutionarily defined significant management units. However, the relative importance of individual patches, while significantly different within a species, were not shared across taxa, meaning that while there are critically important populations within the spatial network of populations, they are not the same set of sites across all species. Overall, multispecies landscape genetic studies are not generally conducted and until they become more mainstream we will not know the extent to which we can classify congruence in the functional response of gene flow among taxa.

### 11.2.4 Synopsis – contemporary processes

The analysis of contemporary genetic processes and the extent to which features of the landscape influence genetic structure provides only a snapshot into potential long-term genetic consequences. Given the large amount of effort required to mount a study on contemporary processes, few studies have looked at temporal stability. One study that did conduct a multiyear analysis was that of Irwin et al. (2003), who examined pollen pool structure in the invasive tree *Albizia julibrissin* (Fabaceae). Across four years of mating events for the same subset of trees in the vicinity of Athens, Georgia, USA, they found significant differences among trees, consistent with spatial restrictions in pollination distance, but almost zero interyear variance in $N_e$. These results suggest that ecological context is perhaps more important to characterizing the relative role individuals play in maintaining population genetic variance and evolutionary potential than interyear variation in mating success, at least as depicted by a few years of data from a long-lived tree.

In general, however, analyses of single or a restricted number of reproductive years only provides a glimpse of factors contributing to the maintenance and distribution of population genetic structure. Generation after generation, the cumulative effects of landscape interference with genetic connectivity can lead to broad-scale genetic substructuring, which itself may also be influenced by landscape features (e.g., Moritz et al. 2009; Lesser & Jackson 2013), a topic addressed in the next section.

## 11.3  HISTORICAL POPULATION GENETIC PROCESSES

Across many mating events, the cumulative effect of landscape features interacting with contemporary gene dispersal results in broad-scale changes in population genetic structure. The characterization of these larger patterns of genetic variation presents several unique challenges that must be considered. First, it is important to realize that increasing the spatial extent of your sampling, even if it is for the purpose of characterizing contemporary processes, necessitates a more focused consideration of how historical processes may have influenced the observed distribution of genetic variance. Even a process such as contemporary pollen pool genetic structure, sampled at large enough scales, will contain among-site variance due to historical demography. If not accounted for, this will introduce bias into derived subsequent biological inferences (e.g., Dyer et al. 2004).

Analysis of large-scale genetic patterns and their association with features of the landscape will also deviate into the realm of phylogeography. In fact, the characterization of "landscape genetics" provided by Manel et al. (2003) is strikingly similar to that for phylogeography, particularly with its foundation in understanding the interactions of dispersal and vicariance as they shape genetic structure. There are methodological changes induced by adopting a phylogeographic approach that are largely not applicable to landscape genetics and should be carefully considered. Knowles and Maddison (2002) suggest that phylogeography is most powerful when examining differences based upon genetic sequence divergence characterized by genealogical histories (e.g., gene trees), which separates it from classical population genetics – a specious argument at best. While technically similar, landscape features large enough to include in coalescent and/or Bayesian phylogeographic analyses such as the historical formation of the Sea of Cortez and its vicariant influence on Sonoran desert biota (e.g., Riddle et al. 2000; Garrick et al. 2009, 2013) are likely not to be the nuanced inferences sought on how intervening habitat modifies connectivity (as shown in Dyer et al. 2010). The analysis of *variants* versus the analysis of *variance* is perhaps a more appropriate and context-specific description of the differences between the two, though these distinctions do not necessarily hold up in all situations. As an example, consider the Sonoran desert cactus *Lophocereus schottii* (Cactaceae). Even though this species has experienced dramatic range contraction and expansion during the last glaciation (Van Devender et al. 1994), the relative number of generations since the last glacial maximum are small enough such that incomplete lineage sorting prevents sequence-based inferences (Hartman et al. 2001), whereas genetic variance decomposition provides more interpretable inferences (Nason et al. 2002; Dyer & Nason 2004).

In this section, examples from the plant literature that focus on quantifying the extent to which landscape features influence the distribution of population genetic variance in a historical context are highlighted. By definition, these studies address processes that have operated over many mating events and will use methodologies that could be considered as "indirect", as previously defined above. The distinction here is that the decomposition of genetic variance is done within the context of spatial and/or ecological separation rather than just among pre-defined demes.

### 11.3.1  Historical processes

A key motivation driving historical landscape genetic analyses is the notion that ongoing climate change may cause species range perturbations at a rate that will erode standing genetic variance. It can also be argued that for many long-lived species, it is critically important to understand connectivity because the rate at which habitat is being changed is sufficiently rapid that an evolutionary selective response is unlikely and the amount of standing genetic variance is all that we have to work with when developing management strategies.

Of primary concern in rapidly changing environments are the abilities of species to keep adequate levels of genetic variation at the front of the expansion. A particular feature of range expansion is the predicted reduction in genetic diversity along the leading edge.

As range expansion occurs, new populations are establishing from existing edge populations, who are often themselves deficient in genetic divergence when compared to similarly sized populations drawn from the core of the species distribution. This pattern of reduced allelic diversity, often referred to as allele surfing, may be strong enough to aid in the identification of alternate Pleistocene refugia, as is the case for the plant *Euphorbia lomelii* (Garrick et al. 2009). In looking at chronosequences of the expanding *Larix decidua* population, Pleuss (2011) was able to show that despite relatively recent expansion, pollen-mediated gene flow was connecting established populations to early successional colonizers. More importantly though, this work showed that in more established populations gene flow was on the order of 2–50 m whereas among individuals on the leading edge it ranged from 115 to 3132 m, a 60-fold difference in maximum dispersal! As investigations of landscape interference proceed into processes that have operated at deeper temporal scales, it is likely that the way in which ecological context influences genetic connectivity may change dramatically (García-Ramos & Rodríguez 2002).

### 11.3.2  Iterative processes

Another challenge associated with examining broad-scale genetic structure is that the larger the spatial extent examined, the more likely that historical phylogeographic variance will contribute to the observed distribution of genetic variation. Of particular concern is the potential for covariance in landscape features and historical demographic perturbations. A classic example is the analysis of genetic structure in *Pinus flexilis* by Latta and Mitton (1997), who found a spatial discontinuity in genetic variance across the slopes of the Rocky Mountains. While the degree of differentiation could be ascribed as a consequence of mountainous vicariance, a broader analysis of the species revealed that the discontinuity was due to secondary contact from alternate refugia rather than ongoing landscape or ecologically induced reductions in gene flow (Mitton et al. 2000).

Regions of secondary contact that impart severe breaks in genetic continuity, such as in *P. flexilis*, are not solely constrained by limiting geographic and topological barriers. The spatial distribution of genetic variation in the invasive C4 grass *Microstegium vimineum* (Poaceae) revealed a contact zone in the James River basin of Virginia, USA (Baker & Dyer 2011), independent of any obvious biogeographic or topographic vicariant feature. In addition to demographic effects, evolutionary constraints can also create coarse-grained spatial genetic variation across a landscape. Polyploidyzation, a major source of both genetic and phylogenetic diversity in plants, has been shown to maintain sympatric populations of cytotypic variants. In *Solidago altissima* (Asteraceae), Halverson et al. (2008) mapped out diploid, tetraploid, and hexaploid variants throughout the midwestern United States, with some local populations having all three ploidy variants. Ignorance of the ploidy level in the plants sampled would have been interpreted as a discontiguous spatial structure. If processes such as historical refugia, range expansions, or sources of vicariance are suggested by paleoecological, paleoclimatic, or other sources, its inclusion can only increase the statistical precision of landscape genetic analyses and subsequent biological inferences.

Dyer et al. (2010) addressed this issue directly when attempting to identify landscape features that impact interpopulation gene flow in the Sonoran desert endemic, *Euphorbia lomelii* (Euphorbiaceae). A previous phylogeographic analysis of this hummingbird pollinated species showed both vicariance and post-Pleistocene range expansion as significant factors influencing the spatial distribution of neutral genetic variation (Garrick et al. 2009). To account for potential bias due to historical demography, an iterative approach was used. The spatial distribution of genetic covariance was first conditioned on phylogeographic history and then subsequently analyzed for the influence of intervening landscape features. The landscape features that best described conditional genetic covariance among populations were those associated with increased overlap in flowering phenology, highlighting the role that pollinator movement has in ensuring genetic connectivity. Perhaps more importantly, though, those data showed that analyses not conditioned on phylogeographic history may lead to incorrect biological inferences in at least two ways. First, similarity in site elevation was indicated as a significant contributor to genetic covariance but fell out of the model when it was conditioned on range expansion. In this landscape (as is common for others), post-Pleistocene range expansion out of refugia is aligned along gradients of increasing elevation, confounding spatial structure and historical processes. Second, even if the correct subset of intervening habitat features has been identified, their relative importance in any model fit to genetic covariance may be influenced by phylogeographic history. For *E. lomelii*, the inclusion of both precipitation and temperature similarities were independent of phylogeography but their relative contribution to describing the spatial distribution of genetic covariance was overestimated by 16% and 30% respectively when appropriate corrections were not made. A valuable extension of the work in *E. lomelii* is the characterization of the relative impact that individual landscape features have on gene flow. Mapped back on to the spatial landscape, an expected gene flow resistance surface was created, serving as a source of subsequent hypothesis generation. It is likely that as landscape sensitive studies increase in frequency, ecologically sensitive resistance maps such as these will find use in the active management conservation programs.

### 11.3.3   Synopsis – historical processes

Plant landscape genetics at deeper historical scales is an exciting area of research, though it is a fine line to walk between addressing landscape genetic hypotheses and ones perhaps more amenable to phylogeography. However, since deeper time processes may leave behind strong spatial and ecologically associated signals, perhaps the separation between phylogeographic and landscape genetic hypotheses should be considered less as a partition of interest and more as an opportunity to perform iterative analyses by conditioning the current distribution of spatial genetic variance on phylogeographic patterns. In the next section, I make the case that in any large-scale landscape genetic studies, and in particular those that seek putatively adaptive inferences, this iterative process becomes paramount.

### 11.4   FUTURE RESEARCH

Landscape genetics, of plants or animals, is a fundamental endeavour for evolutionary biologists that has seen a recent resurgence due in part to both a novel moniker around which studies can be described as well as the continued introduction of more sophisticated treatments of ecological space. Progress in landscape genetic studies of any type are driven by innovations in three separate domains: advances in genetic marker technology, quantification of spatial and ecological context, and continued

development of analytical tools that manipulate and characterize both genetic and spatial data. Since 2005, when 454 Life Sciences launched the first "next-generation DNA sequencer", the promise of next-generation technologies has resulted in a bloom of perspective papers on how complete sequences are going to improve our understanding of almost every aspect of biological analysis including: adaption genomics (Stapley et al. 2010; Davidson 2012), ecological genomics (Elmer & Meyer 2011), conservation (Funk et al. 2012), hybridization (Twyford & Ennos 2012), invasive species management (He et al. 2012), marker development (Davey et al. 2011), phylogeography (Emerson et al. 2010), speciation (Rice et al. 2011), etc. Even with the putatively transformative nature of next-generation technologies, the analysis of how population genetic variation is influenced by landscape features does not automatically become more productive with new or increased numbers of markers (e.g. Cushman 2014). Indeed, it is just as easy, though perhaps more expensive, to implement a poor sampling design using next-generation technologies as it is using current genetic markers. In fact, for many neutral processes, the statistical stability of population genetic parameters is asymptotic in the number of markers used (see, for example, Chakraborty et al. 1988; Boecklen & Howard 1997). As such, having broad coverages is much less statistically important than having large sample sizes in studies of neutral genetic structure, whereas coverage of thousands of loci may be beneficial is when one is interested in identifying potentially adaptive genomic regions. If one knows the genes of interest, the "candidate gene" approach, you do not need a lot of markers. However, if the goal is to identify potential genomic regions under selection, these so-called "genome scan" approaches increase in efficiency with the addition of more scorable markers. This is why the vast majority of gene scan studies to date have been conducted using techniques such as AFLPs, which can produce thousands of markers. Perhaps not surprisingly, some of the most exciting emerging techniques using next-generation marker technology are extending AFLP-like approaches (see Parchman et al. 2012 for one application).

While this chapter focuses on neutral population genetics, attention to covarying processes that influence the spatial distribution of genetic variance are just as salient to analyses of adaptive genetic variance as well (e.g., Prunier et al. 2012; Chapter 9), though rarely are their relative effects factored out. Of particular concern are methods that rely upon either correlations with environmental gradients and/or outlier approaches such as genome scans. While it is true that if positive selection were acting upon a genomic region, allelic variants may exhibit frequency gradients, it is not the only reason; nor is it the most parsimonious one to describe this observed pattern. As shown in the *E. lomelii* example above, entirely neutral genetic processes can lead to spurious correlations with topological and environmental gradients. Genome scans rely upon examining numerous loci (the more the better), looking for those that are significantly more differentiated than the rest of the genome. These outlier loci are thought to be "putatively adaptive", though subsequent sequencing and experimentation are necessary to validate (e.g. Cushman 2014). Both strong spatial phylogeographic patterns and landscape interactions with contemporary landscape features will increase the magnitude of genetic structure across the entire genome, thereby increasing the background structure level and potentially masking outlier loci.

Work in human genetics has shown that once demographic history is factored out, there are very few differences among populations that can be attributed directly to positive selection (e.g., Coop et al. 2009). While a similar level of scrutiny has yet to be applied to plant systems, or any other non-human system for that matter due to the relative depth of genetic resources available, there are several insights that we can harvest from that work to guide experimental design. First, the spatial distribution of adaptive genetic variance is not solely dictated by selective pressures. The spatial distribution of genetic variance, even adaptive variance, is constrained by natural historical and ongoing processes, upon which selection can operate. As a consequence, it is critical that neutral processes be both characterized and iterated out of the analyses prior to searches for adaptive patterns. Second, despite the ubiquity of putatively adaptive genomic regions and their associated linkage neighborhoods, many of the differences found are subtle, exhibiting restricted ranges of differentiation along environmental gradients rather than fixed differences. As a consequence, statistical power, a surrogate for efficient and effective sampling regimes, is going to play a large role in our ability to identify and characterize the presence of adaptation.

One of the most visible characteristics of landscape genetic analyses in the last decade has been the full-scale adoption of isolation-by-resistance (IBR) approaches, stemming in large part from McRae (2006). These IBR approaches, either as least-cost paths (e.g. Cushman et al. 2006) or algorithmic

approaches such as circuitscape (e.g. McRae 2006), extend Wright's classic isolation-by-distance (IBD) paradigm (Wright 1943) and apply it to non-Euclidean physical separation. Given the familiarity with the general approach for most researchers, extending this by analysis along ecological axes has become common, though two factors are limiting. First, for all but the most general large-scale studies, high-quality spatial data are not widely available. There are not many spatially-explicit coverages available to the average researcher at this time. The WorldClim group (Hijmans et al. 2005) continues to provide broad-scale climatic and elevation data found in increasing numbers of manuscripts. Localized spatial data, made available by Federal, State, and local authorities are also beginning to become commonly available. As we look forward, increased specificity will become available through remote sensing technologies such as LiDAR (Hyde et al. 2005) and hyperspectral analysis (Woodcock et al. 1994) that facilitate high-resolution reconstructions of habitat structure. Another limitation in the use of spatially-explicit predictors data is the widespread inability of the average practitioner to physically manipulate spatial data directly with the minimal amount of specialized training. Traditionally, individuals whose skill set included both wet laboratory genetic skills and GIS instruction were exceedingly rare (and rather busy). However, technological changes and the increase in available spatial and genetic quantitative tools such as R (R Core Team 2012), GrassGIS (GRASS Development Core Team 2012), and the movement of ArcGIS (ESRI 2011) from a unix platform to the desktop have supported a population of researchers whose size is ever increasing.

Of the three general components of landscape genetic analysis, genetic data, spatial/ecological data, and the analyses that bind them together, the last has seen the least amount of advancement. Progress in statistical approaches has been more evolutionary than revolutionary, with much focus on refining tools that we are already comfortable using. For example, consider one of the main statistical approaches associated with understanding relationships between genetic and spatial and/or ecological distances, the venerable Mantel test. It was introduced as a way to estimate the degree of correlation between temporal and spatial separation of disease clusters (Mantel 1967). Citations on this approach are growing exponentially in the non-biomedical life sciences as well as in relation to landscape genetics. In the interim, it has been turned into a regression model for partial correlations (Smouse et al. 1986), extended as a non-linear approach (Clarke & Warwick 2001), been compared somewhat favorably to other methods (Borcard et al. 1992; Balkenhol et al. 2009; Cushman & Landguth 2010), criticized for its inefficiency (Dutilleul et al. 2000; Legendre 2000), found to be adequate (Castellano & Balletto 2002), and then subsequently lacking (Legendre & Fortin 2010) in statistical power, and extended into multivariate optimization approaches (Shirk et al. 2010; Cushman et al. 2013). At the end of the day after, however, we are still just asking about a correlation. In a similar way, we have seen refinements on other standard measures such as the multitude of variations for $F_{ST}$ (Weir and Cockerham 1984; Excoffier et al. 1992; Hedrick 2005; Jost 2008), all of which are perhaps just "better hammers" and not new tools.

While our existing tools are becoming more refined, allowing us to make more precise biological inferences, continued refinement of the same tools can only go so far. At some point, we need to evaluate whether we are actually asking the ecological and evolutionary questions we want to ask or are we just addressing the hypotheses that we know how to answer? Kingman's (1982) introduction of coalescent theory revolutionized evolutionary biology and opened up hypotheses and inferences that were not previously addressable using current models (see Nielsen and Beaumont 2009). At the intersection of population genetic and landscape ecology, are there novel tools that will spark innovative approaches to understanding how organisms interact with their environment or are we resigned to only make only incremental refinements on what we already have?

## REFERENCES

Adams, W.T. & Birkes, D.S. (1991). Estimating mating patterns in forest tree populations. In: Fineschi, S., Malvolti, M.E., Cannata, F., & Hattemer, H.H. (eds.), *Biochemical Markers in the Population Genetics of Forest Trees*, pp. 157–72. SPB Academic Publishing, The Hague.

Agnihorti, M.S. & Singh, B.S. (1975) Pollen production and allergenic significance of some grasses around Lucknow. *Journal of Palynology* **11**, 151–4.

Aldrich, P.R., Hamrick, J.L., Chavarriaga, P., & Kochert, G. (1998) Microsatellite analysis of demographic genetic structure in fragmented populations of the tropical tree *Symphonia globulifera*. *Molecular Ecology* **7**, 933–44.

Anderson, J.T., Willis, J.H., & Mitchell-Olds, T. (2011) Evolutionary genetics of plant adaptation. *Trends in Ecology and Evolution* **27**, 258–66.

Austerlitz, F. & Smouse, P.E. (2001) Two-generation analysis of pollen flow across a landscape. II. Relation between $\Phi_{ft}$ pollen dispersal and interfemale distance. *Genetics* **157**, 271–80.

Baker, S.A. & Dyer, R.J. (2011) Invasion genetics of *Microstegium vimineum* (Poaceae) within the James River Basin of Virginia, USA. *Conservation Genetics* **12**, 793–803.

Balkenhol, N., Waits, L.P., & Dezzani, R. (2009) Statistical approaches in landscape genetics: an evaluation of methods for linking landscape and genetic data. *Ecography* **32**, 818–30.

Bazzaz, F.A. (1991) Habitat selection in plants. *The American Naturalist* **137**, S116–S130.

Bera, S.K. (1990) Palynology of *Shorea robusta* (Dipterocarpaceae) in relation to pollen production and dispersal. *Granna* **29**, 251–5.

Boecklen, W.J. & Howard, D.J. (1997) Genetic analysis of hybrid zones: numbers of markers and power of resolution. *Ecology* **78**, 2611–16.

Borcard, D. & Legendre, P. (2002) All-scale spatial analysis of ecological data by means of principal coordinates of neighbour matrices. *Ecological Modeling* **153**, 51–68.

Borcard, D., Legendre, P., & Drapeau, P. (1992) Partialling out the spatial component of ecological variation. *Ecology* **73**, 1045–55.

Burczyk, J. (1998) Mating system variation in a Scots pine clonal seed orchard. *Silvae Genetica* **47**, 155–8.

Campbell, D.R. (1991) Comparing pollen dispersal and gene flow in a natural population. *Evolution* **45**, 1965–8.

Castaing, J.P. & Vergeron, P. (1974) Principes et méthodes d'étude experimentale de la disperson du pollen de pin maritime dans le massif lanais. *Pollen et Spores* **15**, 255–80.

Castellano, S. & Balletto, E. (2002) Is the partial Mantel test inadequate? *International Journal of Organismal Evolution* **56**, 1871–3.

Chakraborty, R., Meagher, T.R., & Smouse, P.E. (1988) Parentage analysis with genetic markers in natural populations. I. The expected proportion of offspring with unambigious paternity. *Genetics* **118**, 527–36.

Clarke, K.R. & Warwick, R.M. (2001) *Changes in Marine Communities: An Approach to Statistical Analysis and Interpretation*, 2nd edition. PRIMER-E, Plymouth.

Coop, G., Pickrell, J.K., Novembre, J., Kudaravalli, S., Li, J., Absher, D., Myers, R.M., Cavalli-Sforza, L.L., Feldman, M.W., & Pritchard, J.K. (2009) The role of geography in human adaptation. *PLoS Genetics* **5**, 1–16.

Coster, S.S. & Kovach, A.I. (2012) Anthropogenic influences on the spatial genetic structure of black bears. *Conservation Genetics* **13**, 1247–57.

Cushman, S.A. (2014) 'Grand challenges in evolutionary and population genetics: the importance of integrating epigenetics, genomics, modeling and experimentation.' *Frontiers in Genetics*, **5**, 197.

Cushman, S.A. & Landguth, E.L. (2010) Spurious correlations and inference in landscape genetics. *Molecular Ecology* **19**, 3592–602.

Cushman, S.A., Wasserman, T.N., Landguth, E.L. & Shirk, A.J. (2013) 'Re-evaluating causal modeling with Mantel tests in landscape genetics.' *Diversity*, **5**, 51–72.

Cushman, S.A. & Lewis, J.S. (2010) Movement behavior explains genetic differentiation in American black bears. *Landscape Ecology* **25**, 1613–25.

Cushman, S.A., McKelvey, K.S., Hayden, J., & Schwartz, M.K. (2006) Gene flow in complex landscapes: testing multiple hypotheses with causal modeling. *The American Naturalist* **168**, 486–99.

Davey, J.W., Hohenlohe, P.A., Etter, P.D., Boone, J.Q., Catchen, J.M., & Blaxter, M.L. (2011) Genome-wide genetic marker discovery and genotyping using next-generation sequencing. *Nature Reviews Genetics* **12**, 499–510.

Davidson, W.S. (2012) Adaptation genomics: next generation sequencing reveals a shared haplotype for rapid early development in geographically and genetically distant populations of rainbow trout. *Molecular Ecology* **21**, 219–22.

Dutilleul, P., Stockwell, J.D., Frigon, D., & Legendre, P. (2000) The Mantel test versus Pearson's correlation analysis: assessment of the differences for biological and environmental studies. *Journal of Agricultural, Biological and Environmental Statistics* **5**, 131–50.

Dyer, R.J. & Nason, J.D. (2004) Population graphs: the graph theoretic shape of genetic structure. *Molecular Ecology* **13**, 1713–27.

Dyer, R.J. & Sork, V.L. (2001) Pollen pool heterogeneity in shortleaf pine, *Pinus echinata* Mill. *Molecular Ecology* **10**, 859–66.

Dyer, R.J., Nason, J.D., & Garrick, R.C. (2010) Landscape modeling of gene flow: improved power using conditional genetic distance derived from the topology of population networks. *Molecular Ecology* **19**, 3746–59.

Dyer, R.J., Westfall, R.D., Sork, V.L., & Smouse, P.E. (2004) Two-generation analysis of pollen flow across a landscape. V: a stepwise approach tor extracting factors contributing to pollen structure. *Heredity* **92**, 204–11.

Dyer, R.J., Chan, D.M., Gardiakos, V.A., & Meadows, C.A. (2012) Pollination graphs: quantifying pollen pool covariance networks and the influence of intervening landscapes on genetic connectivity in the North American understory tree, *Cornus florida* L. *Landscape Ecology* **27**, 239–51.

Elmer, K.R. & Meyer, A. (2011) Adaptation in the age of ecological genomics.' insights from parallelism and convergence. *Trends in Ecology and Evolution* **26**, 298–306.

Emerson, K.J., Merz, C.R., Catchen, J.M., Hohenlohe, P.A., Cresko, W.A., Bradshaw, W.E., & Holzapfel, C.M. (2010) Resolving postglacial phylogeography using high-throughput sequencing. *Proceedings of the National Academy of Sciences of the United States of America* **107**, 16196–200.

Endler, J. (1973) Gene flow and population differentiation. *Science* **19**, 243–50.

Epps, C.W., Wehausen, J.D., Bleich, V.C., Torres, S.G., & Brashares, J.S. (2007) Optimizing dispersal and corridor models using landscape genetics. *Journal of Applied Ecology* **44**, 714–24.

ESRI (2011) ArcGIS Desktop: Release 10. Environmental Systems Research Institute, Redlands, California.

Excoffier, L., Smouse, P.E., & Quattro, J.M. (1992) Analysis of molecular variance inferred from metric distances among DNA haplotypes: application to human mitochondrial DNA restriction data. *Genetics* **131**, 479–91.

Fore, S.A., Hickey, R.J., Vankat, J.L., Guttman, S.I., & Schaefer, R.L. (1992) Genetic structure after forest fragmentation – a landscape ecology perspective on *Acer saccharum*. *Canadian Journal of Botany – Revue Canadienne de Botanique* **70**, 1659–68.

Forget, P.M. & Cuijpers, L. (2008) Survival and scatterhoarding of frugivores-dispersed seeds as a function of forest disturbance. *Biotropica* **40**, 380–5.

Fortuna, M.A., Albaladejo, R.G., Fernández, L., Aparicio, A., & Bascompte, J. (2009) Networks of spatial genetic variation across species. *Proceedings of the National Academy of Sciences* **106**, 19044–9.

Funk, W.C., McKay, J.K., Hohenlohe, P.A., & Allendorf, F.W. (2012) Harnessing genomics for delineating conservation units. *Trends in Ecology and Evolution* **27**, 489–96.

García-Ramos, G. & Rodríguez, D. (2002) Evolutionary speed of species invasions. *Evolution* **56**, 661–8.

Garrick, R.C., Nason, J.D., Meadows, C.A., & Dyer, R.J. (2009) Not just vicariance: phylogeography of a Sonoran Desert euphorb indicates a major role of range expansion along the Baja Peninsula. *Molecular Ecology* **18**, 1916–31.

Garrick, R.C., Nason, J.D., Fernéndez-Manjarrés, J.F., & Dyer, R.J. (2013) Ecological coassociations influence species' responses to past climatic change: an example from a Sonoran Desert bark beetle. *Molecular Ecology* **22**, 3345–61.

Garroway, C.J., Bowman, J., & Wilson, P.J. (2011) Using a genetic network to parameterize a landscape resistance surface for fishers, *Martes pennanti*. *Molecular Ecology* **20**, 3978–88.

Godoy, J.A. & Jordano, P. (2001) Seed dispersal by animals: exact identification of source trees with endocarp DNA microsatellites. *Molecular Ecology* **10**, 2275–83.

Gram, W.K. & Sork, V.L. (2001) Association between environmental and genetic heterogeneity in forest tree populations. *Ecology* **82**, 2012–21.

GRASS Development Core Team (2012) Geographic Resource Analysis Support System (GRASS) Software, Open Source Geospatial Foundation Project,.URL http://grass.osgeo.org.

Halverson, K., Heard, S.B., Nason, J.D., & Stireman, J.O. (2008) Origins, distribution and local co-occurrence of polyploid cytotypes in *Solidago altissima* (Asteraceae). *American Journal of Botany* **95**, 50–8.

Hanski, I. & Simberloff, D. (1997) The metapopulation approach, its history, conceptual domain and application to conservation. *Metapopulation Biology* **124**, 40–4.

Hartmann, S., Nason, J.D., & Bhattacharya, D. (2001) Extensive ribosomal DNA genic variation in the columnar cactus *Lophocereus*. *Journal of Molecular Evolution* **53**, 124–34.

He, R.F., Kim, M.-J., Nelson, W., Balbuena, T.S., Kim, R., Kramer, R., Crow, J.A., May, G.D., Thelen, J.J., Soderlund, C.A., & Gang, D.R. (2012) Next-generation sequencing-based transcriptomic and proteomic analysis of the common reed, *Phragmites australis* (Poaceae), reveals genes involved in invasiveness and rhizome specificity. *American Journal of Botany* **99**, 232–47.

Hedrick, P.W. (2005) A standardized genetic differentiation measure. *Evolution* **59**, 1633–8.

Hijmans, R.J., Cameron, S.E., Parra, J.L., Jones, P.G., & Jarvis, A. (2005) Very high resolution interpolated climate surfaces for global land areas. *International Journal of Climatology* **25**, 1965–78.

Holderegger, R. & Wagner, H.H. (2008) Landscape genetics. *BioScience* **58**, 199–207.

Holderegger, R., Buehler, D., Gugerli, F., & Manel, S. (2010) Landscape genetics of plants. *Trends in Plant Science* **15**, 675–83.

Hyde, P., Dubayah, R., Peterson, B., Blair, J.B., Hofton, M., Hunsaker, C., Knox, R., & Walker, W. (2005) Mapping forest structure for wildlife habitat analysis using waveform lidar: validation of montane ecosystems. *Remote Sensing of Environment* **96**, 427–37.

Irwin, A.J., Hamrick, J.L., Godt, M.J.W., & Smouse, P.E. (2003) A multiyear estimate of the effective pollen donor pool for *Albizia julibrissin*. *Heredity* **90**, 187–94.

Jansen, P.A., Bongers, F., & Hemerik, L. (2004) Seed mass and mast seeding enhance dispersal by a neotropical scatter-hoarding rodent. *Ecological Monographs* **74**, 569–89.

Jordano, P., García, C., Godoy, J.A., & García-Castaño, J.L. (2007) Differential contribution of frugivores to complex seed dispersal patterns. *Proceedings of the National Academy of Sciences of the United States of America* **104**, 3278–82.

Jost, L. (2008) GST and its relatives do not measure differentiation. *Molecular Ecology* **17**, 4015–26.

Kingman, J.F.C. (1982) The coalescent. *Stochastic Processes and Their Applications* **13**, 235–48.

Knight, T.M., McCoy, M.W., Chase, J.M., McCoy, K.A., & Holt, R.D. (2005) Trophic cascades across ecosystems. *Nature* **437**, 880–3.

Knowles, L.L. & Maddison, W.P. (2002) Statistical phylogeography. *Molecular Ecology* **11**, 2623–35.

Kwit, C., Levey, D.J., & Greenberg, C.H. (2004) Contagious seed dispersal beneath heterospecific fruiting trees and its consequences. *Oikos* **107**, 303–8.

Latch, E.K., Boarman, W.I., Walde, A., Fleischer, RC. (2011) Fine-scale analysis reveals cryptic landscape genetic structure in desert tortoises. *PLoS One* **6**, e27794.

Latta, R.G. & Mitton, J.B. (1997) A comparison of population differentiation across four classes of gene marker in limber pine (*Pinus flexilis* James). *Genetics* **146**, 1153–63.

Ledig, F.T. (1992) Human impacts on genetic diversity in forest ecosystems. *Oikos* **63**, 87–108.

Legendre, P. (2000) Comparison of permutation methods for the partial correlation and partial mantel tests. *Journal of Statistical Computation and Simulation* **67**, 37–73.

Legendre, P. & Fortin, M.-J. (2010) Comparison of the Mantel test and alternative approaches for detecting complex multivariate relationships in the spatial analysis of genetic data. *Molecular Ecology Resources* **10**, 831–44.

Lesser, M.R. & Jackson, S.T. (2013) Contributions of long-distance dispersal to population growth in colonising *Pinus ponderosa* populations. *Ecology Letters* **16**, 380–9.

Lichstein, J.W. (2007) Multiple regression on distance matrices: a multivariate spatial analysis tool. *Plant Ecology* **188**, 117–31.

Manel, S., Schwartz, M.K., Luikart, G., & Taberlet, P. (2003) Landscape genetics: combining landscape ecology and population genetics. *Trends in Ecology and Evolution* **18**, 189–97.

Manel, S. & Holderegger, R. (2013) Ten years of landscape genetics. *Trends in Ecology and Evolution* **28**, 614–21.

Manly, B.J.F. (1986) Randomization and regression methods for testing for associations with geographical, environmental and biological distances between populations. *Research in Population Ecology* **28**, 201–18.

Mantel, N. (1967) The detection of disease clustering and a generalized regression approach. *Cancer Research* **27**, 209–20.

Martins, K., Raposo, A., Klimas, C.A., Veasey, E.A., Kainer, K., & Wadt, L.H.O. (2012) Pollen and seed flow patterns of *Crapa guianensis* Aublet (Meliaceae) in two types of Amazonian forests. *Genetics and Molecular Biology* **35**, 818–26.

McRae, B.H. (2006) Isolation by resistance. *Evolution* **608**, 1551–61.

McRae, B.H. & Beier, P. (2007) Circuit theory predicts gene flow in plant and animal populations. *Proceedings of the National Academy of Sciences of the United States of America* **104**, 19885–90.

Meagher, T.R. (1986) Analysis of paternity within a natural population of *Chamaelirium luteum*. *The American Naturalist* **128**, 199–215.

Meagher, T.R. & Thomson, E. (1987) Analysis of parentage for naturally established seedlings of *Chamaelirium luteum* (Liliaceae). *Ecology* **68**, 803–12.

Mitton, J.B. (1992) The dynamic mating systems of conifers. *New Forests* **6**, 197–216.

Mitton, J.B., Kreiser, B.R., & Latta, R.G. (2000) Glacial refugia of limber pine (*Pinus flexilis* James) inferred from the population structure of mitochondrial DNA. *Molecular Ecology* **9**, 91–7.

Molina, R.T., Rodríguez, A.M., Palacios, I.S., & López, F.G. (1996) Pollen production in anemophilus trees. *Grana* **35**, 38–46.

Moritz, C., Hoskin, C.J., MacKenzie, J.B., Phillips, B.L., Tonione, M., Silva, N., van der Wal, J., Williams, S.E., & Graham, C.H. (2009) Identification and dynamics of a cryptic suture zone in tropical rainforest. *Proceedings of the Royal Society B – Biological Sciences* **276**, 1235–44.

Murphy, M.A., Dezzani, R.J., Pilliod, D.S., & Storfer, A. (2010) Landscape genetics of high mountain frog metapopulations. *Molecular Ecology* **19**, 3634–49.

Nason, J.D. & Hamrick, J.L. (1997) Reproductive and genetic consequences of forest fragmentation: two case studies of neotropical canopy trees. *Journal of Heredity* **88**, 264–76.

Nason, J.D., Hamrick, J.L., & Fleming, T.H. (2002) Historical vicariance and postglacial colonization effects on the evolution of genetic structure in Lophocereus, a Sonoran Desert columnar cactus. *Evolution* **56**, 2214–26.

Neigel, J.E. (1997) A comparison of alternative strategies for estimating gene flow from genetic markers. *Annual Review of Ecology and Systematics* **28**, 105–28.

Nielsen, R. & Beaumont, M.A. (2009) Statistical inferences in phylogeography. *Molecular Ecology* **18**, 1034–47.

Okubo, A. & Levin, S.A. (1989) A theoretical framework for data analysis of wind dispersal of seeds and pollen. *Ecology* **70**, 329–38.

Ortego, J., Aguirre, M.P., & Cordero, P.J. (2012) Landscape genetics of a specialized grasshopper inhabiting highly fragmented habitats: a role for spatial scale. *Diversity and Distributions* **18**, 481–92.

Papa, R. & Gepts, P. (2003) Asymmetry of gene flow and differential geographical structure of molecular diversity in wild and domesticated common bean (*Phaseolus vulgaris* L.) from Mesoamerica. *Theoretical and Applied Genetics* **106**, 239–50.

Parchman, T.L., Gompert, Z., Mudge, J., Schilkey, F.D., Benkman, C.W., & Buerkle, C.A. (2012) Genome-wide association genetics of an adaptive trait in lodgepole pine. *Molecular Ecology* **21**, 2991–3005.

Pluess, A.R. (2011) Pursuing glacier retreat: genetic structure of a rapidly expanding *Larix decidua* population. *Molecular Ecology* **20**, 473–85.

Price, M.V. & Waser, N.M. (1982) Experimental studies of pollen carryover: hummingbirds and Ipomopsis aggregata. *Oecologia* **54**, 353–8.

Provan, J., Powell, W., & Hollingsworth, P.M. (2001) Chloroplast microsatellites: new tools for studies in plant ecology and evolution. *Trends in Ecology and Evolution* **16**, 142–7.

Prunier, J., Gerardi, S., Laroche, J., Beaulieu, J. & Bousquet, J. (2012) Parallel and lineage-specific molecular adaptation to climate in boreal black spruce. *Molecular Ecology* **21**, 4270–86.

R Core Team (2012) R: a language and environment for statistical computing. R Foundation for Statistical Computing, Vienna, Austria. URL http://www.R-project.org/.

Rice, A.M., Rudh, A., Ellegren, H., & Qvarnstrom, A. (2011) A guide to the genomics of ecological speciation in natural animal populations. *Ecology Letters* **14**, 9–18.

Riddle, B.R., Hafner, D.J., Alexander, L.F., & Jaeger, J.R. (2000) Cryptic vicariance in the historical assembly of a Baja California Peninsular Desert biota. *Proceedings of the National Academy of Science of the United States of America* **97**, 14438–43.

Rogers, D.L., Millar, C.I., & Westfall, R.D. (1999) Fine-scale genetic structure of whitebark pine (*Pinus albicaulis*): associations with watershed and growth form. *Evolution* **53**, 74–90.

Row, J.R., Gomez, C., Koen, E.L., Bowman, J., Murray, D.L., & Wilson, P.J. (2012) Dispersal promotes high gene flow among Canada lynx populations across mainland North America. *Conservation Genetics* **13**, 1259–68.

Schuster, S.F.W. & Mitton, J.B. (2000) Paternity and gene dispersal in limber pine (*Pinus flexilis* James). *Heredity* **84**, 348–61.

Shirk, A.J. & Cushman, S.A. (2014) 'Spatially-explicit estimation of Wright's neighborhood size in continuous populations.' *Frontiers in Evolutionary and Population Genetics* **2**: 62.

Shirk A., Wallin, D.O., Cushman, S.A., Rice, R.C. & Warheit, C. (2010) 'Inferring landscape effects on gene flow: A new multi-scale model selection framework.' *Molecular Ecology*, **19**, 3603–19.

Slatkin, M. (1985a) Gene flow in natural populations. *Annual Review of Ecology and Systematics* **16**, 393–430.

Slatkin, M. (1985b) Rare alleles as indicators of gene flow. *Evolution* **39**, 53–65.

Smouse, P.E. & Peakall, R. (1999) Spatial autocorrelation analysis of individual multiallele and multilocus genetic structure. *Heredity* **82**, 561–73.

Smouse, P., Long, J.C., & Sokal, R.R. (1986) Multiple regression and correlation extension of the Mantel test of matrix correspondence. *Systematic Zoology* **35**, 627–32.

Smouse, P.E., Dyer, R.J., Westfall, R.D., & Sork, V.L. (2001) Two-generation analysis of pollen flow across a landscape I: Male gamete heterogeneity among females. *Evolution* **55**, 260–71.

Sokal, R.R., Smouse, P.E., & Neel, J.V. (1986) The genetic structure of a tribal population, the Yanomama Indians. XV. Patterns inferred by autocorrelation analysis. *Genetics* **114**, 259–87.

Sorensen, F.C. (1994) Frequency of seedlings from natural self-fertilization in Pacific Northwest Ponderosa pine (*Pinus ponderosa* Dougl. ex Laws). *Silvae Genetica* **43**, 100–8.

Sork, V.L., Campbell, D., Nason, J., & Fernandez, J.F. (1999) Landscape approaches to historical and contemporary gene flow in plants. *Trends in Ecology and Evolution* **14**, 219–23.

Sork, V.L., Davis, F.W., Smouse, P.E., Apsit, V., Dyer, R.J., & Fernandez, J.F. (2002) Pollen movement in declining population of California Valley oak, *Quercus lobata*: Where have all the fathers gone? *Molecular Ecology* **11**, 1657–68.

Spear, S.F., Peterson, C.R., Matocq, M.D., & Storfer, A. (2005) Landscape genetics of the blotched tiger salamander (*Ambystoma tigrinum melanostictum*). *Molecular Ecology* **14**, 2553–64.

Stapley, J., Reger, J., Feulner, P.G.D., Smadja, C., Galindo, J., Ekblom, R., Bennison, C., Ball, A.D., Beckermann, A.P., & Slate, J. (2010) Adaptation genomics: the next generation. *Trends in Ecology and Evolution* **25**, 705–12.

Storfer, A., Murphy, M.A., Spear, S.F., Holderegger, R., & Waits, L.P. (2010) Landscape genetics: Where are we now? *Molecular Ecology* **19**, 3496–514.

Streiff, R., Ducousso, A., Lexer, C., Steinkellner, H., Gloessl, J., & Kremer, A. (1999) Pollen dispersal inferred from paternity analysis in a mixed oak stand of *Quercus robur* L. & *Q. petraea* (Matt.) Liebl. *Molecular Ecology* **8**, 831–41.

Thomson, J.D. (1978) Effects of stand composition on insect visitation in two-species mixtures of *Hieracium*. *American Midland Naturalist* **100**, 431–40.

Twyford, A.D. & Ennos, R.A. (2012) Next-generation hybridization and introgression. *Heredity* **108**, 179–89.

van Devender, T.R., Burgess, T.L., Piper, J.C., & Turner, R.M. (1994) Paleoclimatic implications of holocene plant remains from the Sierra Bacha, Sonora, Mexico. *Quaternary Research* **41**, 99–108.

Wasserman, T.N., Cushman, S.A., Shirk, A.S., Landugth, E.L., & Littell, J.S. (2012) 'Simulating the effects of climate change on population connectivity of American marten (*Martes americana*) in the northern Rocky Mountains, USA.' *Landscape Ecology*, **27**, 211–25.

Watkins, L. & Levin, D.A. (1990) Outcrossing rates as related to plant density in *Phlox drummondii*. *Heredity* **65**, 81–9.

Weir, B.S. & Cockerham, C.C. (1984) Estimating F-statistics for the analysis of population structure. *Evolution* **38**, 1358–70.

Woodcock, C.E., Collins, J.B., Gopal, S., Jakabhazy, V.D., Li, X., Macomber, S., Ryherd, S., Harward, V.J., Levitan, J., Wu, Y.C., & Warbington, R. (1994) Mapping forest vegetation using Landsat™ imagery and a canopy reflectance model. *Remote Sensing of Environment* **50**, 240–54.

Wright, S. (1943) Isolation-by-distance. *Genetics* **28**, 114–38.

Yang, J,. Cushman, S.A., Yang, J., Yang, M. & Bao, T. (2013) 'Effects of climatic gradients on genetic diversification of Caragana on the Ordos Plateau, China.' *Landscape Ecology*, **28**, 1729–41.

Yannic, G., Pellissier, L., Le Corre, M., Dussault, C., Bernatchez, L., & Côté, S.D. (2014) Temporal dynamic habitat suitability predicts genetic relatedness among caribou. *Proceedings of the Royal Society B – Biological Sciences* **218**, 1792.

Young, A.G., Merriam, H.G., & Warwick, S.I. (1993) The effects of forest fragmentation on genetic variation in *Acer saccharum* Marsh. (sugar maple) populations. *Heredity* **71**, 277–89.

Young, A.G., Boyle, T., & Brown, T. (1996) The population genetic consequences of habitat fragmentation for plants. *Trends in Ecology and Evolution* **11**, 413–18.

Zalewski, A., Piertney, S.B., Zalewska, H., & Lambin, X. (2009) Landscape barriers reduce gene flow in an invasive carnivore: geographical and local genetic structure of American mink in Scotland. *Molecular Ecology* **18**, 1601–15.

*Chapter 12*

# APPLICATIONS OF LANDSCAPE GENETICS TO CONNECTIVITY RESEARCH IN TERRESTRIAL ANIMALS

*Lisette P. Waits,[1] Samuel A. Cushman,[2] and Steve F. Spear[3]*

[1]*Fish and Wildlife Sciences, University of Idaho, USA*
[2]*Forest and Woodlands Ecosystems Program, Rocky Mountain Research Station, United States Forest Service, USA*
[3]*The Orianne Society, USA*

## 12.1 INTRODUCTION

This chapter focuses on applications of landscape genetics for understanding connectivity of terrestrial animal populations. We start with a general introduction covering unique characteristics and challenges of the terrestrial study system, followed by an overview of common research questions addressed in terrestrial animals using neutral markers in landscape genetics. These common research questions include detecting and defining barriers, identifying corridors, examining source-sink dynamics, and detecting and predicting animal responses to environmental change. Based on this overview, we highlight limitations of current research approaches and present case studies that have specifically dealt with some of these limitations when testing ecological hypotheses about gene flow in heterogeneous landscapes. Finally, we end the chapter with future directions and current knowledge gaps for terrestrial animal landscape genetic studies.

## 12.2 GENERAL OVERVIEW OF TERRESTRIAL ANIMAL STUDY SYSTEMS AND RESEARCH CHALLENGES

Landscape genetic studies have focused on terrestrial animals more than any other taxonomic group (Storfer et al. 2010). This is not surprising given the inclusion of "land" in the name of the discipline and the usefulness of landscape genetics as a method to infer animal movement, a key process in ecology and conservation. Many aspects of landscape genetic studies are well

*Landscape Genetics: Concepts, Methods, Applications*, First Edition. Edited by Niko Balkenhol, Samuel A. Cushman, Andrew T. Storfer, and Lisette P. Waits.
© 2016 John Wiley & Sons, Ltd. Published 2016 by John Wiley & Sons, Ltd.

suited for applications related to terrestrial animals. Landscape genetics usually requires a model of the continuous environment, and such GIS models are most widely available for terrestrial systems (Chapter 2). This has allowed for a more realistic representation of the heterogeneity present in terrestrial landscapes, whereas it is much more difficult to represent the heterogeneity in aquatic systems using typical remote sensing methods. The ability to incorporate existing GIS data into studies of animal population connectivity has benefited researchers greatly. In addition, landscape genetics has enabled effective research on animals that are secretive or difficult to study using traditional mark-recapture or radio telemetry methods (e.g., Schwartz et al. 2009). Furthermore, landscape genetic studies provide some of the best case studies of how contemporary habitat alteration is affecting population processes such as connectivity in terrestrial species, ranging from insects (e.g., Keyghobadi et al. 2005a), amphibians (e.g., Funk et al. 2005; Murphy et al. 2010a, 2010b; Spear et al. 2012), reptiles (e.g., Stow et al. 2001; Clark et al. 2010), birds (e.g., Pavlacky et al. 2009), and mammals (e.g., Epps et al. 2005; Vignieri et al. 2005; Wasserman et al. 2012).

### 12.2.1 Comparison of terrestrial animal studies with other systems

Landscape genetics research focused on terrestrial animals has a number of unique aspects compared to aquatic systems or terrestrial plant species. The terrestrial landscape can be depicted as a combination of linear elements, mosaics of categorical patch types (Forman 1995), and gradients of continuously varying ecological factors (McGarigal and Cushman 2005; Chapter 2). The patterns of these factors interact with organism behavior to affect population processes, such as mating and dispersal. The functional connectivity of a landscape depends on how the pattern of the environment interacts with behavioral responses and movement abilities of the species to drive population processes.

Terrestrial systems are generally depicted as two-dimensional patterns, which is in contrast to many marine and lacustrine systems, in which hydrodynamic fluxes occur in three dimensions, and riverine systems, which are often depicted as networks of one-dimensional links. This in principle can make analysis of terrestrial landscapes simpler than marine systems and more complicated than riverine networks, simply as a function of the dimensionality of the pattern–process relationships. Temporal dynamics also are very different between typical terrestrial and aquatic systems. Functional connectivity in terrestrial environments is often more constrained, occurs at a finer scale, and changes at a slower rate than connectivity across marine environments (Carr et al. 2003; also see Chapter 13). In many ways, this can make it easier to develop strong statistical models in terrestrial landscape genetic studies if sufficient spatial data are available, as the scale of sampling is more likely to match the scale of the gene flow process. In contrast, studies of seascape genetics often have detected very little genetic structure across broad areas (Selkoe et al. 2010; Schunter et al. 2011), due in part to the combination of long-distance transport of gametes and larva by currents and temporal dynamics that are rapid relative to population turnover. In the terrestrial system, the dynamics of landscape change are generally slower than population processes, enabling the effects of landscape change to be recorded in changes in the genetic structure of populations. However, in many instances rates of change in terrestrial landscapes exceeds the equilibration time of population genetic processes, requiring non-equilibrium methods to analyze current rates of gene flow.

The vast scales, high-frequency dynamics, and typically immense numbers of gametes and larva in marine systems create processes that are essentially "large number problems" in the sense of statistical mechanics and the Boltzmann equation (Boltzmann 1872), in which the overall statistical behavior of the entire system is highly predictable as the outcome of millions of microscale interactions. Terrestrial systems, however, are more often in the range of "middle number problems", where there are often too many interacting individuals and entities to effectively model the mechanistic interactions of the entire system, but too few to average away the individual interactions in a statistical mechanics approach, and dynamics are too slow to collapse them as "noise" and too frequent to ignore and model as a static system. As a result, terrestrial landscape genetic analysis is often a "complex systems" problem, in which researchers attempt to reduce complexity by adopting spatially-explicit approaches to represent the location, movement, and interactions of individuals within populations as functions of the spatial structure of the landscape. This presents a number of conceptual and technical challenges for analysis. For instance, a researcher using resistance modeling (see Chapter 8) typically has to parameterize a multivariate resistance surface in a terrestrial system, whereas following oceanographic currents or stream lengths may be sufficient in an aquatic system (although many aquatic systems do have multiple environmental influences; Chapter 13). Finally, the slower rate of change in functional connectivity in the terrestrial environment introduces the challenge of accounting for temporal lags.

Animals and plants also differ with respect to landscape influence on genetic structure in several ways (also see Chapter 11). The most obvious difference is that individual plants are stationary whereas animals are mobile. The second major difference is that gene flow in terrestrial animals occurs through the transport of gametes by the individual, whereas movement of gametes in plants typically occurs independent of the individual plant through seed and pollen. Therefore, modeling plant gene flow might require a simultaneously modeling of both gametes. On the other hand, the field of adaptive landscape genetics is much more amenable to plants due to their sessile nature (see Chapter 9).

## 12.2.2 Methodological and study design challenges often faced by terrestrial researchers

An important methodological issue that is often faced by researchers investigating landscape genetics of terrestrial animals is whether to sample at an individual or population level (see Chapter 2). Historically, population-level sampling was typical due to its compatibility with population genetic theory, and because groups of individuals could often be tied to discrete terrestrial units, such as ponds or forest patches. However, many species do not occur in discrete population clusters and analysis at the individual level may be more appropriate because artificially imposing clusters may bias results (Schwartz & McKelvey 2009). Recent simulation and empirical work has also shown that population-based analysis may be replaced by individual-based analyses for organisms with patchy distributions, which can allow researchers to increase the number of patches sampled (Prunier et al. 2013). The choice of an individual or population level of analysis has important implications for sampling design and data analysis (Chapter 2 and Chapter 3). The number of terrestrial animal studies using individual-based analyses is increasing since this sampling approach provides a more comprehensive coverage of the landscape and thus better captures landscape heterogeneity. Another important consideration and challenge in sampling design is obtaining a balanced sample of male and female individuals. Multiple studies have demonstrated that habitat use and dispersal patterns can differ by sex and that the more philopatric sex is

often more strongly affected by land conversion, roads, and other forms of habitat fragmentation (e.g., Stow et al. 2001; Proctor et al. 2005; Amos et al. 2014; Paquette et al. 2014; Elliot et al. 2014a, 2014b).

The complex, often indirect, relationship of animal movement with gene flow is another challenge (Bohonak 1999). Animal movement occurs for multiple reasons, many of which are not related to gene flow. Thus, models used to account for landscape influence in a genetic study must recognize that only a subset of movements are relevant when using empirical movement data to identify key landscape variables (see Chapter 8). Furthermore, genetic data may not always be appropriate to direct conservation actions such as movement corridors if the types of movement the actions need to facilitate are not tied to mating or reproduction (i.e. migration). These concerns highlight the importance of understanding the study system and conducting landscape genetic studies with well-designed objectives that can investigate multiple processes.

When analyses are based on individuals, another challenge in many terrestrial animal studies is the fact that most animal species are not stationary, so that the location where an animal is sampled may not represent typical movements or actual places where genetic exchange has occurred (Graves et al. 2012). Thus, locations of genetic samples obtained for terrestrial animals are usually associated with a higher degree of spatial uncertainty than sampling locations for plants. Uncertainty in sampling locations is of the greatest concern in highly heterogeneous landscapes in which organisms might be briefly located or drawn to suboptimal habitats, a challenge that is particularly relevant for wide-ranging species such as large mammals. A simulation study in a binary landscape concluded that, in most cases, location uncertainty was unlikely to significantly alter landscape genetic results (Graves et al. 2012). Of course, the impacts of all of these issues are dependent on the scale of sampling, and on the spatial distribution and dispersal capabilities of the species under study. This is discussed in more detail elsewhere (see Chapter 2), but is an important consideration in many of the case studies we highlight.

## 12.2.3 Research focus of current studies

Most landscape genetic studies of terrestrial animals have focused on assessing the influence of landscape permeability on gene flow using neutral loci. Understanding landscape effects on movement and gene flow is crucial in ecology, evolution, and conservation, and landscape genetic approaches are particularly well suited for understanding effective dispersal (Broquet & Petit 2009; Baguette et al. 2013). Consequently, typical landscape genetics research questions for connectivity of terrestrial animals include detecting barriers, identifying corridors, examining population dynamics, and predicting the response to environmental change. In the following sections, we present an overview of representative studies that have addressed these broad questions. We chose not to conduct a metareview of all landscape genetic papers, as was done in Storfer et al. (2010), but instead highlight examples across multiple taxonomic groups of terrestrial animals.

## 12.3 DETECTING BARRIERS AND DEFINING CORRIDORS

### 12.3.1 Detecting barriers

The detection of landscape barriers using population genetic data has been a long-standing goal in landscape genetics (Storfer et al. 2007, 2010). Multiple studies have evaluated the effectiveness of different methodological approaches for detecting barriers (Landguth et al. 2010;

Safner et al. 2011; Blair et al. 2012) including traditional population genetic metrics (Chapter 3), clustering and assignment methods (Chapter 7), and boundary detection methods (Chapter 3). The most powerful and accurate methods detected in these simulation studies were Bayesian clustering methods (Safner et al. 2011; Blair et al. 2012) and Mantel tests of individual-based genetic distance (Landguth et al. 2010). Overall, these methods were particularly successful when effective population size was low, the species had limited vagility, and the barrier had restricted gene flow for multiple generations.

Landscape genetic studies of terrestrial wildlife have identified a number of both natural and anthropogenic barriers to movement and gene flow (Table 12.1). Across taxonomic groups of species, rivers have frequently been identified as a barrier (Blanchong et al. 2008; Côté et al. 2012; Lugon-Moulin & Hausser 2002; Coulon et al. 2006; Mockford et al. 2007; Cullingham et al. 2009; Robinson et al. 2012). Major ridgelines have been identified as barriers to gene flow in amphibians (Funk et al. 2005; Murphy et al. 2010b), snakes (Manier and Arnold 2006), turtles (Mockford et al. 2007) and small mammals (Zalewski et al. 2009). Conversion of grasslands and forest to agriculture and grazing has been identified as a barrier to gene flow for insects (Marchi et al. 2013), birds (Lindsay

**Table 12.1** A subset of landscape genetic studies that have detected barriers for terrestrial animals.

| Citation | Species | Unit of analysis* | Inference or conclusions |
|---|---|---|---|
| Banks et al. (2005) | *Antechinus agilis* | ind, pop | Pine plantations barrier to gene flow |
| Blanchong et al. (2008) | *Odocoileus virginianus* | pop | River was barrier to gene flow |
| Bush et al. (2011) | *Centrocercus urophasianus* | ind | River and agricultural region were barriers to gene flow |
| Carmichael et al. (2007) | *Canis lupus* | ind, pop | Genetic structure was explained by differences in habitat types |
| Côté et al. (2012) | *Procyon lotor* | pop, ind | River was partial barrier to gene flow |
| Coulon et al. (2006) | *Capreolus capreolus* | ind | Region with canals, highway, and river restricted gene flow |
| Cullingham et al. (2009) | *Procyon lotor* | ind | Rivers are barrier to gene flow and restricts spread of rabies |
| Cushman & Lewis (2010) | *Ursus americanus* | ind | Development, roads, and low forest cover restricted movements |
| Epps et al. (2005) | *Ovis canadensis nelsoni* | pop | Highways, canals and developed areas act as barriers to gene flow |
| Funk et al. (2005) | *Rana luteiventris* | pop | Ridgelines and elevational differences were barriers to gene flow |

**Table 12.1**   *(Continued)*

| Citation | Species | Unit of analysis* | Inference or conclusions |
|---|---|---|---|
| Geffen et al. (2004) | *Canis lupus* | pop | Differences in climate and habitat type restrict gene flow |
| Johansson et al. (2005) | *Rana temporaria* | pop | Agricultural fields were barrier to gene flow. |
| Kuehn et al. (2007) | *Capreolus capreolus* | ind, pop | Transportation network barrier to gene flow |
| Lindsay et al. (2008) | *Dendroica chrysoparia* | pop | Agricultural lands were barrier to gene flow. |
| Lugon-Moulin & Hausser (2002) | *Sorex araneus* | pop | Lowland habitat with rivers and human development restrict gene flow |
| Manier & Arnold (2006) | *Anaxyrus boreas, Thamnophis elegans, Thamnophis sirtalis* | pop | Escarpment was barrier for *T. elegans* and *T. sirtalis* |
| Marchi et al. (2013) | *Bembidion lampros* | pop | Agricultural fields were barrier to gene flow |
| Mockford et al. (2007) | *Emydoidea blandingii* | pop | Mountains and rivers are the strongest barriers to gene flow |
| Murphy et al. (2010b) | *Rana luteiventris* | pop | Gene flow restricted with increased topographic complexity |
| Paquette et al. (2014) | *Procyon lotor* | ind | Agricultural fields were barrier to gene flow for females. |
| Perez-Espona et al. (2008) | *Cervus elaphus* | pop | Greatest barrier was sea lochs and roads, while mountain slopes and forests also restricted gene flow |
| Proctor et al. (2005) | *Ursus arctos* | ind, pop | Highway and associated development barrier to gene flow |
| Radespiel et al. (2008) | *Microcebus ravelobensis* | ind, pop | Open savannah and a major road was a barrier |
| Riley et al. (2006) | *Canis latrans, Lynx rufus* | ind, pop | Highway barrier to gene flow in both species |
| Robinson et al. (2012, 2013) | *Odocoileus virginianus* | ind | Highways and rivers restrict movement of *Odocoileus virginianus* and spread of chronic wasting disease |
| Row et al. (2010) | *Mintoinus gloydi* | ind, pop | Agricultural lands were barrier to gene flow |
| Sacks et al. (2004) | *Canis latrans* | ind | Genetic structure was explained by difference in habitat types |
| Stow et al. (2001) | *Egernia cunninghami* | ind, pop | Forests cleared for grazing were a barrier to gene flow, with females more strongly affected than males |
| van der Wal et al. (2012) | *Cervus canadensis* | ind | Roads were a barrier to gene flow |
| Wasserman et al. (2010) | *Martes americana* | ind | Genetic structure was explained by elevation and hypothesized to relate to differences in vegetation and snowpack. |
| Zalewski et al. (2009) | *Neovison vison* | ind, pop | Mountain range was barrier to gene flow |

*ind, individual; pop, population.

et al. 2008; Bush et al. 2011), amphibians and reptiles (Stow et al. 2001; Johansson et al. 2005; Row et al. 2010), and mammals (Banks et al. 2005; Paquette et al. 2014). Unsuitable natural habitat can also be a significant barrier to gene flow and movement. For example, dry grassland habitat has been shown to fragment salamander populations (Rittenhouse & Semlitsch 2006), and savanna habitats can fragment lemur populations that depend on forested habitat (Radespiel et al. 2008). Anthropogenic landscape features such as roads and other human development have been identified as barriers across several taxonomic groups, including carnivores (Proctor et al. 2005; Riley et al. 2006; Cushman & Lewis 2010), ungulates (Epps et al. 2005; Coulon et al. 2006; Kuehn et al. 2007; Perez-Espona et al. 2008; Robinson et al. 2012) and amphibians (Murphy et al. 2010a). For example, highways were found to be the principle factor driving genetic differentiation in a population of mountain goats (Shirk et al. 2010).

In addition, climate gradients and habitat differences have been shown to be cryptic barriers to gene flow possibly because of species' environmental tolerance limits or natal imprinting (Sacks et al. 2004; Geffen et al. 2004; Carmichael et al. 2007; Wasserman et al. 2010). Cryptic climatic and habitat barriers are of elevated importance because of the difficulty in observing them and the likelihood that global scale changes in climatic and land cover will lead to major shifts in how these factors affect species distribution, abundance, genetic diversity and gene flow (e.g. Wasserman et al. 2012, 2013).

The identification of barriers to gene flow can also provide important information for the management of invasive species and disease in the terrestrial environment (Storfer et al. 2007, 2010; Segelbacher et al. 2010). Landscape genetic models can help predict the geographic nature of disease spread and barriers can quarantine disease within a limited geographic area (Rees et al. 2008; Biek & Real 2010). For example, highways have been shown to restrict the spread of chronic wasting disease in white-tailed deer (*Odocoileus virginianus*; Robinson et al. 2012, 2013) and the spread of bovine tuberculosis in elk (*Cervus Canadensis*; van der Wal et al. 2012). Rivers have been documented as barriers to the spread of chronic wasting disease in white-tailed deer (Blanchong et al. 2008; Robinson et al. 2013) and rabies in raccoons (Cullingham et al. 2009; Côté et al. 2012; Paquette et al. 2014).

### Evaluating and designing corridors

One important, but relatively rare, application of landscape genetic methods is to evaluate the effectiveness of current corridors (Table 12.2). In an early examination

**Table 12.2** Examples of studies that have used landscape genetic approaches to design or evaluate corridors for terrestrial animals.

| Citation | Species | Unit of analysis | Inference or conclusions |
|---|---|---|---|
| Angelone & Holderegger (2009) | *Hyla arborea* | population | Conservation measures have increased connectivity of breeding ponds within valleys, although one valley had greater connectivity |
| Braunisch et al. (2010) | *Tetrao urugallus* | individual | Used landscape genetic associations to design corridors based on groups of "MLP" paths |
| Cushman et al. (2009, 2013) | *Ursus americanus* | individual | Mapped potential corridors using a genetically optimized resistance surface |
| Dixon et al. (2006) | *Ursus americanus* | individual | Population assignments indicated asymmetric migration between two main patches |
| Epps et al. (2007) | *Ovis canadensis* | population | Incorporation of slope and anthropogenic barriers allow modeling of dispersal corridors that matched genetic migration estimates |
| Mech & Hallett (2001) | *Clethrionomys gapperi*, *Peromyscus maniculatus* | population | For *C. gapperi*, highest genetic distance in isolated sites, intermediate in sites connected by corridors, and lowest in continuous forest. No difference for *P. maniculatus* |
| Paetkau et al. (2009) | *Rattus fuscipes*, *Rattus leucopus* | individual | Use and occupation of the corridor was higher for *Rattus fuscipe* than for *Rattus leucopus* and movements were asymmetrical |
| Wells et al. (2009) | *Junonia coenia* | population | Higher genetic diversity and lower $F_{st}$ in connected patches |

of the effectiveness of forest corridors for small mammals, Mech and Hallett (2001) evaluated the genetic connectivity of red-backed vole (*Clethrionomys gapperi*) and deer mouse (*Peromyscus maniculatus*) populations in continuous forest, isolated forest patches, and forest patches connected by forested corridors. They found that genetic distance increased from smallest to largest in contiguous, corridor, and isolated landscapes for the red-backed vole, a closed-canopy specialist, but genetic distances for the deer mouse, a habitat generalist, were not significantly different across treatments. Dixon et al. (2006) used non-invasive genetic sampling of black bears to obtain microsatellite genotypes that could be used to evaluate the effectiveness of a regional corridor in Florida. Using assignment test approaches (Chapter 7), they demonstrated that dispersal and gene flow were occurring primarily south to north within the corridor. A similar approach was used to evaluate whether a newly established corridor in Australia was being utilized by two native small mammals, the bush rat (*Rattus fuscipes*) and the Cape York rat (*Rattus leucopus*) and revealed that use of the corridor was higher for the bush rat than for Cape York rat (Paetkau et al. 2009).

In a study of the declining European tree frog (*Hyla arborea*) in Switzerland, Angelone and Holderegger (2009) evaluated the effectiveness of conservation efforts to establish stepping stones of habitat (i.e., ponds) to provide structural and functional connectivity. The application of assignment test approaches indicated that dispersal and gene flow was occurring among the newly established ponds and establishing a network of interconnected populations. Genetic approaches have also been used to demonstrate the effectiveness of corridors for maintaining genetic diversity, gene flow, and fitness in an open-habitat specialist butterfly, *Junonia coenia* (Wells et al. 2009).

Many review papers have highlighted the potential of landscape genetic methods for designing corridors (Storfer et al. 2007, 2010; Segelbacher et al. 2010; Sork & Waits 2010), but relatively few empirical studies have been conducted (Table 12.2). A study of desert bighorn sheep (*Ovis canadensis nelsoni*) was one of the first to test the effectiveness of different least-cost GIS models and to identify dispersal corridors using the best-fitting models (Epps et al. 2005, 2007). Similar approaches were used to identify optimal corridor regions for black bears (*Ursus americanus*; Cushman et al. 2009, 2013), American marten (*Martes americana*; Wasserman et al. 2013), and wolverine (*Gulo*

*gulo*) in the northern Rocky Mountains of the United States (Schwartz et al. 2009). Another good example of landscape genetic approaches for defining corridors is the work of Braunisch et al. (2010) to quantify landscape permeability for capercaillie in the Black Forest of Germany. The authors calculated pairwise relatedness for 213 individuals and found that relatedness was positively correlated with the proportion of coniferous and mixed forest and negatively correlated to forest edges, roads, settlements, and agricultural land. The authors then used this information to identify the location of corridors crucial to preserving connectivity among habitat patches.

## 12.4  EVALUATING POPULATION DYNAMICS

Understanding source-sink dynamics (Pulliam 1988) and the impact of habitat quality on production and attraction of migrants is important in corridor and reserve design. Landscape genetic approaches have good potential for identifying source and sink habitats by identifying asymmetric gene flow with coalescent approaches (Beerli & Felsenstein 2001) and assignment tests (Paetkau et al. 1995; Wilson & Rannala 2003; Faubet & Gaggiotti 2008; Holderegger & Gugerli 2012). For example, Andreasen et al. (2012) used Bayesian clustering (Chen et al. 2007) and Bayesian assignment test (Faubet & Gaggiotti 2008) approaches to evaluate genetic structure and source-sink dynamics of cougars (*Puma concolor*) in the Great Basin Desert. They detected five populations (south, west, north, east, and central) with asymmetric levels of gene flow, which indicated that the south population was a source while the east, west, and north populations were sinks. The east, west, and north populations had higher levels of hunting and human disturbance while the source area of the south had multiple wildlife refuges and low hunting pressure (Andreasen et al. 2012).

Graph theoretic approaches (Chapter 10) also show great potential for evaluating the importance of individual core areas or nodes to the population viability and connectivity of populations. Using this network-based approach, Garroway et al. (2008) evaluated the influence of landscape characteristics on genetic connectivity and subpopulation level productivity among fisher (*Martes pennanti*) subpopulations sampled from 34 habitat patches in Canada. Fishers in this region were characterized by a higher level of clustering than expected by chance and a short mean path length connecting all pairs of nodes. Nodes with high value for maintaining connectedness in the system were identified and nodes with the greatest snow depth (and thus lower quality habitat) showed less connectedness in the network. These results supported previous work showing that fishers exhibit short dispersal distances, are territorial, and relatively philopatric (Kyle et al. 2001; Koen et al. 2007). From a conservation perspective, this research demonstrated that there was much resiliency to the loss of nodes in the network and suggested that current harvest regimes are unlikely to affect genetic connectivity or induce genetic differentiation.

## 12.5 DETECTING AND PREDICTING THE RESPONSE TO LANDSCAPE CHANGE

Landscape genetics approaches are useful for examining how landscape change has influenced genetic connectivity and how connectivity might change in response to future landscape change (Table 12.3). Long-term viability of populations can be investigated by assessing temporal changes in population genetic structure and genetic diversity that reflect the ability of populations to persist and evolve in response to future changing environments (Frankham et al. 2010; Allendorf et al. 2012). Furthermore, genetic structure is influenced by processes that occur over multiple generations and thus can be used to understand the importance of both historic and contemporary conditions, especially if genetic markers are used that mutate at different rates (see Chapter 3). The other requirements for such an analysis are representations of landscapes from at least two different time points, either a past landscape or a future predicted condition. Thus, there have been two different types of studies investigating landscape change with genetic markers: a retrospective approach in which genetic structure is correlated with both present and past landscape representations and a future predictive approach in which simulations are used to extrapolate current correlations to future landscape change. A third possibility is a "space for time approach" in which

**Table 12.3** Examples of terrestrial wildlife studies that have used landscape genetic approaches to detect and predict response to landscape change.

| Citation | Species | Unit of analysis | Inference or Conclusions |
|---|---|---|---|
| Holzhauer et al. (2006) | *Metrioptera roeseli* | individual | Landscape patterns from 50 years ago explained genetic structure better than current patterns |
| Keyghobadi et al. (2005a, 2005b) | *Parnassius smintheus* | population | Genetic differentiation correlated with contemporary landscape, but genetic diversity correlated with past landscape |
| Murphy et al. (2010a) | *Anaxyrus boreas* | population | Genetic connectivity partially associated with recent fires |
| Pavlacky et al. (2009) | *Orthonyx temmincki* | population | A slightly stronger relationship with contemporary rather than historic landscape; results differed when using different metrics of genetic differentiation |
| Spear & Storfer (2008) | *Ascaphus truei* | population | Timber harvest patterns from 20–40 years were correlated with current estimates of genetic distance |
| Spear et al. (2005) | *Ambystoma tigrinum* | population | Significant correlation of increased gene flow with recently burned forest; suggest fires facilitate connectivity |
| Spear et al. (2012) | *Ascaphus truei* | population | Gene flow among populations in portions of the volcanic zone subject to logging and reforestation more constrained by climatic factors than those in unmanaged area |
| Wasserman et al. (2012, 2013) | *Martes americana* | individual | Climate change scenarios predicted a loss of habitat correlated with genetic connectivity |

landscapes reflecting different conditions at the same time point are compared, but these scenarios are less ideal due to confounding differences between landscapes (McGarigal & Cushman 2002; Short Bull et al. 2011).

It has long been recognized that different aspects of genetic structure can change at different temporal rates (Crow & Aoki 1984), but Keyghobadi et al. (2005b) was one of the first to explicitly address this using landscape genetic methodology. This study used seven microsatellite loci developed for an alpine butterfly (*Parnassius smintheus*) to compare landscape correlations with genetic differentiation and genetic diversity in the Rocky Mountains of Alberta, Canada. The authors used aerial photography to assess landscape conditions in 1952 and 1993 to examine historic and contemporary configurations of meadow and forest. They applied a metric of patch connectivity of both meadows and forests (accounting for dispersal through these two habitat types) as the landscape measure to correlate with aspects of genetic structure. They represented genetic diversity through expected heterozygosity ($H_e$) and genetic differentiation with $G_{ST}$ (see Chapter 3 for detailed descriptions of these metrics). Heterozygosity had a stronger correlation with the historic landscape, whereas $G_{ST}$ had a greater correlation with the contemporary landscape, suggesting that measures of genetic differentiation may be more appropriate tests of contemporary landscape change. Similarly, Zellmer and Knowles (2009) detected a significant association of $F_{ST}$ with contemporary landscape resistance in a wood frog metapopulation after controlling for historic landscapes, but found no association with historic landscapes after controlling for contemporary resistance. However, other studies have demonstrated that patterns of genetic differentiation can have a temporal lag effect. For instance, Holzhauer et al. (2006) studied bush-crickets (*Metrioptera roeseli*) in a rural grassland district of Germany and found that genetic similarity was correlated best with landscape configurations 30–50 years before the study, and that the strength of correlation steadily decreased with more recent landscape configurations. Similarly, Spear and Storfer (2008) found a 20–40 year time lag with a timber harvest date on the Olympic Peninsula of Washington and coastal tailed frog (*Ascaphus truei*) genetic differentiation based on 13 microsatellite loci.

An alternative approach for incorporating landscape change through time is to represent genetic connectivity as different metrics that change at different temporal scales. The advantage of this approach is that it allows the researcher to test the relative influence of historic

and recent genetic connectivity. For instance, Pavlacky et al. (2009) studied logrunner (*Orthonyx temmincki*) connectivity in an Australian subtropical rainforest that had become fragmented due to European colonization. The authors used two types of genetic analysis based on 10 microsatellite loci: a traditional measure of $F_{ST}$ and an estimate of migration using the coalescent approach. Based on $F_{ST}$ analyses, there was similar support for historic and recent landscape composition, although the historic landscape had a slightly higher correlation with forest cover and patch density. In contrast, coalescent estimates of migration had a stronger correlation with the contemporary, fragmented landscape, suggesting it is more reflective of the recent landscape changes that have constrained gene flow. Asymmetric migration observed in logrunners was later associated with the degree of isolation of patches (Pavlacky et al. 2012).

Despite these examples, there are few studies for which researchers have been able to reconstruct past landscapes. Instead, most studies interested in landscape change rely on the assumption that recent, as opposed to historic, landscape changes have caused observed changes in genetic structure. For instance, both Murphy et al. (2010a) and Spear et al. (2005) found that increased gene flow in amphibians was associated with the broad-scale fires in Yellowstone National Park and interpreted this to mean that the recent disturbance had shifted genetic structure. However, recent disturbances may simply be correlated with a spatially associated historic process, resulting in incorrect inferences regarding the process influencing gene flow. One way to address this issue is to study landscapes that have identical or very similar histories but that have had recent landscape changes not dependent on the historic condition. For instance, after the Mount St. Helens eruption, areas in the blast zone that had the same degree of volcanic disturbance were subject to different management regimes. Spear et al. (2012) found that variables associated with gene flow in coastal tailed frogs were different depending on post-eruption management, as populations occurring in areas managed for post-eruption timber were limited by multiple climatic variables and populations in the unmanaged landscape were largely influenced by topography. Since the landscapes before management were highly similar, this study was able to make a stronger inference that the management regime led to differences in genetic structure.

The previous examples all focused on understanding the effect of past landscape change on genetic connectivity. Nonetheless, the current rate of global change demonstrates the need to predict future anthropogenic impacts on genetic connectivity. A series of recent papers have provided examples of predictive landscape genetics. Wasserman et al. (2010) applied an individual-based causal modeling approach to infer relationships between landscape patterns and gene flow processes in a study of the American marten (*Martes americana*) in the United States northern Rocky Mountains. They found that gene flow was a function of elevation, with minimum resistance to gene flow at 1500 m. Alternative hypotheses involving isolation-by-distance, geographical barriers, effects of canopy closure, roads, tree size class, and an empirical habitat model were not supported. Simulation modeling (Chapter 7) was then used to predict the effects of potential climate change on population connectivity and genetic diversity of the American marten population (Wasserman et al. 2012). They then evaluated the influence of five potential future temperature scenarios involving different degrees of warming on population connectivity and genetic diversity (Wasserman et al. 2013). Resistant kernel dispersal models (Compton et al. 2007) were used to assess population connectivity and the CDPOP model (Landguth & Cushman 2010) was used to simulate gene exchange among individual martens under each climate change scenario. Even moderate warming scenarios were shown to result in very large reductions in population connectivity and the size of genetic neighborhoods, with consequent declines in allelic richness and expected heterozygosity. Such predictions can then guide the development of genetic monitoring protocols (Schwartz et al. 2007) that can determine if predictions are accurate and, if not, develop alternative models that better explain the observed genetic changes. However, testing predictions will be difficult for organisms with long generation times and therefore studies on organisms with rapid generations would be most valuable for understanding the potential reliability of predictive landscape genetics.

## 12.6  COMMON LIMITATIONS OF LANDSCAPE GENETIC STUDIES INVOLVING TERRESTRIAL ANIMALS

The preceding overview of landscape genetic studies of terrestrial animals demonstrates how the field of landscape genetics has created a better understanding of the variables that influence connectivity in terrestrial

animal systems and contributed to the conservation and management of many species. However, there are also some important limitations common to many terrestrial landscape studies. The top seven limitations that we identified based on our knowledge of the field and other previous reviews include the following. (1) Few or no hypotheses are explicitly stated and tested in many landscape genetic studies (Storfer et al. 2010; Segelbacher et al. 2010). (2) Most studies evaluate landscape genetic questions and hypotheses in a single landscape and thus provide no replication (Segelbacher et al. 2010). (3) The majority (90%) of studies focus on a single species and thus do not provide insight into the response of multiple species to the same landscape variables (Storfer et al. 2010). (4) Landscape genetic analyses are rarely combined with other data types like field data collected from radiotelemetry or mark-recapture (Spear et al. 2010). (5) Most studies do not evaluate the genetic response to landscape genetic variables at multiple spatial scales (Manel & Holderegger 2013). (6) Landscape genetic researchers rarely conduct uncertainty analyses (i.e., using simulation) (Epperson et al. 2010; Jaquiéry et al. 2011). (7) There can be marked differences in movement behavior (McDevitt et al. 2013; Elliot et al. 2014a) and landscape connectivity (Amos et al. 2014; Elliot et al. 2014b; Paquette et al. 2014) between sexes, size classes, and age classes of the same species, which are highly relevant to ecological understanding and conservation applications, but which have been rarely addressed in landscape genetics research. Below, we highlight groups of terrestrial animal case studies that address many of these limitations while testing ecological hypotheses about gene flow in heterogeneous terrestrial landscapes.

## 12.7  TESTING ECOLOGICAL HYPOTHESES ABOUT GENE FLOW IN HETEROGENEOUS LANDSCAPES

### 12.7.1  Testing landscape resistance hypotheses in black bears

One of the first studies to formally evaluate the support for multiple landscape resistance hypotheses in comparison to null models of isolation-by-distance (IBD) and isolation-by-barriers (IBB) was conducted on the American black bear (*Ursus americanus*). The relative support of more than 100 alternative hypotheses about the effects of landscape features such as roads, elevation gradients, forest cover, and topographical slope on genetic differentiation was tested in an American black bear population in northern Idaho, USA (Cushman et al. 2006). Follow-up papers evaluated the methods employed (Landguth & Cushman 2010; Cushman & Landguth 2010a, 2010b), tested the results with independent radiotelemetry data (Cushman & Lewis 2010), and spatially replicated the study to evaluate whether generalities in relationships between bear genetic differentiation and landscape features held across a number of study areas (Short Bull et al. 2011). Testing multiple competing resistance hypotheses (Cushman et al. 2006), verifying the high performance of the analytical methods employed (Cushman & Landguth 2010a), evaluating the predicted resistance maps using independent movement data (Cushman & Lewis 2010), and confirming these relationships across a large replicated study gave the researchers high confidence that forest cover, human development, roads, and elevation drive landscape resistance for black bears in the northern Rocky Mountains.

### 12.7.2 Comparative landscape genetics studies

One promising and relatively unexplored area of research is comparative landscape genetics of multiple species in a shared landscape (Table 12.4). Evaluating genetic diversity and population structure of multiple species in the same landscape can increase our ability to identify similarities and differences among species in response to a common set of landscape features. In one of the first and most comprehensive comparative landscape genetic studies, Manier and Arnold (2006) evaluated potential barriers and ecological correlates of several population genetic parameters for three interacting species, the western toad (*Anaxyrus boreas*), a terrestrial garter snake (*Thamnophis elegans*), and a common garter snake (*Thamnophis sirtalis*), in a 1000 km² landscape of northern California, US. They found the same landscape barrier in snakes, a 300 m escarpment formed by a series of block faults, but no evidence of this barrier in the western toad. They identified different variables, as well as some commonalities, that influenced migration of the three species including the presence of conspecifics (*T. elegans*, *T. sirtalis*), the presence of competing species (*T. elegans*), the presence of predators (*A. boreas*), geographic distance (*T. elegans*, *A. boreas*), elevation (*T. elegans*, *A.*

*boreas*), and pond depth (*T. elegans*, *A. boreas*). When evaluating landscape variables that explained genetic distance ($F_{st}$), geographic distance was significant for all species while elevation also was significant for *T. elegans* and the presence of *T. elegans* and differences in pond perimeter were important for *T. sirtalis*. This study provides important insight into the microevolutionary processes that influence the population genetic structure for a group of interacting species and is one of the only studies to demonstrate the impacts of interspecific competition and predation on migration or gene flow.

Goldberg and Waits (2010) also conducted a comparative landscape genetics study by evaluating landscape genetic patterns of the co-distributed Columbia spotted frog (*Rana luteiventris*) and long-toed salamander (*Ambystoma macrodactylum*) in a 213 km² study area in northern Idaho. These species are phylogenetically, physiologically, and behaviorally distinct, but share many of the same breeding ponds in this region (Goldberg & Waits 2009). The authors evaluated (i) whether genetic patterns in this area were structured by distances along drainages at a broad scale or by watershed at a fine scale and (ii) what combinations of landscape features were most likely to predict variation in genetic distances between breeding locations for each species. Forested land cover was hypothesized to be associated with lower landscape resistance values for both species to minimize water loss. The riparian watershed network in this landscape was not found to be associated with gene flow for either species, and thus terrestrial movements across the landscape were likely to be driving gene flow patterns. For both species, urban and rurally developed land cover types provided the highest landscape resistances but the species differed in their response to other landscape variables. Resistance values for long-toed salamanders followed a moisture gradient where forest provided the least resistance, whereas agriculture and shrub/clearcut provided the least resistance for Columbia spotted frogs.

In a comparative landscape genetic study of woodland-dependent birds, Harrisson et al. (2012) evaluated the impacts of habitat loss and fragmentation on four bird species sampled at over 55 sites across 12 replicated landscape blocks (100 km² each) in southern Australia. They sampled a gradient of woodland tree cover ranging from 11% to 78% and conducted analyses at multiple spatial scales including within landscape, between landscapes, and with all landscapes combined. They collected nuclear DNA data from 6–16 neutral loci for three species known to decline with loss of forest habitat (decliners), the eastern yellow robin (*Eopsaltria australis*), weebill (*Smicrornis brevirostris*), and spotted pardalote (*Pardalotus punctatus*), and one tolerant species, the striated pardalote (*Pardalotus striatus*). The authors found that genetic diversity and effective population size decreased with decreasing forest cover for the decliner species but not the tolerant species. Also for two decliner species, the robin and weebill, the relatedness of individuals increased in more fragmented habitats and an isolation-by-distance effect of increased relatedness at smaller spatial scales was detected for the robin species, the least mobile decliner. However, no significant genetic differentiation was detected for any species despite simulation analyses that indicated sufficient time had passed since fragmentation to create detectable genetic structure under scenarios with moderate restrictions of gene flow. In the same study system, Amos et al. (2014) expanded analyses to 10 bird species and demonstrated sex-specific responses to habitat fragmentation for four species with the philopatric sex more affected by fragmentation in three of the four species. Overall, these studies provide a good example of how species and sexes may respond differently to habitat loss and fragmentation and provides evidence that some species of birds can maintain gene flow and connectivity in the face of major habitat alteration.

Other comparative landscape genetic studies have examined the effects of roadways and urbanization on the genetic structure of multiple terrestrial animals. For example, Riley et al. (2006) used radiotelemetry data and nuclear DNA microsatellite data to demonstrate that a freeway in southern California, US, was a partial barrier to both bobcats (*Lynx rufus*) and coyotes (*Canis latrans*). Delaney et al. (2010) investigated the genetic effects of urban fragmentation on three lizards, the side-blotched lizard (*Uta stansburiana*), western skink (*Plestiodon skiltonianus*), and western fence lizard (*Sceloporus occidentalis*), and one bird, the wrentit (*Chamaea fasciata*) in southern California, and found that the greatest restriction to gene flow occurred across the widest and oldest expanses of urban area, which included a major highway. Also, Frantz et al. (2012) found that a motorway in Belgium created a barrier for red deer (*Cervus elaphus*) but not wild boars (*Sus scrofa*). In a study of sympatric newt species in Greece, gene flow of the species with a terrestrial form of dispersal (*Lissotriton vulgaris*) was restricted by roadways while roads had no effect on genetic structure of the newt species (*Triturus*

*macedonicus*) with an aquatic mode of dispersal (Sotiropoulos et al. 2013).

### 12.7.3 Scale dependence and threshold effects in landscape genetics

The effect of scale on ecological processes is a centrally important question (Wiens 1989; Levin 1992) because of the fundamental dependencies between scale and pattern and between pattern and process (Wu & Loucks 1995; Chapter 2). Landscape genetics explicitly focuses on how relationships between genetic processes and landscape patterns vary in heterogeneous environments in relation to spatial and temporal scales. Extensive work has focused on how apparent relationships between landscape patterns and species occurrence change with scale of observation (e.g., Tischendorf 2001; Thompson & McGarigal 2002; Cushman 2006; Corry & Nassauer 2005; Cushman et al. 2008). These studies have clearly shown that the drivers of an ecological process may each act at unique scales in space and time (Wiens 1989). It is critically important to correctly match the scale of each driving variable to the response process.

Despite the central importance of scale dependence in landscape genetics very few studies have explicitly focused on scale issues in landscape genetics (Anderson et al. 2010; Table 12.4). One of the earliest examples of a study that assessed scale dependence in landscape genetics was conducted by Cushman and Landguth (2010b), who showed using simulations that the strength of landscape-genetic relationships was highly dependent on matching the grain, extent, and thematic resolution of the analysis with that of the underlying biological process. An empirical analysis that provided a similar insight was conducted by Angelone et al. (2011). They examined the effects of landscape elements on genetic differentiation at three distance classes reflecting varying frequencies of European tree frog (*Hyla arborea*) movement, and found that different variables (rivers, geographic distance, roads, and forest cover) were associated with gene flow at different distance classes ranging from 0 to 8 km, indicating scale dependence of the relationship between frog dispersal and landscape structure. Similarly, in a landscape genetics study of boreal toads (*Bufo boreas*) in Yellowstone National Park, Murphy et al. (2010a) found that landscape variables were significantly associated with genetic differentiation at different spatial

**Table 12.4** Examples of landscape genetic studies that tested ecological hypotheses for terrestrial animals by (a) applying a comparative landscape genetics framework and (b) assessing scale dependence and threshold effects.

**(a) Comparative landscape genetic studies**

| Citation | Species | Unit of analysis | Inference or conclusions |
|---|---|---|---|
| Amos et al. (2014) | 10 woodland dependent birds | population | Four of the least mobile species impacted by forest loss and fragmentation and philopatric sexes more affected |
| Delaney et al. (2010) | *Uta stansburiana, Plestiodon skiltonianus, Sceloporus occidentalis, Chamaea fasciata* | indvidual, population | For all species, gene flow most restricted across the widest and oldest expanses of urban area, which included a major highway |
| Frantz et al. (2012) | *Cervus elaphus, Sus scrofa* | individual | Motorway created barrier for *C. elaphus* but not *S. scrofa* |
| Goldberg & Waits (2010) | *Ambystoma macrodactylum, Rana luteiventris* | population | Development was least permeable land cover for both species and movement was not restricted to riparian corridors. Agriculture, shrub, and forest cover had different levels of permeability for each species |
| Harrisson et al. (2012) | *Eopsaltria australis, Smicrornis brevirostris, Pardalotus punctatus, Pardalotus striatus* | population | Genetic diversity and effective population size decreased with decreasing forest cover for all species except *P. striatus* |
| Manier & Arnold (2006) | *Anaxyrus boreas, Thamnophis elegans, Thamnophis sirtalis* | population | Landscape variables that restricted gene flow varied across species; *T. elegans* and *T. sirtalis* had the same natural landscape barrier |
| Riley et al. (2006) | *Lynx rufus, Canis latrans* | individual | Highway partial barrier to gene flow in both species, stronger effect for bobcats |
| Sotiropoulos et al. (2013) | *Lissotriton vulgaris, Triturus macedonicus* | individual, population | *L. vulgaris* showed more differentiation than *T. macedonicus*; roads restricted gene flow in *L. vulgaris* but not *T. macedonis* |

**(b) Assessing scale dependence and threshold effects**

| | | | |
|---|---|---|---|
| Angelone et al. (2011) | *Hyla arborea* | population | Different landscape variables (rivers, forest cover, geographic distance, and roads) were associated with gene flow at different distance classes ranging from 0 to 8 km |
| Balkenhol et al. (2013) | *Marmosops incanus* | population | Genetic diversity was significantly lower in landscape with 31% forest cover compared to landscapes with 49% and 86%; genetic structure was significantly lower in the landscape with 86% forest cover |

*(continued)*

**Table 12.4** *(Continued)*

**(b) Assessing scale dependence and threshold effects**

| | | | |
|---|---|---|---|
| Cushman & Landguth (2010b) | *Ursus americanus* | individual | Simulation study showed that that the strength of landscape-genetic relationships was highly dependent on matching the grain, extent, and thematic resolution of the analysis with that of the underlying biological process |
| Galpern et al. (2012) | *Rangifer tarandus caribou* | individual | Using multiple spatial grains can reveal landscape influences on genetic structure that may be overlooked with a single grain, and coarsening the grain of landcover data may be appropriate for highly mobile species |
| Keller et al. (2013) | *Stethophyma grossum* | population | Showed that the highest model fits were found when restricting landscape genetic analysis to smaller scales (0–3 km) and neighboring populations |
| Lange et al. (2010) | *Metrioptera roeselii, Pholidoptera griseoaptera* | population | Threshold effects on genetic diversity and structure detected for *P. griseoaptera* when suitable habitat dropped below 20% |
| Murphy et al. (2010a) | *Bufo boreas* | population | Landscape variables were significantly influenced by genetic structure at different spatial scales ranging from 60 to 960 m |

scales ranging from 60 m to 960 m. The optimal cross-scale model included metrics measured at fine scales (canopy, elevation relief ratio), mid-scale (impervious surfaces), broad scale (ridges), and across multiple scales (growing-season precipitation and slope temperature–moisture). However, cross-scale models only improved the percentage of variation explained by 3–5%.

Other recent studies have also demonstrated the importance of considering spatial scale and network topology when conducting landscape genetic analyses. For example, Galpern et al. (2012) used a novel approach based on patch-based landscape graphs to develop resistance surfaces with multiple spatial grains to evaluate landscape-genetic relationships among woodland caribou (*Rangifer tarandus caribou*). Their results showed that using multiple spatial grains can reveal landscape influences on genetic structure that may be overlooked with a single grain, and that coarsening the grain of landcover data may improve landscape genetic inferences for highly mobile species. In a

similar study, Keller et al. (2013) studied landscape-genetic relationships among wetland grasshopper (*Stethophyma grossum*) populations in a fragmented agricultural landscape and evaluated results across multiple spatial scales and network topologies. They found that the highest model fits were obtained when restricting landscape genetic analysis to smaller scales (0–3 km) and neighboring populations (Keller et al. 2013).

One key question in landscape and conservation genetics is whether there are landscape-scale threshold effects of habitat loss or fragmentation that lead to major increases in genetic structure and losses of genetic variation. So far, only a few studies have attempted to address this research question (Table 12.4). Lange et al. (2010) evaluated threshold effects of habitat fragmentation for a grassland species of bush cricket, *Metrioptera roeselii*, and a forest-edge species, *Pholidoptera griseoaptera*. No threshold effects were detected for the grassland species, but for the forest-edge species genetic differentiation ($G_{ST}$, Jost's $D$) increased substantially and genetic diversity was significantly lower when the amount of suitable habitat dropped below 20% in the very high fragmentation class. In a similar study, Balkenhol et al. (2013) evaluated the effects of varying levels of forest fragmentation on genetic diversity and structure in a marsupial forest-specialist species (*Marmosops incanus*) of the Brazilian Atlantic forest and detected threshold effects. For example, genetic diversity was significantly lower in the landscape with 31% forest cover compared to landscapes with 49% and 86% forest cover and genetic structure was significantly lower in the landscape with 86% forest cover compared to the more fragmented landscapes.

## 12.8 KNOWLEDGE GAPS AND FUTURE DIRECTIONS

Landscape genetic studies published to date have been primarily descriptive and few have been hypothesis-driven (Segelbacher et al. 2010; Storfer et al. 2010). This limits the ability of landscape genetics to provide broader insights into important ecological and evolutionary questions and makes these studies highly vulnerable to inferential errors when only one or a few models are tested (Cushman & Landguth 2010a). Thus, it is important to develop explicit hypotheses in the early stages of a landscape genetics study so that the sampling design can be optimized to improve the inferential power and to also consider the use of controlled experiments for testing hypotheses. Another important challenge in testing landscape genetic hypotheses is differentiating the impacts of historic and recent landscape change on genetic diversity and structure.

One of the factors currently limiting our understanding of landscape-genetic relationships in terrestrial animals is the relatively limited taxonomic and geographic scope of current research. Several reviews of the effects of habitat heterogeneity and fragmentation over the past 10–15 years have demonstrated a bias toward temperate forest ecosystems in developed countries and vertebrates, especially birds and mammals, as focal species (McGarigal & Cushman 2002; Storfer et al. 2010; Zeller et al. 2012; Manel & Holderegger 2013). However, these chosen study systems may not be the most tractable to landscape genetic questions. For example, more studies could focus on small mammals, like rodents, given the relative sensitivity of these species to recent landscape change because of relatively short generation times and limited dispersal distances

(e.g. Landguth et al. 2010). Additionally, invertebrate taxa provide a good study system because they are more sensitive to climate change and other environmental perturbations and tend to be easily manipulated in controlled experimental studies (e.g. McGarigal & Cushman 2002; Prather et al. 2013). While applied studies of single species of conservation interest are important and will continue and be useful to those focal systems, more research is needed on model organisms and multispecies systems to address central questions in landscape genetics.

There has been extensive debate and considerable disagreement about the best analytical methods for evaluating relationships between landscape structure and gene flow processes. Naturally, the choice of method will depend on the nature of the question and will differ for studies aiming at identifying the drivers of long-term patterns in gene flow than those focused on measuring current gene flow, and likewise will differ between these and studies addressing the effects of landscape factors on adaptive genetic variation. Some generalities, however, have emerged, such as the importance of verifying the performance of statistical methods using simulation modeling (e.g., Cushman & Landguth 2010a; Jacquiéry et al. 2011; Shirk et al. 2012; Graves et al. 2013), developing generalized frameworks to compete multiple alternative hypotheses in a unified framework (e.g., Cushman et al. 2006; Wasserman et al. 2010), and the importance of employing analytical and statistical approaches that enable effective optimization of predictions of pattern–process relationships (e.g., Wang et al. 2009; Shirk et al. 2010; Cushman et al. 2014). These are large and abiding challenges that will remain the focus of research in landscape genetics for years to come. Chapter 5 of this book presents a comprehensive framework that enables delineation of different approaches of landscape genetic analysis based on the scope and objectives of analysis.

Another key for landscape genetics to successfully address many ecological hypotheses is to move to a more process-based approach when designing studies. One of the major criticisms of current correlative analytical approaches is that they do not directly model key factors driving gene flow, such as inter- and intraspecific competition, mating system, or dispersal patterns (Graves et al. 2013). To date, a vast majority of studies have used unvalidated expert opinion in the absence of empirical data on movement or gene flow to parameterize resistance surfaces (Spear et al. 2010;

Zeller et al. 2012), leading to a high degree of uncertainty in predictions. While there are a number of formal methods to assess variability of expert opinion and combine and weight expert rankings, which improve its implementation (e.g., Burgman 2004; Okoli & Pawlowski 2004), it is desirable to utilize empirical data wherever possible. In some instances where it would be excessively expensive or time consuming to gather empirical data to parameterize resistance surfaces and when there is considerable urgency, such as when dealing with endangered species in a rapidly changing landscape, careful utilization of methods to integrate expert knowledge may be appropriate.

Dispersal can be incorporated by linking individual behavior and fine-scale movement data to gene flow (e.g., Cushman & Lewis 2010). With the increasing availability of GPS telemetry for small organisms, a more nuanced understanding of how individual movement changes genetic patterns is possible and can serve as model inputs (e.g., Elliot et al. 2014b). The availability of fine-scale movement data can also allow for a better classification of different types of movement (Zeller et al. 2014), which can be used to focus on the movements relevant to gene flow. Further discussion of these issues is presented in Chapter 8.

While modeling dispersal is becoming more feasible, processes such as inheritance and mating are still difficult to study in nature, particularly in vertebrates. A solution to this is the linkage of empirical landscape genetics modeling with individual-based simulation of gene flow in complex landscapes (e.g., Landguth & Cushman 2010; Shirk et al. 2012; see Chapter 6). Simulation modeling can easily incorporate a number of processes and the mechanisms by which a landscape resists gene flow can be inferred by evaluating the relationship between landscape models and an observed pattern of genetic isolation. Furthermore, simulation modeling allows for more robust sensitivity analyses that can elucidate which factors are likely to most influence genetic structure, and important processes can then be focal points for future empirical studies. If applied to a variety of organisms and study systems, the simulation approach can also identify general patterns that will be relevant across landscape genetic studies. For example, simulations could assess how different mating systems affect landscape genetic results if organisms with similar dispersal tendencies are modeled.

Ultimately, responses to global change are one of the most important avenues for research for landscape genetics and many biological disciplines (Manel & Holderegger 2013). Thus, simulation modeling along with empirical genetic work across a variety of current landscape conditions using both neutral and adaptive markers (see Chapter 9) is critical to predict how species will respond to climate change. Such studies are in their infancy, but a more rigorous framework is needed to advance predictive landscape genetics so that it can truly inform future actions to address global change.

In closing, we recommend a broadening of the scope of landscape genetics to address a wider taxonomic range across a more diverse set of ecological systems. Particular focus should be given to invertebrate taxa and rodents, which provide experimentally tractable study species, with short generation times and limited dispersal ability, making them ideal for landscape genetic studies. Also, there is great value in directly incorporating important processes into landscape genetic models and in synergistically combining empirical analysis and simulation modeling. We believe that these research areas will pay rapid dividends in increasing our knowledge of landscape genetic processes for terrestrial animal species and their influences on populations, species, and communities.

## REFERENCES

Allendorf, F.W., Luikart, G.H., & Aitken, S.N. (2012) *Conservation and the Genetics of Populations*. John Wiley & Sons.

Amos, J.N., Harrisson, K.A., Radford, J.Q., White, M., Newell, G., Mac Nally, Snuucks, P., & Pavlova, A. (2014) Species- and sex-specific connectivity effects of habitat fragmentation in a suite of woodland birds. *Ecology* **95**, 1556–68.

Anderson, C.D., Epperson, B.K., Fortin, M.-J., Holderegger, R., James, P.M.A., Rosenberg, M.S., Scribner, K.T., & Spear, S. (2010) Considering spatial and temporal scale in landscape-genetic studies of gene flow. *Molecular Ecology* **19**, 3565–75.

Andreasen, A.M., Stewart, K.M., Longland, W.S., Beckmann, J.P., & Forister, M.L. (2012) Identification of source-sink dynamics in mountain lions of the Great Basin. *Molecular Ecology* **21**, 5689–701.

Angelone, S. & Holderegger, R. (2009) Population genetics suggests effectiveness of habitat connectivity measures for the European tree frog in Switzerland. *Journal of Applied Ecology* **46**, 879–87.

Angelone, S., Kienast, F., & Holderegger, R. (2011) Where movement happens: scale-dependent landscape effects on genetic differentiation in the European tree frog. *Ecography* **34**, 714–22.

Baguette, M., Blanchet, S., Legrand, D., Stevens, V.M., & Turlure, C. (2013) Individual dispersal, landscape connectivity and ecological networks. *Biological Reviews* **88**, 310–26.

Balkenhol, N., Pardini, R., Cornelius, C., Fernandes, F., & Sommer, S. (2013) Landscape-level comparison of genetic diversity and differentiation in a small mammal inhabiting different fragmented landscapes of the Brazilian Atlantic Forest. *Conservation Genetics* **14**, 355–67.

Banks, S.C., Lindenmayer, D.B., Ward, S.J., & Taylor, A.C. (2005) The effects of habitat fragmentation via forestry plantation establishment on spatial genotypic structure in the small marsupial carnivore, *Antechinus agilis*. *Molecular Ecology* **14**, pp. 1667–80.

Beerli, P. & Felsenstein, J. (2001) 'Maximum likelihood estimation of a migration matrix and effective population sizes in n subpopulations by using a coalescent approach', *Proceedings of the National Academy of Sciences* **98**, 4563–8.

Biek, R. & Real, L.A. (2010) The landscape genetics of infectious disease emergence and spread. *Molecular Ecology* **19**, 3515–31.

Blair, C., Weigel, D.E., Balazik, M., Keeley, A.T.H., Walker, F.M., Landguth, E.L., Cushman, S.A., Murphy, M., Waits, L.P., & Balkenhol, N. (2012) A simulation-based evaluation of methods for inferring linear barriers to gene flow. *Molecular Ecology Resources* **12**, 822–33.

Blanchong, J.A., Samuel, M.D., Scribner, K.T., Weckworth, B.V., Langenberg, J.A., & Filcek, K.B. (2008) Landscape genetics and the spatial distribution of chronic wasting disease. *Biology Letters* **4**, 130–3.

Bohonak, A.J. (1999) Dispersal, gene flow and population structure. *The Quarterly Review of Biology* **74**, 21–45.

Boltzman, L. (1872) Weitere Studien über das Wärmegleichgewicht unter Gasmolekülen. *Wiener Berichte* **66**, 275–370.

Braunisch, V., Segelbacher, G., & Hirzel, A.H. (2010) Modeling functional landscape connectivity from genetic population structure: a new spatially explicit approach. *Molecular Ecology* **19**, 3664–78.

Broquet, T. & Petit, E.J. (2009) Molecular estimation of dispersal for ecology and population genetics. *Annual Review of Ecology, Evolution and Systematics* **40**, 193–216.

Burgman, M.A. (2004) Expert frailties in conservation risk assessment and listing decisions. In: P. Hutchings, D. Lunney, & C. Dickman (eds.), *Threatened Species Legislation: Is It Just an Act?* Royal Zoological Society of New South Wales, Mosman, NSW.

Bush, K.L., Dyte, C.K., Moynahan, B.J., Aldridge, C.L., Sauls, H.S., Battazzo, A.M., Walker, B.L., Doherty, K.E., Tack, J., Carlson, J., Eslinger, D., Nicholson, J., Boyce, M.S., Naugle, D.E., Paszkowski, C.A., & Coltman, D.W. (2011) Population structure and genetic diversity of greater sage-grouse (*Centrocercus urophasianus*) in fragmented landscapes at the northern edge of their range. *Conservation Genetics* **12**, 527–42.

Carmichael, L.E., Krizan, J., Nagy, J.A., Fuglei, E., Dumond, M., Johnson, D., Veitch, A., Berteaux, D., & Strobeck, C. (2007) Historical and ecological determinants of genetic structure in arctic canids. *Molecular Ecology* **16**, 3466–83.

Carr, M.H., Neigel, J.E., Estes, J.A., Andelman, S., Warner, R.R., & Largier, J.L. (2003) Comparing marine and terrestrial ecosystems: implications for the design of coastal marine reserves. *Ecological Applications* **13**, 90–107.

Chen, C., Durand, E., Forbes, F., & François, O. (2007) Bayesian clustering algorithms ascertaining spatial population structure: a new computer program and a comparison study. *Molecular Ecology Notes* **7**, 747–56.

Clark, R.W., Brown, W.S., Stechert, R., & Zamudio, K.R. (2010) Roads, interrupted dispersal and genetic diversity in timber rattlesnakes. *Conservation Biology* **24**, 1059–69.

Compton, B.W., McGarigal, K., Cushman, S.A., & Gamble, L.R. (2007) A resistant-kernel model of connectivity for amphibians that breed in vernal pools. *Conservation Biology* **21**, 788–99.

Corry, R.C. & Nassauer, J.I. (2005) Limitations of using landscape pattern indices to evaluate the ecological consequences of alternative plans and designs. *Landscape and Urban Planning* **72**, 265–80.

Côté, H., Garant, D., Robert, K., Mainguy, J., & Pelletier, F. (2012) Genetic structure and rabies spread potential in raccoons: the role of landscape barriers and sex-biased dispersal. *Evolutionary Applications* **5**, 393–404.

Coulon, A., Guillot, G., Cosson, J.-F., Angibault, J.M.A., Aulagnier, S., Cargnelutti, B., Galan, M., & Hewison, A.J.M. (2006) Genetic structure is influenced by landscape features: empirical evidence from a roe deer population. *Molecular Ecology* **15**, 1669–79.

Crow, J.F. & Aoki, K. (1984) Group selection for a polygenic behavioral trait: estimating the degree of population subdivision. *Proceedings of the National Academy of Sciences of the United States of America* **81**, 6073–7.

Cullingham, C.I., Kyle, C.J., Pond, B.A., Rees, E.E., & White, B.N. (2009) Differential permeability of rivers to raccoon gene flow corresponds to rabies incidence in Ontario, Canada. *Molecular Ecology* **18**, 43–53.

Cushman, S.A. (2006) Effects of habitat loss and fragmentation on amphibians: a review and prospectus. *Biological Conservation* **128**, 231–40.

Cushman, S.A. & Landguth, E.L. (2010a) Spurious correlations and inference in landscape genetics. *Molecular Ecology* **19**, 3592–602.

Cushman, S.A. & Landguth, E.L. (2010b) Scale dependent inference in landscape genetics. *Landscape Ecology* **25**, 967–79.

Cushman, S.A. & Lewis, J.S. (2010) Movement behavior explains genetic differentiation in American black bears. *Landscape Ecology* **25**, 1613–25.

Cushman, S.A., McKelvey, K.S., Hayden, J., & Schwartz, M.K. (2006) Gene flow in complex landscapes: testing multiple hypotheses with causal modeling. *The American Naturalist* **168**, 486–99.

Cushman, S.A., McGarigal, K., & Neel, M.C. (2008) Parsimony in landscape metrics: strength, universality and consistency. *Ecological Indicators* **8**, 691–703.

Cushman, S.A., McKelvey, K.S., & Schwartz, M.K. (2009) Use of empirically derived source-destination models to map regional conservation corridors. *Conservation Biology* **23**, 368–76.

Cushman, S.A., Lewis, J.S., & Landguth, E.L. (2013) Evaluating the intersection of a regional wildlife connectivity network with highways. *Movement Ecology* **1**, 1–11.

Cushman, S.A., Wasserman, T.N., Landguth, E.L., & Shirk, A.J. (2014) Re-evaluating causal modeling with Mantel tests in landscape genetics. *Diversity* **5**, 51–72.

Delaney, K.S., Riley, S.P.D., & Fisher, R.N. (2010) A rapid, strong and convergent genetic response to urban habitat fragmentation in four divergent and widespread vertebrates. *PLoS One* **5**, e12767.

Dixon, J.D., Oli, M.K., Wooten, M.C., Eason, T.H., McCown, J.W., & Paetkau, D. (2006) Effectiveness of a regional corridor in connecting two Florida black bear populations. *Conservation Biology* **20**, 155–62.

Elliot, N.B., Cushman, S.A., Loveridge, A.J., Mtare, G., & Macdonald, D.W. (2014a) Movements vary according to dispersal stage, group size and rainfall: the case of the African lion. *Ecology* **95**, 2860–9.

Elliot, N.B., Cushman, S.A., Macdonald, D.W., & Loveridge, A.J. (2014b) The devil is in the dispersers: predictions of landscape connectivity change with demography. *Journal of Applied Ecology* **51**, 1169–78.

Epperson, B.K., McRae, B.H., Scribner, K., Cushman, S.A., Rosenberg, M.S., Fortin, M.-J., James, P.M.A., Murphy, M., Manel, S., Legendre, P., & Dale, M.R.T. (2010) Utility of computer simulations in landscape genetics. *Molecular Ecology* **19**, 3549–64.

Epps, C.W., Palsbøll, P.J., Wehausen, J.D., Roderick, G.K., Ramey, R.R., & McCullough, D.R. (2005) Highways block gene flow and cause a rapid decline in genetic diversity of desert bighorn sheep. *Ecology Letters* **8**, 1029–38.

Epps, C.W., Wehausen, J.D., Bleich, V.C., Torres, S.G., & Brashares, J.S. (2007) Optimizing dispersal and corridor models using landscape genetics. *Journal of Applied Ecology* **44**, 714–24.

Faubet, P. & Gaggiotti, O.E. (2008) A new Bayesian method to identify the environmental factors that influence recent migration. *Genetics* **178**, 1491–504.

Forman, R.T.T. (1995) *Land Mosaics: The Ecology of Landscapes and Regions.* Cambridge University Press, Cambridge.

Frankham, R., Ballou, J.D., and Briscoe, D.A. (2010) *Introduction to Conservation Genetics*, 2nd edition. Cambridge, UK: Cambridge University Press.

Frantz, A.C., Bertouille, S., Eloy, M.C., Licoppe, A., Chaumont, F., & Flamand, M.C. (2012) Comparative landscape genetic analyses show a Belgian motorway to be a gene flow barrier for red deer (*Cervus elaphus*) but not wild boars (*Sus scrofa*). *Molecular Ecology* **21**, 3445–57.

Funk, W.C., Blouin, M.S., Corn, P.S., Maxell, B.A., Pilliod, D.S., Amish, S., & Allendorf, F.W. (2005) Population structure of Columbia spotted frogs (*Rana luteiventris*) is strongly affected by the landscape. *Molecular Ecology.* **14**, 483–96.

Galpern, P., Manseau, M., & Wilson, P. (2012) Grains of connectivity: analysis at multiple spatial scales in landscape genetics. *Molecular Ecology* **21**, 3996–4009.

Garroway, C.J., Bowman, J., Carr, D., & Wilson, P.J. (2008) Applications of graph theory to landscape genetics. *Evolutionary Applications* **1**, 620–30.

Geffen, E., Anderson, M.J., & Wayne, R.K. (2004) Climate and habitat barriers to dispersal in the highly mobile grey wolf. *Molecular Ecology* **13**, 2481–90.

Goldberg, C.S. & Waits, L.P. (2009) Using habitat models to determine conservation priorities for pond-breeding amphibians in a privately-owned landscape of northern Idaho, USA. *Biological Conservation* **142**, 1096–104.

Goldberg, C.S. & Waits, L.P. (2010) Comparative landscape genetics of two pond-breeding amphibian species in a highly modified agricultural landscape. *Molecular Ecology* **19**, 3650–63.

Graves, T.A., Wasserman, T.N., Ribeiro, M.C., Landguth, E.L., Spear, S.F., Balkenhol, N., Higgins, C.B., Fortin, M.-J., Cushman, S.A., & Waits, L.P. (2012) The influence of landscape characteristics and home-range size on the quantification of landscape–genetics relationships. *Landscape Ecology* **27**, 253–66.

Graves, T.A., Beier, P., & Royle, J.A. (2013) Current approaches using genetic distances produce poor estimates of landscape resistance to interindividual dispersal. *Molecular Ecology* **22**, 3888–903.

Harrisson, K., Pavlova, A., Amos, J., Takeuchi, N., Lill, A., Radford, J., & Sunnucks, P. (2012) Fine-scale effects of habitat loss and fragmentation despite large-scale gene flow for some regionally declining woodland bird species. *Landscape Ecology* **27**, 813–27.

Holderegger, R. & Gugerli, F. (2012) Where do you come from, where do you go? Directional migration rates in landscape genetics. *Molecular Ecology* **21**, 5640–2.

Holzhauer, S.I.J., Ekschmitt, K., Sander, A.-C., Dauber, J., & Wolters, V. (2006) Effect of historic landscape change on the genetic structure of the bush-cricket *Metrioptera roeseli*. *Landscape Ecology* **21**, 891–9.

Jaquiéry, J., Broquet, T., Hirzel, A.H., Yearsley, J., & Perrin, N. (2011) Inferring landscape effects on dispersal from genetic distances: how far can we go? *Molecular Ecology* **20**, 692–705.

Johansson, M., Primmer, C.R., Sahlsten, J. & Merilä, J. (2005) The influence of landscape structure on occurrence, abundance and genetic diversity of the common frog, *Rana temporaria*. *Global Change Biology* **11**, 1664–79.

Keller, D., Holderegger, R., & van Strien, M.J. (2013) Spatial scale affects landscape genetic analysis of a wetland grasshopper. *Molecular Ecology* **22**, 2467–82.

Keyghobadi, N., Roland, J., & Strobeck, C. (2005a) Genetic differentiation and gene flow among populations of the alpine butterfly, *Parnassius smintheus,* vary with landscape connectivity. *Molecular Ecology* **14**, 1897–909.

Keyghobadi, N., Roland, J., Matter, S.F., & Strobeck, C. (2005b) Among- and within-patch components of genetic diversity respond at different rates to habitat fragmentation: an empirical demonstration. *Proceedings of the Royal Society B: Biological Sciences* **272**, 553–60.

Koen, E.L., Bowman, J., & Findlay, C.S. (2007) Fisher survival in Eastern Ontario. *The Journal of Wildlife Management* **71**, 1214–19.

Kuehn, R., Hindenlang, K.E., Holzgang, O., Senn, J., Stoeckle, B., & Sperisen, C. (2007) Genetic effect of transportation infrastructure on roe deer populations (*Capreolus capreolus*). *Journal of Heredity* **98**, 13–22.

Kyle, C.J., Robitaille, J.F., & Strobeck, C. (2001) Genetic variation and structure of fisher (*Martes pennanti*) populations across North America. *Molecular Ecology* **10**, 2341–7.

Landguth, E.L. & Cushman, S.A. (2010) cdpop: a spatially explicit cost distance population genetics program. *Molecular Ecology Resources* **10**, 156–61.

Landguth, E.L., Cushman, S.A., Schwartz, M.K., McKelvey, K.S., Murphy, M., & Luikart, G. (2010) Quantifying the lag time to detect barriers in landscape genetics. *Molecular Ecology* **19**, 4179–91.

Lange, R., Durka, W., Holzhauer, S.I.J., Wolters, V., & Diekötter, T. (2010) Differential threshold effects of habitat fragmentation on gene flow in two widespread species of bush crickets. *Molecular Ecology* **19**, 4936–48.

Levin, S.A. (1992) The problem of pattern and scale in ecology: the Robert H. MacArthur award lecture. *Ecology* **73**, 1943–67.

Lindsay, D.L., Barr, K.R., Lance, R.F., Tweddale, S.A., Hayden, T.J., & Leberg, P.L. (2008) Habitat fragmentation and genetic diversity of an endangered, migratory songbird, the golden-cheeked warbler (*Dendroica chrysoparia*). *Molecular Ecology* **17**, 2122–33.

Lugon-Moulin, N. & Hausser, J. (2002) Phylogeographical structure, postglacial recolonization and barriers to gene flow in the distinctive Valais chromosome race of the common shrew (*Sorex araneus*). *Molecular Ecology* **11**, 785–94.

Manel, S. & Holderegger, R. (2013) Ten years of landscape genetics. *Trends in Ecology and Evolution* **28**, 614–21.

Manier, M.K. & Arnold, S.J. (2006) Ecological correlates of population genetic structure: a comparative approach using a vertebrate metacommunity. *Proceedings of the Royal Society B: Biological Sciences* **273**, 3001–9.

Marchi, C., Andersen, L.W., & Loeschcke, V. (2013) Effects of land management strategies on the dispersal pattern of a beneficial arthropod. *PLoS One* **8**, e66208.

McDevitt, A.D., Oliver, M.K., Piertney, S.B., Szafrańska, P.A., Konarzewski, M. & Zub, K. (2013) Individual variation in dispersal associated with phenotype influences fine-scale genetic structure in weasels. *Conservation Genetics* **14**, 499–509.

McGarigal, K. & Cushman, S.A. (2002) Comparative evaluation of experimental approaches to the study of habitat fragmentation effects. *Ecological Applications* **12**, 335–45.

McGarigal, K. & Cushman, S.A. (2005) The gradient concept of landscape structure. In: J. Wiens & M. Moss (eds.), *Issues and Perspectives in Landscape Ecology.* Cambridge University Press, Cambridge, pp. 112–19.

Mech, S.G. & Hallett, J.G. (2001) Evaluating the effectiveness of corridors: a genetic approach. *Conservation Biology* **15**, 467–74.

Mockford, S.W., Herman, T.B., Snyder, M., & Wright, J.M. (2007) Conservation genetics of Blanding's turtle and its application in the identification of evolutionarily significant units. *Conservation Genetics* **8**, 209–19.

Murphy, M.A., Evans, J.S., & Storfer, A. (2010a) Quantifying *Bufo boreas* connectivity in Yellowstone National Park with landscape genetics. *Ecology* **91**, 252–61.

Murphy, M.A., Dezzani, R.J., Pilliod, D.S., & Storfer, A. (2010b) Landscape genetics of high mountain frog metapopulations. *Molecular Ecology* **19**, 3634–49.

Okoli, C. & Pawlowski, S.D. (2004) The Delphi method as a research tool: an example, design considerations and applications. *Information & Management* **42**, 15–29.

Paetkau, D., Calvert, W., Stirling, I., & Strobeck, C. (1995) Microsatellite analysis of population structure in Canadian polar bears. *Molecular Ecology* **4**, 347–54.

Paetkau, D., Vázquez-Domínguez, E., Tucker, N.I.J., & Moritz, C. (2009) Monitoring movement into and through a newly planted rainforest corridor using genetic analysis of natal origin. *Ecological Management and Restoration* **10**, 210–16.

Paquette, S.R., Talbot, B., Garant, D., Mainguy, J., & Pelletier, F. (2014) Modeling the dispersal of the two main hosts of the raccoon rabies variant in heterogeneous environments with landscape genetics. *Evolutionary Applications* **7**, 734–49.

Pavlacky, D.C., Goldizen, A.W., Prentis, P.J., Nicholls, J.A., & Lowe, A.J. (2009) A landscape genetics approach for quantifying the relative influence of historic and contemporary habitat heterogeneity on the genetic connectivity of a rainforest bird. *Molecular Ecology* **18**, 2945–60.

Pavlacky, D.C., Possingham, H.P., Lowe, A.J., Prentis, P.J., Green, D.J., & Goldizen, A.W. (2012) Anthropogenic landscape change promotes asymmetric dispersal and limits regional patch occupancy in a spatially structured bird population. *Journal of Animal Ecology* **81**, 940–52.

Pérez-Espona, S., Pérez-Barbería, F.J., Mcleod, J.E., Jiggins, C.D., Gordon, I.J., & Pemberton, J.M. (2008) Landscape features affect gene flow of Scottish Highland red deer (*Cervus elaphus*). *Molecular Ecology* **17**, 981–96.

Prather, C.M., Pelini, S.L., Laws, A., Rivest, E., Woltz, M., Bloch, C.P., Del Toro, I., Ho, C.K., Kominoski, J., Scott-Newbold, T.A., Parsons, S., & Joern, A. (2013) Invertebrates, ecosystem services and climate change. *Biological Reviews* **88**, 327–48.

Proctor, M.F., McLellan, B.N., Strobeck, C., & Barclay, R.M.R. (2005) Genetic analysis reveals demographic fragmentation of grizzly bears yielding vulnerably small populations. *Proceedings of the Royal Society B: Biological Sciences* **272**, 2409–16.

Prunier, J.G., Kaufmann, B., Fenet, S., Picard, D., Pompanon, F., Joly, P., & Lena, J.P. (2013) Optimizing the trade-off between spatial and genetic sampling efforts in patchy populations: towards a better assessment of functional connectivity using an individual-based sampling scheme. *Molecular Ecology* **22**, 5516–30.

Pulliam, H.R. (1988) Sources, sinks and population regulation. *The American Naturalist* **132**, 652–61.

Radespiel, U., Rakotondravony, R., & Chikhi, L. (2008) Natural and anthropogenic determinants of genetic structure in the largest remaining population of the endangered golden-brown mouse lemur, *Microcebus ravelobensis*. *American Journal of Primatology* **70**, 860–70.

Rees, E.E., Pond, B.A., Cullingham, C.I., Tinline, R., Ball, D., Kyle, C.J., & White, B.N. (2008) Assessing a landscape barrier using genetic simulation modeling: implications for raccoon rabies management. *Preventive Veterinary Medicine* **86**, 107–23.

Riley, S.P.D., Pollinger, J.P., Sauvajot, R.M., York, E.C., Bromley, C., Fuller, T.K., & Wayne, R.K. (2006) FAST-TRACK: a southern California freeway is a physical and social barrier to gene flow in carnivores. *Molecular Ecology* **15**, 1733–41.

Rittenhouse, T.A.G. & Semlitsch, R.D. (2006) Grasslands as movement barriers for a forest-associated salamander: migration behavior of adult and juvenile salamanders at a distinct habitat edge. *Biological Conservation* **131**, 14–22.

Robinson, S.J., Samuel, M.D., Lopez, D.L., & Shelton, P. (2012) The walk is never random: subtle landscape effects shape gene flow in a continuous white-tailed deer population in the Midwestern United States. *Molecular Ecology* **21**, 4190–205.

Robinson, S.J., Samuel, M.D., Rolley, R.E., & Shelton, P. (2013) Using landscape epidemiological models to understand the distribution of chronic wasting disease in the Midwestern USA. *Landscape Ecology* **28**, no. 10, pp. 1923–1935.

Row, J.R., Blouin-Demers, G., & Lougheed, S.C. (2010) Habitat distribution influences dispersal and fine-scale genetic population structure of eastern foxsnakes (*Mintonius gloydi*) across a fragmented landscape. *Molecular Ecology* **19**, 5157–71.

Sacks, B.N., Brown, S.K., & Ernest, H.B. (2004) Population structure of California coyotes corresponds to habitat-specific breaks and illuminates species history. *Molecular Ecology* **13**, 1265–75.

Safner, T., Miller, M.P., McRae, B.H., Fortin, M.-J., & Manel, S. (2011) Comparison of Bayesian clustering and edge detection methods for inferring boundaries in landscape genetics. *International Journal of Molecular Sciences* **12**, 865–89.

Schunter, C., Carreras-Carbonell, J., Planes, S., Sala, E., Ballesteros, E., Zabala, M., Harmelin, J.-G., Harmelin-Vivien, M., MacPherson, and Pascual, M. (2011) Genetic connectivity patterns in an endangered species: the dusky grouper (Epinephelus marginatus). *Journal of Experimental Marine Biology and Ecology* **401**, 126–33.

Schwartz, M.K. & McKelvey, K.S. (2009) Why sampling scheme matters: the effect of sampling scheme on landscape genetic results. *Conservation Genetics* **10**, 441–52.

Schwartz, M.K., Luikart, G., & Waples, R.S. (2007) Genetic monitoring as a promising tool for conservation and management. *Trends in Ecology and Evolution* **22**, 25–33.

Schwartz, M.K., Copeland, J.P., Anderson, N.J., Squires, J.R., Inman, R.M., McKelvey, K.S., Pilgrim, K.L., Waits, L.P., & Cushman, S.A. (2009) Wolverine gene flow across a narrow climatic niche. *Ecology* **90**, 3222–32.

Segelbacher, G., Cushman, S.A., Epperson, B.K., Fortin, M.-J., François, O., Hardy, O.J., Holderegger, R., Taberlet, P., Waits, L.P., & Manel, S. (2010) Applications of landscape genetics in conservation biology: concepts and challenges. *Conservation Genetics* **11**, 375–85.

Selkoe, K.A., Watson, J.R., White, C., Horin, T.B., Iacchei, M., Mitarai, S., Siegel, D.A., Gaines, S.D., & Toonen, R.J. (2010) Taking the chaos out of genetic patchiness: seascape genetics reveals ecological and oceanographic drivers of genetic patterns in three temperate reef species. *Molecular Ecology* **19**, 3708–26.

Shirk, A.J., Wallin, D.O., Cushman, S.A., Rice, C.G., & Warheit, K.I. (2010) Inferring landscape effects on gene flow: a new model selection framework. *Molecular Ecology* **19**, 3603–19.

Shirk, A.J., Cushman, S.A., & Landguth, E.L. (2012) Simulating pattern–process relationships to validate landscape genetic models. *International Journal of Ecology* **2012**, 539109.

Short Bull, R.A., Cushman, S.A., Mace, R., Chilton, T., Kendall, K.C., Landguth, E.L., Schwartz, M.K., McKelvey, K., Allendorf, F.W., & Luikart, G. (2011) Why replication is important in landscape genetics: American black bear in the Rocky Mountains. *Molecular Ecology* **20**, 1092–107.

Sork, V.L. & Waits, L.P. (2010) Contributions of landscape genetics – approaches, insights and future potential. *Molecular Ecology* **19**, 3489–95.

Sotiropoulos, K., Eleftherakos, K., Tsaparis, D., Kasapidis, P., Giokas, S., Legakis, A., & Kotoulas, G. (2013) Fine scale spatial genetic structure of two syntopic newts across a network of ponds: implications for conservation. *Conservation Genetics* **14**, 385–400.

Spear, S.F. & Storfer, A. (2008) Landscape genetic structure of coastal tailed frogs (*Ascaphus truei*) in protected vs. managed forests. *Molecular Ecology* **17**, 4642–56.

Spear, S.F., Peterson, C.R., Matocq, M.D., & Storfer, A. (2005) Landscape genetics of the blotched tiger salamander (*Ambystoma tigrinum melanostictum*). *Molecular Ecology* **14**, 2553–64.

Spear, S.F., Balkenhol, N., Fortin, M.-J., McRae, B.H., & Scribner, K.T. (2010) Use of resistance surfaces for landscape genetic studies: considerations for parameterization and analysis. *Molecular Ecology* **19**, 3576–91.

Spear, S.F., Crisafulli, C.M., & Storfer, A. (2012) Genetic structure among coastal tailed frog populations at Mount St. Helens is moderated by post-disturbance management. *Ecological Applications* **22**, 856–69.

Storfer, A., Murphy, M.A., Evans, J.S., Goldberg, C.S., Robinson, S., Spear, S.F., Dezzani, R., Delmelle, E., Vierling, L., & Waits, L.P. (2007) Putting the "landscape" in landscape genetics. *Heredity* **98**, 128–42.

Storfer, A., Murphy, M.A., Spear, S.F., Holderegger, R., & Waits, L.P. (2010) Landscape genetics: where are we now? *Molecular Ecology* **19**, 3496–514.

Stow, A.J., Sunnucks, P., Briscoe, D.A., & Gardner, M.G. (2001) The impact of habitat fragmentation on dispersal of Cunningham's skink (*Egernia cunninghami*): evidence from allelic and genotypic analyses of microsatellites. *Molecular Ecology* **10**, 867–78.

Thompson, C.M. & McGarigal, K. (2002) The influence of research scale on bald eagle habitat selection along the lower Hudson River, New York (USA). *Landscape Ecology* **17**, 569–86.

Tischendorf, L. (2001) Can landscape indices predict ecological processes consistently? *Landscape Ecology* **16**, 235–54.

van der Wal, E., Paquet, P.C., & Andrés, J.A. (2012) Influence of landscape and social interactions on transmission of disease in a social cervid. *Molecular Ecology* **21**, pp. 1271–82.

Vignieri, S.N. (2005) 'Streams over mountains: influence of riparian connectivity on gene flow in the Pacific jumping mouse (*Zapus trinotatus*)', *Molecular Ecology*, vol. **14**, no. 7, pp. 1925–37.

Wang, I.J., Savage, W.K. & Bradley Shaffer, H. (2009) 'Landscape genetics and least-cost path analysis reveal unexpected dispersal routes in the California tiger salamander (*Ambystoma californiense*)', *Molecular Ecology*, vol. **18**, no. 7, pp. 1365–74.

Wasserman, T.N., Cushman, S.A., Schwartz, M.K., & Wallin, D.O. (2010) Spatial scaling and multi-model inference in landscape genetics: *Martes americana* in northern Idaho. *Landscape Ecology* **25**, 1601–12.

Wasserman, T.N., Cushman, S.A., Shirk, A.S., Landguth, E.L., & Littell, J.S. (2012) Simulating the effects of climate change on population connectivity of American marten (*Martes americana*) in the northern Rocky Mountains, USA. *Landscape Ecology* **27**, 211–25.

Wasserman, T.N., Cushman, S.A., Littell, J.S., Shirk, A.J., & Landguth, E.L. (2013) Population connectivity and genetic diversity of American marten (*Martes americana*) in the United States northern Rocky Mountains in a climate change context. *Conservation Genetics* **14**, 529–41.

Wells, C.N., Williams, R.S., Walker, G.L., & Haddad, N.M. (2009) Effects of corridors on genetics of a butterfly in a landscape experiment. *Southeastern Naturalist* **8**, 709–22.

Wiens, J.A. (1989) Spatial scaling in ecology. *Functional Ecology* **3**, 385–97.

Wilson, G.A. & Rannala, B. (2003) Bayesian inference of recent migration rates using multilocus genotypes. *Genetics* **163**, 1177–91.

Wu, J. & Loucks, O.L. (1995) From balance of nature to hierarchical patch dynamics: a paradigm shift in ecology. *The Quarterly Review of Biology* **70**, 439–66.

Zalewski, A., Piertney, S.B., Zalewska, H., & Lambin, X. (2009) Landscape barriers reduce gene flow in an invasive carnivore: geographical and local genetic structure of American mink in Scotland. *Molecular Ecology* **18**, 1601–15.

Zeller, K.A., McGarigal, K., Beier, P., Cushman, S.A., Vickers, T.W., & Boyce, W.M. (2014) Sensitivity of landscape resistance estimates based on point selection functions to scale and behavioral state: pumas as a case study. *Landscape Ecology* **29**, 541–57.

Zeller, K.A., McGarigal, K., & Whiteley, A.R. (2012) Estimating landscape resistance to movement: a review. *Landscape Ecology* **27**, 777–97.

Zellmer, A.J. & Knowles, L.L. (2009) Disentangling the effects of historic vs. contemporary landscape structure on population genetic divergence. *Molecular Ecology* **18**, 3593–602.

*Chapter 13*

# WATERSCAPE GENETICS – APPLICATIONS OF LANDSCAPE GENETICS TO RIVERS, LAKES, AND SEAS

*Kimberly A. Selkoe,[1] Kim T. Scribner,[2] and Heather M. Galindo[3]*

[1]*National Center for Ecological Analysis and Synthesis (NCEAS), University of California Santa Barbara, USA & Hawaii Institute of Marine Biology, University of Hawaii, USA*
[2]*Department of Fisheries and Wildlife & Department of Zoology, Michigan State University, USA*
[3]*University of Washington Bothell, USA*

## 13.1 INTRODUCTION

Aquatic environments dominate the planet, are critical to the planet's ecological functioning and climate and are important sources of ecosystem services and biodiversity. Yet aquatic landscape genetic studies are underrepresented in the literature (Postel & Carpenter 1997; Wilson & Carpenter 1999; Zedler & Kercher 2005; Esselman et al. 2011). Storfer et al. (2010) found that only 15% of landscape genetic studies were conducted in freshwater and 6% in marine habitats. The integration of ecological concepts with aquatic population genetic data has been challenging due to a combination of disparate training for scientists, lack of communication among disciplines, and both technical and conceptual difficulties in

collecting appropriate multiscale data to quantify how **waterscape features** influence genetics (Storfer et al. 2007; Balkenhol et al. 2009; Wagner & Fortin 2013).

However, even before the term landscape genetics was coined, interdisciplinary data sets combining spatial environmental and genetic information contributed to basic understanding of marine and freshwater population patterns (e.g., the relationship of biogeographic and phylogeographic patterns and processes: Avise 1992; Bernatchez & Wilson 1998) and made important management contributions (e.g., Berst & Simon 1981; Bermingham & Avise 1986; Allendorf et al. 1987; Allendorf & Waples 1996; Hendry & Stearns 2004). This early work also highlighted particular challenges of applying simplistic population genetic models to the

*Landscape Genetics: Concepts, Methods, Applications*, First Edition. Edited by Niko Balkenhol, Samuel A. Cushman, Andrew T. Storfer, and Lisette P. Waits.
© 2016 John Wiley & Sons, Ltd. Published 2016 by John Wiley & Sons, Ltd.

great biological diversity and spatial complexity of aquatic populations, which is caused by extreme or variable effective population size and genetic diversity, **anisotropic** and episodic gene flow that create non-equilibrium conditions, as well as low genetic differentiation in marine systems (e.g., Waples 1990; Hedgecock et al. 1992; Whitlock & McCauley 1999; Grosberg & Cunningham 2001). In all cases, new developments in conceptual and statistical landscape genetic approaches have opened up new understanding of the permeability of aquatic habitats to dispersal and gene flow, and promise new routes to overcoming particular challenges to synthesizing waterscape genetic data.

In this chapter we start by describing the dominant waterscape features of marine and freshwater environments and key biological characteristics of aquatic organisms that are most relevant to the design and analysis of waterscape genetic studies. Then we provide examples from the literature that represent a range of research questions and methodological approaches to analyses of aquatic genetic data, with a focus on applications of the emerging analytical tools described in earlier chapters. Information on sources for obtaining spatial aquatic landscape data and software can be found in Box 13.1. Finally, we suggest several future directions for research.

---

**Box 13.1**  An introductory guide to spatial aquatic landscape data and software for geneticists

Considerable advances in GIS system capabilities have increased the amount and diversity of data that are accessible to landscape geneticists, facilitating characterizations of aquatic landscape features. Remotely sensed aquatic data are frequently used to characterize environmental, structure, and biotic features associated with genetic sampling sites (or co-located), to estimate environmental niche (e.g., temperature, precipitation, hydrogeomorphology) and to model or predict species distributions and species richness. Spatially-explicit data are also increasingly available regarding alterations of aquatic landscapes associated with human use, for example, the extent of urban development (e.g., amount of impervious surfaces that affect runoff and waste), agriculture use (e.g., nutrient and pesticide input), and impoundments (e.g., affecting the degree of connectivity, mean and variance in stream temperatures, dissolved organic carbon (DOC), stream velocity, and turbidity).

Several analytical tools are widely used with aquatic landscape data and species presence–absence data that utilize remotely sensed or empirical field data to model species niche and to predict the effects of human disturbance on measures of biodiversity. Examples of statistical approaches include classification and regression trees (CARTs), general linear models (GLMs), and multivariate adaptive regression splines (MARSs). Alternative approaches have utilized presence-only observations and other analytical methods that are implemented in ecological niche models, such as BIOCLIM, DOMAIN, ESRI, MaxEnt, and GARP.

**Aquatic data resources for landscape geneticists (assembled in 2011)**
- Global Biodiversity Information Facility (GBIF) – http://www.gbif.org
- WorldClim GIS data – http://www.worldclim.org
- Hydro 1K – http://eros.usgs.gov/
- National Water Quality Assessment Program (NAWQA) – http://water.usgs.gov/nawqa/
- USGS National Water Information System (NWIS) – http://water.usgs.gov/
- FishBase – http://www.fishbase.org/search.php
- VertNet (FishNet, HerpNET, MaNIS, ORNIS) – http://vertnet.org/index.php
- AmphibiaWeb – http://amphibiaweb.org/
- Aquatic GAP – http://gapanalysis.usgs.gov/aquatic-gap/
- Natural Resources Conservation Service (NRCS) Geospatial Data Gateway – http://datagateway.nrcs.usda.gov/
- US Fish and Wildlife Service (USFWS) – http://www.fws.gov/GIS/index.htm
- National Environmental Satellite, Data, and Information Service – http://www.nesdis.noaa.gov/index.html
- Tropicos (plants) – http://www.tropicos.org
- OBIS (marine life) – http://www.iobis.org
- NOAA National Centers for Environmental Information – http://www.nodc.noaa.gov/

- Human Impacts to Marine Ecosystems Project – http://www.nceas.ucsb.edu/globalmarine/impacts
- SeaWiFS Project – http://oceancolor.gsfc.nasa.gov/SeaWiFS/
- Landsat – http://landsat.gsfc.nasa.gov/
- National Data Buoy Center – http://www.ndbc.noaa.gov/
- National Weather Service Climate Prediction Center – http://www.cpc.ncep.noaa.gov/data/indices/
- Pacific Decadal Oscillation – http://jisao.washington.edu/pdo/
- North Pacific Gyre Oscillation – http://www.o3d.org/npgo/
- National Biodiversity Network Gateway – http://data.nbn.org.uk/
- Mapping European Seabed Habitats – http://www.searchmesh.net/
- Joint Nature Conservation Committee – http://jncc.defra.gov.uk/page-2117
- NOAA Office for Coastal Management – http://www.csc.noaa.gov/digitalcoast/data/benthiccover
- PISCO – http://www.piscoweb.org/data/access-and-applications
- UNEP-WCMC – http://www.unep-wcmc.org/
- Ocean Health Index – http://www.oceanhealthindex.org

### National freshwater efforts that use remotely sensed landcover and land-use data plus analytical tools for the management of fisheries resources in the United States

- US Army Corps of Engineers National Inventory of Dams – http://geo.usace.army.mil/nwpdocs/metadata/DISTRICT.htm
- US Environmental Protection Agency (USEPA) Regional Environmental Monitoring and Assessment Program www.epa.gov/emap/remap/index.html
- US Environmental Protection Agency and US Geological Survey (USEPA and USGS). National hydrography dataset plus, NHDPlus – www.horizon-systems.com/nhdplus/
- US Geological Survey (USGS) – Active mines and mineral processing plants in the United States http://tin.er.usgs.gov/metadata/mineplant.faq.html
- US Environmental Protection Agency and US Geological Survey (USEPA and USGS) National Elevation Dataset – http://ned.usgs.gov/
- US Environmental Protection Agency and US Geological Survey (USEPA and USGS) Multi-Resolution Land Characteristics Consortium (MRLC) – National Land Cover Database (NLCD) – www.mrlc.gov/nlcd_multizone_map.php

### Examples of national efforts to apply GIS and landscape data to fisheries resources in freshwater landscapes in the United States

- A nationwide assessment of the status of fish habitats in freshwater systems throughout the United States: National Fish Habitat Action Plan – www.fishhabitat.org
- National effort to characterize stream vulnerability to climate and land use changes to predict fish changes in habitat at multiple spatial scales – http://fishhabclimate.org
- VAST – Brenden et al. (2008) – valley segment affinity search technique – valley segments have been advocated as a focal spatial scale for river assessment and monitoring
- FLoWS – functional linage of watersheds and streams – Theobald et al. (2005) – tool to spatially represent stream environmental covariates and hydrologic features

### Sources for further information

Bolstad, P. (2008) GIS Fundamentals: A First Text on Geographic Information Systems, 3rd edition. Eider Press.
  Longley, P.A., Goodchild, M.F., Maguire, D.J., & Rhind, D.W. (2005) *Geographic Information Systems and Science*, 2nd edition. John Wiley & Sons, Inc.
  Carlos A. Furuti – http://www.progonos.com/furuti/MapProj/Normal/TOC/cartTOC.html.
  ESRI – http://training.esri.com/acb2000/showdetl.cfm?DID=6andProduct_ID=826.

## 13.2 UNDERSTANDING MARINE AND FRESHWATER ENVIRONMENTS

In contrast to terrestrial species, aquatic organisms must often exert more energy to stay in place than to move (Mann & Lazier 2006). Consequently, dispersal and migration are ubiquitous features of aquatic life. However, characteristics of marine and freshwater environments differ considerably, as do the phenotypic and genetic adaptations of organisms inhabiting these environments. For example, contemporary census and effective population sizes are typically smaller in freshwater than marine systems, increasing the likelihood of local extinction via stochastic demographic and environmental events (Lande 1988), whereas large population sizes in marine species limit the effects of drift and genetic differentiation. The spatial and temporal resolution (grain) and the size (extent) of system dynamics and species ranges are often larger in the marine realm, where boundaries with ecological relevance are also harder to ascertain.

The scale and scope of human impacts on abundance and connectivity also differ among terrestrial, freshwater, and marine systems. Terrestrial landscape genetics is often motivated by a need to protect remaining dispersal corridors or mitigate increasing human land development. While habitat fragmentation and loss can be dominant types of anthropogenic stress in riverine freshwater and coastal systems like seagrass habitats, they are not in offshore marine systems. Our present ability to successfully manipulate aqueous dispersal processes to create "dispersal corridors" is limited, particularly in the marine realm. Waterscape genetic studies often seek to relate scales and patterns of dispersal to environmental drivers, motivated by management decisions about where to preserve or create habitats for connectivity. Nevertheless, waterscape genetic analyses are badly needed to investigate causal relationships among measures of human disturbance, the abilities of aquatic organisms to respond via adaptation or relocation across aquatic landscapes, and to identify drivers of aquatic biodiversity.

### 13.2.1 Waterscape features of marine systems

The majority of marine species begin life as minute larvae that are adapted for dispersal in ocean currents over days or months (Scheltema 1971; Strathmann 1985; Pechenik 1999). Water can be a source of diffusion and dilution, spreading nutrients, gametes, larvae, or prey over large distances. However, physical and biological forces often interact to *advect* (i.e., push) and aggregate these same elements, creating distinct habitat patches and temporal pulses of nutrients, food, temperature, and larval density, all which effect survival during transport (Siegel et al. 2003; Mann & Lazier 2006; Woodson & McManus 2007; Marshall et al. 2010). The clumped nature of larval distributions constrains the trajectories and **dispersal kernel** produced by a group of larvae spawned at a single time and place, but not always in a predictable or consistent way (Mitarai et al. 2008). Genetic approaches are commonly used to estimate dispersal patterns in marine species or to test predictions about transport mechanisms in the sea that are otherwise hard to quantify (Palumbi & Pinsky 2014; Selkoe et al. 2008; Selkoe & Toonen 2011; Riginos & Liggins 2013; Liggins et al. 2013).

The importance of high small-scale variation in dispersal rates and pathways to longer-term gene flow estimates is still unclear, but may act like frequent recolonization, preventing equilibrium between rates of genetic drift and migration (Mitarai et al. 2009; Watson et al. 2012). For some species, frequent output and high abundance of larvae may smooth over this unpredictable and irregular small-scale patchiness. For others, it is possible that so few larval cohorts survive that **sweepstake reproductive success** ensues, depressing effective population size and disrupting genetic equilibrium (reviewed in Hedgecock & Pudovkin 2011). Effective population size has been shown to typically be many orders of magnitude lower than census size, supporting theories that variance in fecundity is high (Hauser & Carvalho 2008). Importantly, variation in census population size can also be extreme over both space and time and grossly violate assumptions about uniform population size in population genetic models (Sagarin et al. 2006), so independent estimates of population densities and effective population sizes should also be incorporated into analyses.

At certain spatiotemporal scales, discontinuities in the fluid environment and the shifting "migration corridors" they form constitute significant barriers, yet data sets used to quantify these invisible sources of waterscape resistance are challenging to generate. Robust data on locations of stationary and mobile **eddies** and **jets** are only recently accumulating, enabling delineation of the ocean into areas with low mixing, high mixing, and unidirectional flow (Fig. 13.1). Even more difficult to quantify are the locations and effects of smaller-scale eddies and other forms of **turbulence** that are key features associated with larval transport (Siegel et al. 2003; Pineda et al. 2007; Mitarai et al. 2008). Furthermore, temporal variation in larger-scale **hydrographic regimes** can be seasonal, annual, decadal, or storm-based, with complete flow reversals that often provide important means for highly mobile larvae to return to natal habitats or to colonize areas that seem otherwise unreachable (Zacherl et al. 2003; Woodson & McManus 2007; Morgan & Fisher 2010). Over longer time scales, current patterns shape spatial configurations and scale of habitat patches (e.g., biogenic reefs), possibly creating co-variation between habitat- and transport-based influences on spatial genetic structure. On geologic time scales, the signatures of past glaciation events and dramatic changes in sealevel rise can still be seen in extant marine spatial genetic structure (Hellberg et al. 2001; Marko 2004; Hickerson & Cunningham 2005; Maggs et al. 2008). Landscape genetic approaches hold promise to address unknowns about how short-term events influence long-term patterns and how historical events influence present-day patterns (Marko & Hart 2011).

For the majority of marine species with pelagic larvae, realized migration rates are influenced not just by the physical transport in currents but potentially also processes related to how these larvae reach offshore flows, behave during transport, access suitable settlement habitats, and then recruit to a population (Fig. 13.2). The net effect of these often overlooked "details" is to decouple dispersal trajectories from physical transport processes, but their roles as waterscape effects on gene flow are poorly understood. More sophisticated spatial data on

waterscape features such as **tidal bores** (which deliver larvae to inshore habitats; Pineda et al. 2010), **upwelling/relaxation cycles** (which enable larvae to return to the natal habitat; Graham & Largier 1997; Morgan & Fisher 2010), soundscapes (which allow larvae to navigate to settlement habitats; Leis et al. 2011), and community composition (which influences density-dependent recruitment; Schmitt & Holbrook 2007) are needed. In sum, waterscape genetic analyses hold great promise to help elucidate the basic controls on marine population dynamics and dispersal, but will require creative and potentially data-demanding combinations of hydrodynamic, habitat, and biological data.

Marine dispersal is not always dominated by the larval life stage. While highly migratory species, including sharks, marine mammals, and many commercially exploited fish species, may seem to have no physical dispersal barriers in the ocean, many show spatial genetic structure due to site-associated spawning, breeding or kin aggregation (e.g., Jørgensen et al. 2005; Duncan et al. 2006; Kershaw & Rosenbaum 2014). Both migratory and otherwise sedentary species can show remarkably long spawning migrations, daily or yearly, disconnecting spatial patterns of genetic structure from the distribution of adults over the habitat and necessitating tailored approaches to testing dispersal influences on genetic patterns.

### 13.2.2  Features of freshwater systems

In contrast to marine systems, the geography of freshwater systems, inclusive of both **lentic** and **lotic** types, often has strong hierarchical organization (Fausch et al. 2002; Soranno et al. 2010; Figs 13.3 and 13.4). In the small-scale **tributary** networks of linear

**Fig. 13.1** "Perpetual Ocean," a visualization of the world's surface ocean currents. Width of lines indicates speed of flow. The brightest features are large-scale stable and more stationary mean flow, which tends to be laminar (e.g., the Gulf Stream running around Florida and up the east coast of North America). Only large-scale eddies are depicted here, as they are fairly stable and slow moving (even when fast-spinning). Small-scale fleeting eddies are important transport features that are not visible here. The many thin, dash-size lines indicate areas of the ocean with slow flow, which can create areas "cut off" from mixing, such as the panhandle of Florida. (Picture from NASA/Reuters. See animation at http://www.nasa.gov/topics/earth/features/perpetual-ocean.html.) (For a color version of this figure, please refer to the color plates section.)

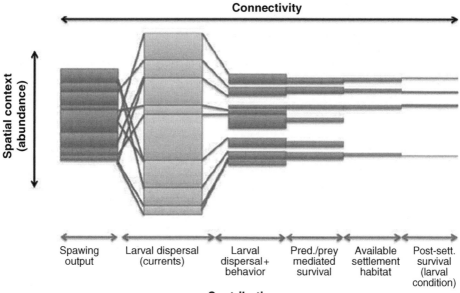

**Fig. 13.2** A representation of the combined processes that act to determine the spatial connections among populations via larval dispersal and survival. The spatial context is depicted by a series of populations (each represented by a single band on the left). The spawning output of those populations is dependent on the size, fecundity, etc., of each population; hence the relative size of each band indicates spatial variability in reproductive output. The larval physical dispersal process (***advection*** and diffusion) acts to mix the young (the crossing lines to different bars), expand the spatial distribution of the larvae (expanded size of the bands), and consequently dilute the concentration (lighter color). With the inclusion of larval behaviors (e.g., vertical migration, horizontal swimming), the diffusive nature of the physical processes may be counteracted through biophysical interactions; hence the constriction of the spatial extent and increase in density (smaller bands and darker color). In addition to physical dispersal, various biological and physical factors operate to extract a spatially varying loss (mortality), as depicted by a continual reduction in available survivors (smaller bands) due to predator/prey interactions, availability of settlement habitat, and post-settlement survivorship, in part driven by the larval condition at the time of settlement (e.g., carryover effects). Comparison of the two extremes of this figure (the source and receiving locations) provides an idea of the pattern of connectivity or the scale of successful dispersal. The steps between these two ends provide an overview of the processes that contribute to the dispersal kernel. (Reprinted from Cowen and Sponaugle 2009 with permission from Annual Reviews, Inc.)

(e.g., riverine) settings, scales, rates, and directionality of connectivity are influenced by stream flow, volume, gradient, and dynamic interactions within ***riparian***, floodplain, and subsurface ***groundwater*** areas, including episodic flooding events (Benda et al. 2004; Stanford et al. 2005). Long- and short-term interactions of climate, floods and succession produce a shifting habitat mosaic (Ward & Stanford 1995). At the scale of ***stream reaches*** or valley segments, habitat is affected by the superficial geology (e.g., soil type, permeability, depth, parent substrate) and by landuse within adjoining aquatic habitats (e.g., sediment, nutrient, and pollutant inputs). Productivity and diversity are affected by organic matter, nutrients, and upwelling

(Franken et al. 2001). Individual wetlands, lakes, tributary networks, or stream reaches are physically connected either directly within a larger matrix of surface water or hydrologically through subsurface flow of groundwater in the ***hyphoretic zone*** (Fig. 13.4C and D; Benda et al. 2004; Stanford et al. 2005; Brenden et al. 2008; Boulton et al. 2010). At the largest hierarchical scale are watersheds (i.e., drainage basins), which vary in the composition and fluxes of organisms, materials, and energy among habitats, creating unique functional dynamics (Fig. 13.3; Schlosser & Angermeier 1995; Wiens 1999). Drainage networks are a strong force of hierarchical genetic structuring for many aquatic species (Hughes et al. 2009) and sometimes

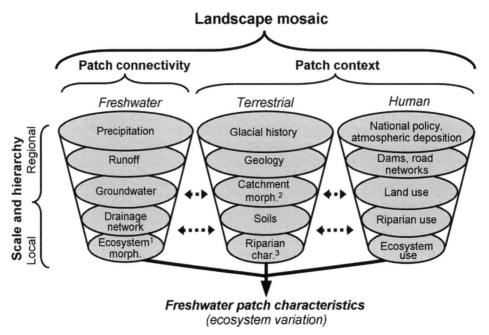

**Fig. 13.3** Components of freshwater limnological landscapes. (Reproduced from Soranno et al. 2010 with permission from the American Institute of Biological Sciences.)

historical delineations show stronger genetic signatures than those at the present day (Mock et al. 2010; Tamkee et al. 2010; Bezault et al. 2011; Cook et al. 2011; Loxterman & Keeley 2012).

Similar to marine habitats, large lakes such as the North American Great Lakes are characterized by spatial heterogeneity in bathometry, currents, temperature, and levels of primary productivity impacting permeability of lake environments to dispersing individuals. Depth and temperature preferences can act to isolate populations (e.g., Berst and Spangler 1973; Fabrizio et al. 2000). Least-cost paths based on species depth limitations and current patterns have been used to adjust shoreline distances for spatial genetic studies (e.g., Homola et al. 2012). Lakescape genetic studies have linked genetic similarity of shoreline populations to prevailing currents and seasonal spawning temperature regimes (e.g., genetic isolation by spawning time; see Hendry & Day 2005; Scribner & Stott, unpublished data).

Many organisms use specific, and often spatially separated, habitats during different seasons and ontogenetic stages (Fig. 13.5). Given the temporally and spatially dynamic nature of freshwater aquatic systems, species' access to resources or areas of suitable habitat required to meet life requirements necessitates a high degree of movement, often including large areas over the course of a year (Hitt & Angermeier 2008). For example, many **anadromous** or **adfluvial** species spawning areas are often some distance from habitats required for larval and juvenile rearing, which are in turn different from areas occupied by adults (Duong et al. 2011). Semi-aquatic organisms (e.g., insects, amphibians, and turtles) often travel between networks of permanent bodies of water, ephemeral wetlands, and adjacent terrestrial landscapes that are differentially permeable to dispersal (Morreale et al. 1984; Semlitsch 1998; Calhoun et al. 2003; Roe et al. 2009). For example, some turtles and amphibians make extensive terrestrial movements in search of mates (Morreale et al. 1984), ephemeral resources in seasonally flooded wetlands (Bodie & Semlitsch 2000), and overwintering sites (Buhlmann et al. 2009). The extent and frequency of movements in aquatic and terrestrial habitats used to complete life functions have been shown to influence levels of spatial genetic structure among populations

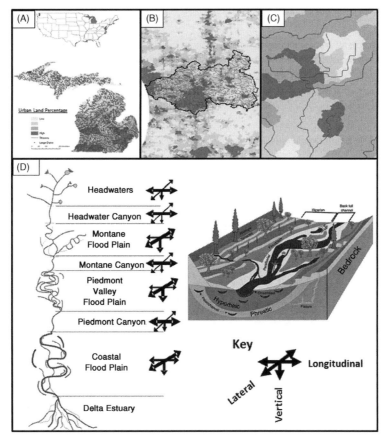

**Fig. 13.4** (A–C) Hierarchical representation of freshwater landscapes from regional to local scales including superficial (subsurface) sources of connectivity for aquatic organisms. (Figure provided by Dana Infante and Kyle Herreman.) (D) Landscape map highlighting hydrogeomorphologic features and habitat types, and with inset showing hydrodynamic features that make up flow. (From Stanford et al. 2005, Stanford 2006.) (For a color version of this figure, please refer to the color plates section.)

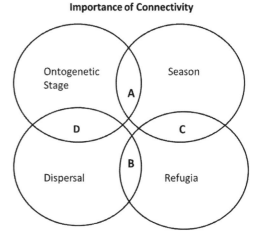

**Fig. 13.5** Freshwater organisms must have access to resources to meet life requirements during all times of the year and across all ontogenetic stages.: (A) spawning migration, (B) recolonization following disturbance (metapopulation level), (C) seasonal shifts in response to changing thermal regimes, and (D) ontogenetic shifts in forage needs.

(Semlitsch & Bodie 2003; Scribner et al. 1986, 2000). Furthermore, "atypical" drought, flooding, and other fluctuations in habitat availability or quality can result in episodic but high rates of gene flow, and can dramatically alter effective population size and levels of genetic diversity, especially in isolated and ephemeral ponds (e.g., Kennett & Georges 1990; Scribner et al. 1995; Huey et al. 2011). Capturing these events in landscape genetic analyses requires sampling at appropriate time and space scales, and is increasingly critical as the frequency and amplitude of water level fluctuations continue to intensify with climate variability and change (Faulks et al. 2010). Sensitivity of freshwater systems to change means that current distributions may be relatively new and vicariance may result in non-intuitive patterns (e.g., drainages in close proximity with historical but not current day connectivity), confounding landscape genetic analysis. For example, the high levels of genetic diversity observed for current fish populations in the Great Lakes basin are attributed to admixture from four glacial refugia (Hubbs & Lagler 1964; Bailey & Smith 1981; Underhill 1986). Despite concordant spatial distributions across species, variation in dispersal and other ecological traits created different responses to landscape features of the Great Lakes region, resulting in distinct connectivity levels among refugia following glacial

retreat (Taylor & Bentzen 1993; Lafontaine & Dodson 1997; Bernatchez & Wilson 1998; Turgeon et al. 1999; Short & Caterino 2009).

Land use changes in riparian areas can have strong impacts on connectivity, for both semi- and fully aquatic species. Dams, roads, and culverts prevent movement and change connectivity (Arthington et al. 2006; Faulks et al. 2011). Human development along waterways causes changes in water temperature and chemistry (Arthington et al. 2006) or loss of connectivity (Blum et al. 2012). Forestry practices affect stream temperatures via sedimentation, impacting movements of coldwater stream fishes such as western cutthroat trout (Valdal & Quinn 2011). Increased use of ground water for human use in agriculture and for drinking water affects stream hydrology, relative contributions of groundwater to surface water and seasonal water levels and temperatures that can have dramatic effects on stream community composition, population abundance, movement and gene flow. Landscape data collected at intermediate and large landscape scales (>1 km) have been widely used in the United States to assess effects of anthropogenic stressors on measures of ecological condition and diversity in aquatic systems studies but have been under-utilized in genetic studies (Fig. 13.6; Angermeier et al. 2002; Allan 2004; Wang et al. 2006; Vörösmarty et al. 2010; Esselman et al. 2011).

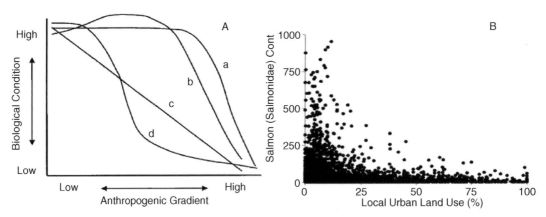

**Fig. 13.6** (A) Hypothetical relationship depicting possible responses of stream biological condition (species richness, assemblage similarity, population size or genetic diversity) to a gradient of increasing environmental stress. Possible responses include (a) non-linear response occurring in the high range of the gradient, (b) subsidy-stress response, (c) linear response, and (d) nonlinear (threshold) response occurring in the low range of the gradient. Curves (a) versus (d) indicate low versus high sensitivity to a stressor. (Modified from Allan 2004 with permission from Annual Reviews Inc.) (B) Salmonid fish abundance response to local urban land use for the Eastern United States. (Provided by Dan Wesley and Dana Infante.)

## 13.3 TYPICAL RESEARCH QUESTIONS AND APPROACHES

### 13.3.1 Drivers of spatial genetic structure in aquatic environments

Early waterscape genetic studies have sought to compare and quantify influences of physical and biotic features on spatial genetic patterns, including hydrogeomorphology, coastal upwelling, gradients in salinity and temperature, river slope and tributary scaling, habitat configuration, dams, and water pollution (e.g., Costello et al. 2003; Jørgensen et al. 2005; Hemmer-Hansen et al. 2007; Guy et al. 2008; Puritz & Toonen 2011). Below we present some examples of testing metrics of dispersal distance, environmental correlates, historical events, and other drivers of spatial genetic structure that reflect unique characteristics of waterscape genetics. We begin with reviewing particular challenges and insights associated with testing the seemingly simple expectation of isolation-by-distance (IBD) in marine and freshwater systems before covering more complex study designs.

### 13.3.2 Isolation by Euclidean distance in marine systems

One of the compelling challenges in marine ecology is determining the extent to which populations are demographically "open" or "closed", in other words, do local populations largely replenish themselves or are they strongly dependent on immigrants for replenishment (Swearer et al. 2002)? Despite conceptual differences between demographic and genetic independence (Lowe & Allendorf 2010), early studies addressing this question focused on how genetic differentiation (e.g., $F_{ST}$) tended to be low in marine systems (e.g., Waples 1998), supporting the widely held view through the 1990s that long-distance dispersal was widespread. However, low $F_{ST}$ can result simply from large effective population size and/or stepping stone dispersal at small scales (Guillot et al. 2009). Newer analytical tools can be used to quantify the relative effects of drift and gene flow, and landscape genetic approaches can help confirm whether very low $F_{ST}$ values are due to dispersal processes. Current thinking is that larvae often avoid long-distance dispersal via behavior and timing, but infrequent long-distance dispersal events still occur (e.g., Treml et al. 2012). A landscape genetics framework will be crucial to

determining the factors, such as storm events, that affect rates of successful long-distance dispersal.

A new generation of isolation-by-distance studies that estimate spatial scales of mean dispersal at small scales have been critical to this paradigm shift from viewing marine populations as open to closed (e.g., Puebla et al. 2009; Pinsky et al. 2010). Previously, many studies failed to find support for IBD, leading to the widely held belief that stepping stone dispersal and drift-migration equilibrium are uncommon in marine systems, and that long-distance dispersal results in a weak population genetic structure (reviewed in Grosberg & Cunningham 2001). However, repeated failure to reject the null hypothesis of panmixia or to detect IBD may have resulted from inadequate sample sizes (i.e., low statistical power; Waples 1998; Taylor et al. 2000) and inappropriate sampling design due to poor understanding of the relevant "local" scale (reviewed in Selkoe & Toonen 2011).

Individual-level genetic analysis based on continuously distributed, geo-referenced samples is often a more powerful and appropriate sampling design for landscape genetic analysis than population-based analyses (Manel et al. 2003; Landguth et al. 2010), but it is rarely used in marine systems. Where used, it has led to powerful insights. Puebla et al. (2012) sampled individuals in a continuous fashion for five coral reef fishes characterized by moderate or high dispersal potential based on larval duration (ranging from 13 to 94 days). IBD was observed at each reef. The slopes of the IBD linear models were combined with estimates of density so that slopes could be converted to mean-squared parent–offspring distance (i.e., mean dispersal distance; Rousset 2004). The work suggested mean dispersal distances of just 7–42 km for the fishes, many of which were thought to be high-dispersal and panmictic, based on previous Caribbean-wide mtDNA studies (Shulman & Bermingham 1995; Purcell et al. 2006). Regardless of whether the predictor variable is Euclidean distance or a landscape factor, proper study scale and replication are critical to robust model testing.

### 13.3.3 Isolation by Euclidean distance in freshwater systems

In contrast to marine systems, the difficulty of large-scale dispersal across distinct watersheds in freshwater systems leads to strong genetic structuring. Instead of struggling to detect IBD where it exists, in many freshwater cases strong IBD relationships can mask the importance of other factors that affect spatial genetic structure. The hierarchical structuring of freshwater communities (e.g., Fig. 13.4) and theoretical studies suggest that IBD should be more common in one-dimensional linear systems such as rivers than in two-dimensional systems such as freshwater lakes or ocean systems (e.g., Slatkin & Maddison 1990; Slatkin 1993). However, meta-analyses conducted by Crispo and Hendry (2005) found limited support for this theoretic expectation.

The bounded spatial structure and branching topologies of **dendritic** ecological networks such as streams affect population distributions and movement behaviors (Grant et al. 2007) that in turn influence IBD. Dendritic landscapes are structured hierarchically by elevation and therefore gene flow is typically asymmetric. Levels of genetic diversity and patterns of gene flow often differ markedly from upstream to downstream. Downstream reaches concentrate and propagate influence from spatially expansive upstream areas (Morrisey & de Kerckhove 2009).

As on land, temporal dynamism and spatial complexity of freshwater systems may disrupt IBD. For example, Koizumi et al. (2006) used a decomposed pairwise regression model to estimate the relative strengths of drift and gene flow affecting each stream population of dolly varden charr (*Salvelinus malma*). A strong pattern of IBD existed for most populations but outlier populations showed confounding effects of anthropogenic factors associated with physical barriers and population size. Other factors confounding IBD relationships in freshwater systems include stocking with hatchery fish (Ward 2006), and population position (e.g., **headwater** versus main stem or river mouth; Hughes et al. 2009).

Many freshwater systems, particularly those in high latitudes, are comparatively recently occupied following glacial retreat, such that colonization timing, routes, and distribution of glacial refugia may influence IBD (Castric et al. 2001; Castric & Bernatchez 2003; Costello et al. 2003). Historical hydrological structure dominates current day genetic relationships of brook charr (*Salvelinus fontinalis*) in New Foundland, Canada (Poissant et al. 2005). A sliding-window approach to IBD analysis, for which IBD slope is calculated with only a few adjacent populations, revealed weakening IBD at northern latitudes for brook charr sampled along the entire North West Atlantic coast, likely due to the more recent colonization in the north following glaciation (Castric &

Bernatchez 2003). This study highlights the importance of spatially replicating estimates of IBD slope – or any version of a spatial genetic structure linear model – if they are to be accurately estimated.

Importantly, a survey of anadromous and marine fish species found that the slope of IBD relationships varied as a function of the life history traits of different species (Bradbury & Bentzen 2007). For example, larger body size was an indicator of a dispersal phenotype, straying of hatchery fish following introduction contributes to gene flow among populations that can alter IBD relationships (Ward 2006), and dams and other barriers can disrupt IBD (Hughes et al. 2009). Connectivity is also affected by population position in the stream network (e.g., headwater versus main stem or river mouth). Species inhabiting upper reaches of streams are generally habitat specialists that rely on comparatively cooler waters, whereas generalist species are more common in lower sections of streams (Hughes et al. 2009); these traits are expected to impact IBD and influence landscape drivers of genetic patterns.

### 13.3.4 Isolation by hydrodynamic distance in marine systems

There is great interest in using **hydrodynamic connectivity** estimates and **biophysical models** to estimate marine larval connectivity (reviewed by Werner et al. 2007; Miller 2007; Cowen & Sponaugle 2009). Comparing models outputs to cross-validate gene flow and ocean flow models is fraught with pitfalls (reviewed in Liggins et al. 2013). Cross-validation studies require comparable metrics, and have sometimes created an "oceanographic" metric of dispersal distance based on simulation approaches similar in concept to least-cost path analysis, but with varying construction, as elaborated below.

In the approach of White et al. (2010), simulated "virtual larvae" were "released" into computer-modeled ocean currents for the Santa Barbara Channel. The larval release dates and duration of the study species, a predatory whelk in rocky reef and kelp forest habitat, were input into the model. The frequencies of transport among all coastal grid cells in the region were averaged across all particle release dates to calculate dispersal probabilities. A multigenerational stepping stone dispersal process was simulated with a Markov chain approach to generate steady-state long-term connectivity probabilities between all cells. These probabilities were then transformed to relative alongshore distances by using a model of dispersal in a simulated two-dimensional flow field along a linear coastline, which linearized their distribution to fit the IBD framework. This estimate of oceanographic distance explained a third of the variation in spatial genetic structure, compared to near zero using either Euclidean distance or the single generational version of the oceanographic distance. The study highlights the role of multigenerational stepping stone dispersal in maintaining genetic affinities across large distances in marine systems.

An alternative approach to generating oceanographic distance from flow simulations that does not require knowledge of the length of larval or spore life stages is to use mean transit time (in days) for a particle to travel from a source to destination site, instead of frequency or probability of exchange (Alberto et al. 2011). A study characterizing kelp genetic structure found that the shorter of the two transit times between a pair of sites explained more variance in genetic differentiation than mean transport time or Euclidean distance (Alberto et al. 2011), suggesting that dispersal or recruitment success may decline with dispersal distance and skew the dispersal kernel toward small distances.

Several studies used exploratory approaches to identify combinations of oceanographic and biological predictor variables that improved correlation between ocean flow and gene flow estimates. This approach can generate hypotheses about phenomena impacting the genetic patterns that are not captured by the hydrodynamic simulations, and suggest improvements to oceanographic simulations to capture dispersal processes with more realism. For instance, Galindo et al. (2010) used a detailed bio-physical oceanographic model of larval transport to investigate the maintenance of a cline in allele frequencies in the acorn barnacle (*Balanus glandula*) across central California. They combined an ocean circulation model that included nutrient availability with a species-specific larval development model that represented variation in larval trajectories and survival based on temperature, food availability, and depth distribution. Simulated connectivity matrices were used to generate predictions of spatial genetic patterns, to determine whether the observed genetic cline could be reproduced. Only by introducing a forcing variable (e.g., selection, increased local retention, or northward biased transport) could the cline be replicated.

In a study conducted at a larger scale across the Caribbean, Foster et al. (2012) applied a coupled bio-oceanographic simulation model of larval dispersal and a matrix-based population genetic projection model to investigate correlations with empirical spatial genetic structure for a reef building coral. The largest scale breaks were well replicated by the model, but the smaller-scale variation in connectivity between these breaks differed, perhaps due to omission of seasonal barriers caused by freshwater input from the model.

To date, cross-validation studies show varying concordance between ocean flow and gene flow estimates, ranging from moderate (e.g., Kool et al. 2010; White et al. 2010; Alberto et al. 2011; Coleman et al. 2011; Schunter et al. 2011) to weak correlation (e.g., Gilg & Hilbish 2003; Galindo et al. 2010; Selkoe et al. 2010; Rivera et al. 2011; Knutsen et al. 2012). It is possible that fine-scale hydrodynamic modeling needs improvement to be rigorously compared with local-scale genetic patterns, but comparing patterns from the two approaches may be fruitful at larger scales. However, without accounting for the processes flanking the dispersal phase (e.g., production and departure of larvae and their eventual recruitment success) in concert with hydrodynamic transport models, cross-validation of ocean flow and gene flow estimates may never find high correlations (Bowler & Benton 2005; Clobert et al. 2009; Selkoe & Toonen 2011; Faurby & Barber 2012). Historical effects on gene flow may also need to be accounted for (e.g. Dyer et al. 2010; Marko & Hart 2011).

Importantly, if single generational dispersal for many coastal marine species is often under 10–20 km, as results from Puebla et al. (2012) suggest, regional oceanographic models of average currents may be less relevant to understanding common dispersal events, especially for species with larval swimming behavior that can resist dispersal. Instead, current models may be more relevant to characterizing gene flow that occurs over several generations in a stepping stone manner or rare long-distance dispersal that results from maximum pelagic larval duration (Woodson & McManus 2007). Increasingly, studies are recognizing the need to consider distinct influences on minimum, mean, and maximum larval duration or dispersal distance separately (e.g., Weersing & Toonen 2009; Alberto et al. 2011; Faurby & Barber 2012). However, exploratory approaches using complex and more "realistic" models of transport must consider that accumulated error over multiple parameters may overwhelm accuracy and spurious effects can dominate correlative results (e.g. Cushman & Landguth 2010).

Nevertheless, there is still much insight yet to be gained on dispersal processes and the cross-validation approach holds much promise for making valuable contributions (Broquet & Petit 2009).

### 13.3.5   Isolation by hydrologic distance in freshwater systems

In freshwater systems, connectivity mediated by flow of water is commonly called ***hydrologic connectivity***, whereas marine studies use the term hydrodynamic connectivity (Pringle 2003; Rodriguez-Iturbe et al. 2009; see Fig. 13.4D). Hydrological models have been widely integrated into landscape-scale research and management (e.g., Wiley et al. 2010; Freeman et al. 2012), including modeling of material and organism transport processes and dispersal (Shen et al. 2010). Modeling studies by Morrissey and de Kerckhove (2009) found that asymmetric (downstream) migration and dendritic metapopulation structure resulted in higher equilibrium levels of genetic diversity than in simpler settings (e.g., symmetrical migration and linear array). Lower population genetic diversity is typically found in headwater areas than in intermediate and lower stream reaches (e.g., Columbia spotted frogs, *Rana luteiventris*; Funk et al. 2005). Ozerov et al. (2012) found significant associations between stream gradient, stream carrying capacity, and availability of lake (juvenile rearing) habitat to spatial genetic structure of Russian populations of ***catadromous*** Atlantic salmon (*Salmo salar*). Empirical studies of asymmetrical migration in dendritic stream systems shows that it is strongly affected by an organism's dispersal ability (Winemiller & Rose 1992). Genetic data compiled by Hänfling and Weetman (2006) documented high levels of differentiation at small scales (<0.5 km) in river sculpin (*Cottus gobio*) that also reflected a source-sink structure, evidenced by signatures of asymmetric gene flow, whereby small upstream populations were characterized by high net emigration and larger downstream populations were characterized by high net immigration rates. Given the complex ecology and heterogeneity of hydrology within and among systems (reviewed in Grant et al. 2007) and importance of other riverscape features associated with human development (Arthington et al. 2006), a landscape genetic approach can be profitably employed to understand processes affecting spatial genetic structure in stream systems.

In lake systems, current patterns and general wind direction are important, particularly for benthic invertebrates and demersal fish species with planktonic larval stages (Bradbury & Snelgrove 2001). There are several studies of Great Lakes fishes that use biophysical models, simulating the dispersal of larvae based on numerically generated flow models and larval traits related to buoyancy, feeding, and swimming (Beletsky et al. 2007; Zhao et al. 2009; Humphrey et al. 2012). Lake hydrological data have been useful to disentangle the effects of natural and anthropogenically mediated dispersal of invasive species into the Great Lakes at different spatial scales. Darling and Folino-Rorem (2009), studying the non-native hydrozoan Cordylophora, and Bronnenhuber et al. (2011), studying invasive round gobys (*Neogobius melanostomus*) in the Great Lakes and their tributaries, found that secondary range expansion was associated with dispersal mediated by hydrological features. However, studies that directly compare modeled connectivity based on hydrodynamics to gene flow patterns using landscape genetic techniques are far less common for freshwater lake systems than marine systems so far.

### 13.3.6   Site-based factors and alternative model testing

As mentioned above, dispersal pathways alone often do not explain observed spatial genetic structure and studies are increasingly considering site-specific environmental factors (Saura & Rubio 2010; Luque et al. 2012). For steam-dwelling amphibians, exit from and entry into populations occurs over land and can be related to the environmental profiles of sites such as elevation, solar radiation, temperature, and moisture levels (e.g., Spear & Storfer 2008; Murphy et al. 2010). For freshwater fishes, the elevation and position of sites up or downstream can influence migration rates (Hughes et al. 2009). For marine systems, current speeds or flushing rates at sites can impact migration rates (Woodson & McManus 2007).

Some of these site-associated features are likely to act more directly on $N_e$ than on the migration rate. Spear and Storfer (2008) used spatial regression to quantify the influence of terrestrial landscape features between aquatic breeding habitats on connectivity in the coastal tailed frog (*Ascaphus truei*). The authors found that elevation and amount of solar radiation (associated with unforested habitats) explained levels of differentiation among populations and between forested and non-forested (harvested) regions for this stream-dwelling amphibian. Moore et al. (2011) used a landscape approach to quantify the effects of landscape features on gene flow in the boreal toad (*Bufo boreas*) in Alaska (USA). The authors constructed resistance surfaces using least-cost paths and circuit theory and found that saltwater was a significant physiological barrier to boreal toad gene flow. Murphy et al. (2010) measured habitat permeability, topographic relief, and ground temperature/moisture conditions as ecological processes that could affect population connectivity in the boreal toad (*Bufo boreas*). Models that included temperature and moisture regimes were generally more predictive of interpopulation genetic distance than were models with topology and habitat features. The authors found that predictive power of features depended on the spatial scale of analysis. Variables quantifying the degree of habitat permeability were comparatively more important at shorter interpopulation differences while topography was important among populations separated over larger distances. The effects of temperature and moisture were important predictors of spatial genetic structure at all spatial scales.

Puritz and Toonen (2011) found samples of a sea star (*Patiria miniata*) showed that genetic differentiation of both microsatellite and mitochondrial markers increased close to offshore storm water and wastewater discharge sites in the Southern California Bight. The study teased out effects of natural freshwater plumes from current day point-source wastewater impacts by first fitting a regression line to river mouth distance and then fitting the residual genetic variance to the wastewater driver. The authors posit that high larval mortality near the pollution sources reduced genetic diversity (rarefied allelic richness) observed there. Without invoking spatial variation in the dispersal rate or distance, frequent mortality events at these polluted sites could result in few successful settlement events, increasing genetic drift near outfalls. In another Southern California nearshore study, Selkoe et al. (2010) found that habitat patch size negatively correlated with genetic diversity and differentiation for three predators found in kelp forests. Large kelp forests tend to be older and home to older populations that may intensify either density-dependent recruitment or sweepstakes reproductive success, which could dampen genetic diversity. Site-based effects on genetic differentiation indicate roles for ecological factors driving connectivity (Berkeley et al. 2004) and links between genetics, demographic connectivity (Lowe &

Allendorf 2010), and habitat quality, a topic almost completely unexplored in marine studies. However, the relationships between genetic and ecological patterns remain speculative without more detailed comparison and simulations to test the likelihood of alternative processes (e.g., effects of effective population size, $N_e$, versus migration rate, $m$, the two components of gene flow).

Both studies above used alternative model-testing frameworks to evaluate candidate environmental drivers. Puritz and Toonen (2011) used multiple regression on distance matrices (MRM) (Lichstein 2006) and Selkoe et al. (2010) used a generalized linear model with Toeplitz covariance structure to address non-independence of pair-wise matrix data. However, if site-based environmental or ecological factors can be compared to site-based calculations of genetic differentiation (e.g., local $F_{ST}$; Foll & Gaggiotti 2008), there is less need for complex treatments of pair-wise data and alternative model testing is more straightforward (Wagner & Fortin 2013).

Gaggiotti et al. (2009) used a creative approach to alternative model testing to disentangle demographic history, temperature and/or salinity, and migratory behavior effects on spatial genetic structure based on microsatellite analysis of Atlantic herring (*Clupea harengus*). Demographic history (i.e., a possible eastward population expansion) was represented by longitude and distance to a possible zone of secondary contact. Ten separate potential environmental drivers were tested against local $F_{ST}$ estimates in a modified version of the software GESTE (Foll & Gaggiotti 2008), which uses a logistic-regression model with separate locus and population effects. Ten factors and interactions led to >1000 alternative models, but by dropping the five factors associated with the lowest posterior probabilities and rerunning the smaller set of 32 alternative models, spurious models were avoided. The results suggested selection related to salinity at spawning sites and feeding migrations largely drove spatial genetic structure, with little evidence for lingering signals of demographic history. The authors highlight the approach's utility for resolving joint influences of selective and neutral genomic regions on spatial genetic structure, especially for marine species with a large effective population size, which expands the effect of selection relative to drift (Allendorf et al. 2008). White et al. (2010) used a similar approach to compare seascape drivers of neutral and adaptive divergence with GESTE for the Roundnose Grenadier (*Coryphaenoides rupestris*).

## 13.4 APPLICATIONS OF LANDSCAPE GENETIC APPROACHES

### 13.4.1 Population genetic simulations

An increasing number of studies use population genetic simulations to gain insight into complex spatial genetic structures (Epperson et al. 2010; Chapter 6), including coalescent-based approaches and forward-time individual based models (e.g., Galindo et al. 2006). Simulation programs have been developed specifically for freshwater systems (e.g., demographic-based coalescent simulations as implemented in Aquasplatche (Neuenschwander 2006), and individual-based spatially-explicit simulations implemented in CDFish enable modeling the two-phased life histories of many aquatic species (Landguth et al. 2011). We focus here on examples demonstrating how simulations can help disentangle sampling effects and scale dependencies, multiple dispersal pathways/strategies, and historical influences on gene flow.

There are many examples of scale-dependency in demographic and ecological phenomena in aquatic species, and genetics are no exception. Berry et al. (2012a) found small-scale spatial autocorrelation of genotypes at the juvenile (but not adult) stage but apparent panmixia at broad scales for an Australian reef fish. This study is one example of a not uncommon observation by marine geneticists that small-scale genetic structuring exceeds large-scale structuring; the same pattern has been reported for birds (e.g., Finnegan et al. 2013) and for stream trout that show kin aggregation (Anderson & Dunham 2008). Determining which sampling, landscape, and biological phenomena produce these unintuitive results is a challenge for the field. Berry et al. (2012b) applied a simulation approach using the software IBDsim (Leblois et al. 2009) and a hydrodynamic model to test potential roles of adult and larval dispersal in generating the pattern. They point out that although panmixia normally suggests management as a single stock, contrary evidence for sedentary adults and fine-scale structure may nevertheless make local populations sensitive to overharvest.

One study of habitat effects on salmon density demonstrated how the scale over which observations are made affects conclusions about relationships between landscape and spatial structure (Burnett et al. 2007). This study provides a useful example of using mixed-scale models to consider spatial scale explicitly in landscape genetic analysis and also highlights the roles

of scale in study design and interpretation (Anderson et al. 2010). Kanno et al. (2011) designed a multiscale sampling protocol for brook trout with individual-level continuous sampling replicated within population-level samples to test influences of a diversity of riverscape attributes. The study used both IBD testing and two clustering methods (the Bayesian clustering method implemented in STRUCTURE, Pritchard et al. 2000, and discriminant analysis of principal components, Jombart et al. 2009). The results revealed heirarchical population structuring, with fine-scale continuous gradients of spatial genetic structure influenced by stream grade and temperature and larger-scale discrete genetic clusters separated by dams and waterfalls. Mullen et al. (2010) found a similar scale-dependency dictated by stream genetic architecture for a salamander.

Simulation is also an effective way to set bounds on the extent and type of dispersal driving gene flow (Faurby & Barber 2012). Some aquatic species (e.g., invertebrates) have the potential to disperse both through river networks and over land. Similarly, it is often unclear what the roles of adult and larval dispersal are for gene flow in marine species (Berry et al. 2012b). Simulation is also a good way of exploring the role of asymmetric exchange. Chaput-Bardy et al. (2009) used individual-based models to quantify the relative contributions of upstream versus downstream movement on gene flow under different dispersal scenarios that reflected different species' ecologies. Knutsen et al. (2012) used population genetic simulation to test alternative hypotheses about the roles of complex ocean currents and diffusive, stepping stone dispersal in generating an observed increase in genetic structure at the range edges compared to the center.

Patterns of spatial genetic structure do not always just reflect the balance of drift and migration – history can play an important role, especially for species with large effective size and/or limited gene flow. Marko and Hart (2011) emphasized the complementarity of simulation approaches (e.g., IMa, SimCoal, PopABC, msBayes, Approximate Bayesian Computation methods, and others) for any empirical study to jointly estimate time since divergence, migration rate, mutation rate, and population size instead of assuming that genetic differentiation reflects gene flow (also see Hart & Marko 2010). For instance, McGovern et al. (2010) used six nuclear sequences and mtDNA with the software IMa for the bat star (*Patiria miniata*) and the frilled dog whelk (*Nucella lamellosa*) to examine whether the location of a major region of genetic discontinuity was spatially concordant between species and whether the break indicated either ongoing non-equilibrium or lack of gene flow. Limborg et al. (2012) used $\theta_{ST}$, $F_{ST}$, and $R_{ST}$ to compare the influences of drift and mutation separately in generating the stock structure of European sprat (*Sprattus sprattus* L.), allowing inference on the age of cladogenic events leading to the stock boundaries. This study's use of the genetic structure simulator PowSim (Ryman & Palm 2006) also demonstrated the importance of power analysis in evaluating empirical results on genetic boundaries, some of which may be too weak to detect with statistical significance given the limitations of sampling design.

### 13.4.2 Graph theoretic network approaches

Despite the intuitive fit of graph theory as a hypothesis-testing framework in both dendric freshwater and three-dimensional marine systems, graph theoretic approaches have been underutilized in aquatic settings. The majority of studies employing graph, gravity, or network-based approaches in freshwater systems focused on amphibians (e.g., Murphy et al. 2010; Chapter 10). By representing connectivity as a network, dispersal corridors, barriers, and asymmetries can be effectively visualized to lend insight into the ways biological and genetic factors interact with the waterscape to influence scales and patterns of gene flow or community similarity (Treml et al. 2007; Moalic et al. 2012). Barriers are represented by the lack of edges and resulting graph topology can be compared to networks built from other types of connectivity data or other species' genetic data. For example, Kininmonth et al. (2010) compared a genetic network based on $F_{ST}$ values of populations of a coral directly with an analogous connectivity network created from hydrodynamic data, with a focus on how to characterize and overcome genetic sampling gaps.

Network metrics can reveal structuring where traditional approaches fail. Oceanic zooplankton have long been assumed to have little means to maintain neutral spatial genetic structure in the face of current systems. However, Goetze (2011) used microsatellite analysis of a widespread copepod to reveal a network structure with five subgraphs, revealing a surprising degree of structuring and insights on the nature of transition zones. Moalic et al. (2011) represented the genetic relationship of sister species of brown algae with networks of individuals. This approach allowed testing alternative hypotheses about why a handful of individuals were found to have polymorphism shared between species: ongoing speciation or present-day secondary contact via hybridization. In contrast to Bayesian clustering, which simply identified admixed individuals (albeit with slightly better accuracy), the network analysis revealed these individuals to be nodes of interspecies gene flow between the two species and enabled effective visualization of the role of geography in this hybridization phenomenon.

### 13.4.3 Assignment and clustering

Graph theory is highly complementary to the many other techniques for barrier and cluster analyses, many of which have strong statistical frameworks but also tend to require large amounts of data and strong spatial genetic structure to be appropriately applied. Because clustering and assignment techniques often isolate the most recent migration events, they can inform pressing management decisions requiring information on current-day demographic connectivity that is otherwise difficult to estimate (e.g., Andrews et al. 2010; Mokhtar-Jamaï et al. 2011; Amaral et al. 2012; Horne et al. 2012; Soria et al. 2012; Chapter 7). Because there are so many approaches to identification of barriers and clustering, many studies employ multiple approaches to increase statistical rigor of inferences of descriptive patterns and to quantify associations supporting identified underlying mechanisms.

A multifaceted study of stock structure, effective population sizes and seascape associations of white hake (*Urophycis tenuis*) in the northwest Atlantic is a good example of how combining multiple tools can improve clustering results (Roy et al. 2012). This highly exploited species was assumed to be panmictic due to a 3-month pelagic larval period, high abundance, and extremely high fecundity. Nine Canadian management zones were thus based on hydrographic features thought to be possible dispersal barriers. Using nearly two thousand georeferenced individuals genotyped at 16 microsatellites, the authors first applied a Structurama routine (Huelsenbeck & Andolfatto 2007), which searches for the most probable number of genetic clusters ($K$). They used the resulting $K = 3$ in a set of STRUCTURE runs and amalgamated results across runs into a single estimate of population assignment using CLUMPP (Jakobsson & Rosenberg 2007).

Admixed individuals could then be excluded from estimation of short-term and long-term $N_e$ for each population using temporal replication and coalescent simulation (IMa2; Hey 2010). These three stocks showed differing spatial ranges, $N_e$ estimates, and diversity, and differences among stocks accounted for 3.6% of genetic variance in an Analysis of Molecular Variance (AMOVA), among the highest observed for large-ranged, abundant marine species, and a striking contradiction of the previously assumed panmixia. Lastly, important effects of seascape features (e.g., latitude, depth, salinity) and ecological factors (e.g., spawning time) on this stock structure were identified with the software BIMR (Faubet & Gaggiotti 2008) and individual-based modeling of group membership. These data are crucial to improving the management of this depleted species and suggest specific ways to create new boundaries based on seascape features.

### 13.4.4   Adaptation, selection, and genomics

One focus of genomic-level analysis of spatial structure is finding alleles under environmental selection to determine the functional roles of genetic variation and the forces shaping local adaptation. For example, management of salmonids has recently sought to better grasp the extent of adaptive diversity behind an observed "portfolio effect" (Hilborn et al. 2003; Schindler et al. 2010), in which phenotypic variation among spawning populations is linked to differential responses to environmental variation. This response diversity leads to spatially decoupled population dynamics, such that individual populations fluctuate substantially over time. Nevertheless, in aggregate at the regional scale populations are much more stable. Species distributions are often determined by thermal tolerance (Brannon et al. 2004) and genetic markers from (or linked to) functional genes have been used to address questions involving effects of environmental variables on salmon spatial genetic structure (Teel et al. 2011). For example, Narum et al. (2010) found a number of candidate SNP candidate markers that revealed a strong pattern of "isolation-by-temperature" between mountain and desert populations of redband trout (*Onchorhynchus mykiss gairdneri*). In a follow-up study, Narum et al. (2013) used a common garden experiment to study the response of individuals of different genotypes obtained from climatically highly differentiated habitats (mountain and desert) at heat shock genes and documented significant differences in survival.

There is also increasing evidence for human-induced evolutionary changes in many species (e.g., Stockwell et al. 2003; Parmesan 2006; Kuparinen and Merilä 2007; Reusch and Wood 2007; Hendry et al. 2008; Allendorf et al. 2008) that warrant further landscape genetic investigation. This is particularly true in freshwater systems where loss and homogenization of habitats is a primary factor associated with the incidence of hybridization in fishes (Scribner et al. 2001).

There is still little understanding of the extent of local adaptation in marine species with extensive dispersal capacity and/or large $N_e$. Hice et al. (2012) used common garden experiments along a latitudinal gradient to evaluate whether latitudinal variation in three distinct phenotypic traits correlated with spatial gradients in environmental factors for the Atlantic silverside (*Menidia menidia*), a nearshore fish with very low neutral spatial genetic structure. Spatial autocorrelation analysis revealed high spatial correlations in allele frequencies over short geographic distances. Breakpoint regression identified specific points at which the relationship of traits to latitude changed in association with biogeographical boundaries. Results indicated that each trait had a unique spatial genetic structure, signaling the complexity of characterizing structure at a genome level.

Despite tendencies for low levels of spatial genetic structure at neutral loci, selection can be strong when drift is weak, creating a complex picture of spatial genetic structure at the genomic scale (Allendorf et al. 2010). Increasing numbers of studies are using genome scans to find loci under selection and explore this question. Nielsen et al. (2009) used BAYESCAN on a panel of 98 gene-associated SNPs to detect signatures of selection in Atlantic cod (*Gadus morhua*). This study was one of the first to consider spatial scale and geographic replication in testing for outlier loci. They then used the program SAM (Joost et al. 2008), which finds candidate loci under selection via exploratory testing of correlations with environmental data, a strategy that is still largely unvetted. Most loci were significantly correlated with several temperature, salinity, and geographic gradients, suggesting either weak selection or landscape drivers of neutral spatial genetic structure.

As more studies use genomic scale sampling to investigate spatial genetic structure of neural and adaptive loci simultaneously, the reality that many loci will fall into a "gray area" on a gradient between neutral and weak selective behavior may bring challenges in interpreting the mechanism for environmental influence. For instance, the cod study showed that global $F_{ST}$ estimates from the neutral SNPs were lower than most microsatellite-based estimates, perhaps suggesting that microsatellites are frequently subject to hitchhiking effects or that many of the putatively neutral SNP markers are actually under balancing selection.

The environment dictates how genotypes map to phenotypes. One of the goals in genomics and phenomics is to better understand the interplay and impact of genetic, epigenetic, and transcriptional variation on phenotypic diversity. Poikilothermic vertebrates and northern freshwater fishes in particular (Robinson & Schluter 2000) are ideally suited for studies of the effects of natural and anthropogenically mediated selection on adaptive genetic variation because of the importance of environmentally mediated divergent selection on heritable phenotypic traits (Schluter 1996).

## 13.5 FUTURE DIRECTIONS: KNOWLEDGE GAPS, RESEARCH CHALLENGES, AND LIMITATIONS

The collection of studies in this chapter demonstrate that with proper sampling and rigorous analysis, insights can be gained on the scales and patterns of population structuring, dispersal, and the suite of physical, biological, and historical drivers of those patterns at both neutral and adaptive loci. While there is need for a continuation of this trajectory, there are noteworthy avenues for future research that would fill gaps in understanding particular to marine and freshwater ecology.

### 13.5.1 Quantifying and accounting for species' traits

A challenge shared by marine and freshwater genetics is the need to account for the complex processes that influence the dispersal of organisms, such as individual behavior and inter- and intraspecific interactions that affect the costs and benefits of dispersal. Unlike in terrestrial systems, dispersal pathways are fluid, shifting, and hard to measure, and there are distinct physical challenges and constraints to entering and exiting

the flowing transport pathways. Some of these may largely be mediated by organismal behavior, such as the location and timing of spawning (e.g., timed to outgoing tides or positioned midstream in a river) and chemical sensory systems' cuing behavior. Probabilities of dispersal from natal areas will depend in part on an organism's reactions to conditions at the natal site and encountered during dispersal (condition dependence; Ronce 2007), as well as sex, age and physiological condition (Bowler & Benton 2005), and intraspecific competition, kin competition, and mate choice (Clobert et al. 2009). Interspecific interactions often affect occupancy of habitats and can affect the propensity of individuals to disperse.

### 13.5.2 Comparative waterscape genetics across species

Multispecies genetic analysis is needed to determine the shared and divergent landscape influences among co-distributed species (Storfer et al. 2010; Selkoe et al. 2010; Carpenter et al. 2011; Selkoe et al. 2014), especially in marine systems where it is still unclear to what extent physical features constrain dispersal pathways across divergent organisms. Graph theoretic congruence networks may be a powerful way to uncover multispecies genetic connectivity, elucidating "metawebs" (i.e., a "collection of networks of genetic variation of all species within a community"; Fortuna et al. 2009).

### 13.5.3 Comparing genetic and ecological connectivity

There are direct linkages between landscape genetics and existing and emerging "metacommunity" statistics (e.g., Leibold et al. 2010; Peres-Neto et al. 2012; Mouquet et al. 2012). Relating structural connectivity of aquatic habitats to functional connectivity at species and community levels is needed to protect metacommunity dynamics and biodiversity in managed ecosystems (Esselman et al. 2011; Perkin & Gido 2012). Waterscapes affect the species composition of aquatic communities and metacommunities (Leibold et al. 2004). Understanding differences in genetic signatures among species that have different habitat requirements or that are differentially sensitive to environmental and human stressors would represent a considerable advance over current analyses using other surrogates of biodiversity (e.g., Allan 2004). For instance, metapopulation models predict landscape-dependent dispersal and extinction rates (Elkin & Possingham 2008). Metagenomic analyses of microbial community composition and metabolic profiles have revealed strong correlates with spatial patterns of human impacts on coral reefs and other urbanized coastal zones (Gianoulis et al. 2009; McDole et al. 2012). These studies emphasize ways that variation in community composition can be related to spatial and environmental effects, but fail to incorporate concepts of environmental resistance to dispersal that landscape genetic analyses can provide.

### 13.5.4  Applied problems

The footprint of human disturbance is widespread in aquatic habitats across the globe: freshwater resources are being depleted (Poff 2009), riparian and estuarine habitat destruction continues (Sweeney et al. 2004), freshwater and estuarine extinction and nascent marine extinction are increasing (Ricciardi & Rasmussen 1999), and climate change is accelerating, all with effects on both biodiversity and human ecosystem services. Streams are important sentinels and integrators of terrestrial and atmospheric processes and of the impacts of human disturbance (Williamson et al. 2008). Climatic warming can affect timing and length of optimal periods for growth within and across life history transitions (Bradshaw & Holzapfel 2008) and disrupt developmental selection regimes that have created and maintain the phenotypic diversity of numerous inhabitants of aquatic ecosystems (Stefan et al. 2001).

In the marine realm, warming can bring sudden decline in coral reef communities, and ocean acidification and sea level rise impact a multitude of marine taxa and habitat types (Harley et al. 2006; Hoegh-Guldberg et al. 2007). Continuing work to catalogue and synthesize spatial data on human impacts is central to waterscape genetics as the dominance of human influence on connectivity increases (e.g., Halpern et al. 2008; Wang et al. 2008; Host et al. 2011). The advances in applications of genetic tools and the ability to quantify landscape features at multiple spatial scales in aquatic systems promise to help us understand how to manage human uses and influences on aquatic systems in ways that best preserve connectivity and resilience at both population and community levels.

## ACKNOWLEDGMENTS

Funding was provided to K. Selkoe by the National Science Foundation (BioOCE Award Number 1260169) and the National Oceanic and Atmospheric Administration (NMSP MOA#2005-008/66882). Funding was provided to K. Scribner through the Partnership for Ecosystem Research and Management (PERM) program between the Fisheries Division of the Michigan Department of Natural Resources and the Department of Fisheries and Wildlife at Michigan State University. The authors thank D. Infante, P. Sorrano, K. Cheruvelil, and members of their laboratories for discussions of ideas. Members of the Infante and Scribner Labs reviewed earlier drafts of the manuscript. D. Infante, D. Wesley, and K. Herreman provided data and assisted with the preparation of the figures.

## REFERENCES

Alberto, F., et al. (2011) Isolation by oceanographic distance explains genetic structure for *Macrocystis pyrifera* in the Santa Barbara Channel. *Molecular Ecology.* **20**, 2543–54.

Allan, J. (2004) Landscapes and riverscapes: the influence of land use on stream ecosystems. *Annual Review of Ecology, Evolution, and Systematics* **35**, 257–84.

Allendorf, F. & Waples, R. (1996) Conservation and genetics of salmonid fishes. In: Avise, J. & Hamrick, J. (eds.), *Conservation Genetics: Case Histories from Nature.* Chapman and Hall, New York.

Allendorf, F., Ryman, N., & Utter, F. (1987) Genetics and fishery management: past, present, and future. In: Ryman, N. & Utter, F. (eds.), *Population Genetics and Fishery Management.* University of Washington Press, Seattle, WA.

Allendorf, F., et al. (2008) Genetic effects of harvest on wild animal populations. *Trends in Ecology and Evolution* **23**, 327–37.

Allendorf, F., Hohenlohe, P., & Luikart, G. (2010) Genomics and the future of conservation genetics. *Nature Reviews: Genetics* **11**, 697–709.

Amaral, A., et al. (2012) Seascape genetics of a globally distributed, highly mobile marine mammal: the short-beaked common dolphin (genus *Delphinus*). *PloS One* **7**, e31482.

Anderson, E. & Dunham, K. (2008) The influence of family groups on inferences made with the program STRUCTURE. *Molecular Ecology Resources* **8**, 1219–29.

Anderson, C., et al. (2010) Considering spatial and temporal scale in landscape-genetic studies of gene flow. *Molecular Ecology* **19**, 3565–75.

Andrews, K., et al. (2010) Rolling stones and stable homes: social structure, habitat diversity and population genetics of the Hawaiian spinner dolphin (*Stenella longirostris*). *Molecular Ecology* **19**, 732–48.

Angermeier, P., Krueger, K., & Dolloff, C. (2002) Discontinuity in stream-fish distributions: implications for assessing and predicting species occurrence. In: Scott, J., et al. (eds.), *Predicting Species Occurrences: Issues of Accuracy and Scale.* Island Press, Covelo, CA.

Arthington, A., et al. (2006) The challenge of providing environmental flow rules to sustain river ecosystems. *Ecological Applications*, **16**, 1311–18.

Avise, J. (1992) Molecular population structure and the biogeographic history of a regional fauna: a case history with lessons for conservation biology. *OIKOS*, **63**, 62–76.

Bailey, R. & Smith, G. (1981) Origin and geography of the fish fauna of the Laurentian Great Lakes Basin. *Canadian Journal of Fisheries and Aquatic Sciences* **38**, 1539–61.

Balkenhol, N., et al. (2009) Identifying future research needs in landscape genetics: where to from here? *Landscape Ecology* **24**, 455–63.

Beletsky, D., et al. (2007) Biophysical model of larval yellow perch advection and settlement in Lake Michigan. *Journal of Great Lakes Research* **33**, 842–66.

Benda, L., et al. (2004) Confluence effects in rivers: interactions of basin scale, network geometry, and disturbance regimes. *Water Resources Research* **40**, W05402.

Berkeley, S., et al. (2004) Fisheries sustainability via protection of age structure and spatial distribution of fish populations. *Fisheries* **29**, 23–32.

Bermingham, E. & Avise, J. (1986) Molecular zoogeography of freshwater fishes in the southeastern United States. *Genetics* **113**, 939–65.

Bernatchez, L. & Wilson, C. (1998) Comparative phylogeography of Nearctic and Palearctic fishes. *Molecular Ecology* **7**, 431–52.

Berry, O., England, P., Fairclough, D., et al. (2012a) Microsatellite DNA analysis and hydrodynamic modelling reveal the extent of larval transport and gene flow between management zones in an exploited marine fish (*Glaucosoma hebraicum*). *Fisheries Oceanography* **21**, 243–54.

Berry, O., England, P., Marriott, R., et al. (2012b) Understanding age-specific dispersal in fishes through hydrodynamic modelling, genetic simulations and microsatellite DNA analysis. *Molecular Ecology* **21**, 2145–59.

Berst, A. & Simon, R. (1981) Introduction to the Proceedings of the 1980 Stock Concept International Symposium (STOCS). *Canadian Journal of Fisheries and Aquatic Sciences* **38**, 1457–8.

Berst, A. & Spangler, G. (1973) Lake Huron: the ecology of the fish community and man's effects on it. Great Lakes Fishery Commission, Ann Arbor, MI.

Bezault, E., et al. (2011) Spatial and temporal variation in population genetic structure of wild Nile tilapia (*Oreochromis niloticus*) across Africa. *BMC Genetics* **12**, 102.

Blum, M., et al. (2012) Genetic diversity and species diversity of stream fishes covary across a land-use gradient. *Oecologia* **168**, 83–95.

Bodie, R. & Semlitsch, R. (2000) Spatial and temporal use of floodplain habitats by lentic and lotic species of aquatic turtles. *Oecologia* **122**, 138–46.

Boulton, A., et al. (2010) Ecology and management of the hyporheic zone: stream-groundwater interactions of running waters and their floodplains. *Journal of the North American Benthological Society* **29**, 26–40.

Bowler, D. & Benton, T. (2005) Causes and consequences of animal dispersal strategies: relating individual behaviour to spatial dynamics. *Biological reviews of the Cambridge Philosophical Society*, **80**, 205–25.

Bradbury, I. & Bentzen, P. (2007) Non-linear genetic isolation by distance: implications for dispersal estimation in anadromous and marine fish populations. *Marine Ecology Progress Series* **340**, 245–57.

Bradbury, I.R. & Snelgrove, P.V.R. (2001) Contrasting larval transport in demersal fish and benthic invertebrates: the roles of behaviour and advective processes in determining spatial pattern. *Canadian Journal of Fisheries and Aquatic Sciences* **58**, 811–23.

Bradshaw, W.E. & Holzapfel, C.M. (2008) Genetic response to rapid climate change: it's seasonal timing that matters. *Molecular Ecology* **17**, 157–66.

Brannon, E.L., et al. (2004) Population structure of Columbia River Basin Chinook salmon and steelhead trout. *Research in Fisheries Science* **12**, 99–232.

Brenden, T.O., Wang, L., & Seelbach, P.W. (2008) A River Valley segment classification of Michigan streams based on fish and physical attributes. *Transactions of the American Fisheries Society* **137**, 1621–36.

Bronnenhuber, J.E., et al. (2011) Dispersal strategies, secondary range expansion and invasion genetics of the non-indigenous round goby, *Neogobius melanostomus*, in Great Lakes tributaries. *Molecular Ecology* **20**, 1845–59.

Broquet, T. & Petit, E.J. (2009) Molecular estimation of dispersal for ecology and population genetics. *Annual Review of Ecology, Evolution, and Systematics* **40**, 193–216.

Buhlmann, K.A., et al. (2009) Ecology of chicken turtles (*Deirochelys reticularia*) in a seasonal wetland ecosystem: exploiting resource and refuge environments. *Herpetologica* **65**, 39–53.

Burnett, K.M., et al. (2007) Distribution of salmon-habitat potential relative to landscape characteristics and implications for conservation. *Ecological applications: a publication of the Ecological Society of America* **17**, 66–80.

Calhoun, A.J.K., et al. (2003) Evaluating vernal pools as a basis for conservation strategies: a Maine case study. *Wetlands* **23**, 70–81.

Carpenter, K.E., et al. (2011) Comparative phylogeography of the Coral Triangle and implications for marine management. *Journal of Marine Biology* **2011**, 1–14.

Castric, V. & Bernatchez, L. (2003) The rise and fall of isolation by distance in the anadromous Brook Charr (*Salvelinus fontinalis* Mitchill). *Genetics* **163**, 983–96.

Castric, V., Bonney, F., & Bernatchez, L. (2001) Landscape structure and hierarchical genetic diversity in the brook charr, *Salvelinus fontinalis*. *Evolution* **55**, 1016–28.

Chaput-Bardy, A., et al. (2009) Modelling the effect of in-stream and overland dispersal on gene flow in river networks. *Ecological Modelling* **220**, 3589–8.

Clobert, J., et al. (2009) Informed dispersal, heterogeneity in animal dispersal syndromes and the dynamics of spatially structured populations. *Ecology Letters* **12**, 197–209.

Coleman, M.A., et al. (2011) Variation in the strength of continental boundary currents determines continent-wide connectivity in kelp. *Journal of Ecology* **99**, 1026–32.

Cook, B.D., et al. (2011) Landscape genetic analysis of the tropical freshwater fish *Mogurnda mogurnda* (Eleotridae) in a monsoonal river basin: importance of hydrographic factors and population history. *Freshwater Biology* **56**, 812–27.

Costello, A.B., et al. (2003) The influence of history and contemporary stream hydrology on the evolution of genetic diversity within species: an examination of microsatellite DNA variation in bull trout, *Salvelinus confluentus* (Pisces: Salmonidae). *Evolution* **57**, 328–44.

Cowen, R.K. & Sponaugle, S. (2009) Larval dispersal and marine population connectivity. *Annual Review of Marine Science* **1**, 443–66.

Crispo, E. & Hendry, A.P. (2005) Does time since colonization influence isolation by distance? A meta-analysis. *Conservation Genetics* **6**, 665–82.

Cushman, S.A., & Landguth E.L. (2010) Spurious correlations and inference in landscape genetics. *Molecular Ecology* **19**, 3592–602.

Darling, J.A. & Folino-Rorem, N.C. (2009) Genetic analysis across different spatial scales reveals multiple dispersal mechanisms for the invasive hydrozoan Cordylophora in the Great Lakes. *Molecular Ecology* **18**, 4827–40.

Duncan, K.M., et al. (2006) Global phylogeography of the scalloped hammerhead shark (*Sphyrna lewini*). *Molecular Ecology* **15**, 2239–51.

Duong, T.Y., et al. (2011) Relative larval loss among females during dispersal of Lake Sturgeon (*Acipenser fulvescens*). *Environmental Biology of Fishes* **91**, 459–69.

Dyer, R.J., Nason J.D., & Garrick, R.C. (2010) Landscape modelling of gene flow: improved power using conditional genetic distance derived from the topology of population graphs. *Molecular Ecology* **19**, 3746–59.

Elkin, C.M. & Possingham, H. (2008) The role of landscape-dependent disturbance and dispersal in metapopulation persistence. *The American Naturalist* **172**, 563–75.

Epperson, B.K., et al. (2010) Utility of computer simulations in landscape genetics. *Molecular Ecology* **19**, 3549–64.

Esselman, P.C., et al. (2011) An index of cumulative disturbance to river fish habitats of the Conterminous United States from landscape anthropogenic activities. *Ecological Restoration* **29**, 133–51.

Fabrizio, M.C., Raz, J., & Bandekar, R.R. (2000) Using linear models with correlated errors to analyze changes in abundance of Lake Michigan fishes: 1973–1992. *Canadian Journal of Fisheries and Aquatic Sciences* **57**, 775–88.

Faubet, P. & Gaggiotti, O.E. (2008) A new Bayesian method to identify the environmental factors that influence recent migration. *Genetics* **178**, 1491–504.

Faulks, L.K., Gilligan, D.M., & Beheregaray, L.B. (2010) Islands of water in a sea of dry land: hydrological regime predicts genetic diversity and dispersal in a widespread fish from Australia's arid zone, the golden perch (*Macquaria ambigua*). *Molecular Ecology* **19**, 4723–37.

Faulks, L.K., Gilligan, D.M., & Beheregaray, L.B. (2011) The role of anthropogenic vs. natural in-stream structures in determining connectivity and genetic diversity in an endangered freshwater fish, Macquarie perch (*Macquaria australasica*). *Evolutionary Applications* **4**, 589–601.

Faurby, S. & Barber, P.H. (2012) Theoretical limits to the correlation between pelagic larval duration and population genetic structure. *Molecular Ecology* **21**, 3419–32.

Fausch, K.D., et al. (2002) Landscapes to riverscapes: bridging the gap between research and conservation of stream fishes. *BioScience* **52**, 483.

Finnegan, L., et al. (2013) Fine-scale analysis reveals cryptic patterns of genetic structure in Canada geese. *The Condor* **115**, 738–49.

Foll, M. & Gaggiotti, O. (2008) A genome-scan method to identify selected loci appropriate for both dominant and codominant markers: a Bayesian perspective. *Genetics*, **180**, 977–93.

Fortuna, M.A., et al. (2009) Networks of spatial genetic variation across species. *Proceedings of the National Academy of Sciences of the United States of America*. **106**, 19044–9.

Foster, N.L., et al. (2012) Connectivity of Caribbean coral populations: complementary insights from empirical and modelled gene flow. *Molecular Ecology* **21**, 1143–57.

Franken, R.J.M., Storey, R.G., & Williams, D.D. (2001) Biological, chemical and physical characteristics of downwelling and upwelling zones in the hyporheic zone of a north-temperate stream. *Hydrobiologia* **444**, 183–95.

Freeman, M.C., et al. (2012) Linking river management to species conservation using dynamic landscape-scale models. *River Research and Applications* **7**, 906–18.

Funk, W.C., et al. (2005) Population structure of Columbia spotted frogs (*Rana luteiventris*) is strongly affected by the landscape. *Molecular Ecology* **14**, 483–96.

Gaggiotti, O.E., et al. (2009) Disentangling the effects of evolutionary, demographic, and environmental factors influencing genetic structure of natural populations: Atlantic herring as a case study. *Evolution; International Journal of Organic Evolution* **63**, 2939–51.

Galindo, H.M., Olson, D.B., & Palumbi, S.R. (2006) Seascape genetics: a coupled oceanographic-genetic model predicts population structure of Caribbean corals. *Current Biology* **16**, 1622–6.

Galindo, H.M., et al. (2010) Seascape genetics along a steep cline: using genetic patterns to test predictions of marine larval dispersal. *Molecular Ecology* **19**, 3692–707.

Gianoulis, T.A., et al. (2009) Quantifying environmental adaptation of metabolic pathways in metagenomics. *Proceedings of the National Academy of Sciences* **106**, 1374–79.

Gilg, M.R. & Hilbish, T.J. (2003) The geography of marine larval dispersal: coupling genetics with fine-scale physical oceanography. *Ecology (Washington DC)* **84**, 2989–98.

Goetze, E. (2011) Population differentiation in the open sea: insights from the pelagic copepod *Pleuromamma xiphias*. *Integrative and Comparative Biology* **51**, 580–97.

Graham, W.M. & Largier, J.L. (1997) Upwelling shadows as nearshore retention sites: the example of northern Monterey Bay. *Continental Shelf Research* **17**, 509–32.

Grant, E.H.C., Lowe, W.H., & Fagan, W.F. (2007) Living in the branches: population dynamics and ecological processes in dendritic networks. *Ecology Letters* **10**, 165–75.

Grosberg, R. & Cunningham, C,W. (2001) Genetic structure in the sea: from populations to communities. In: Bertness, M.D., Gaines, S., & Hay, M. (eds.), *Marine Community Ecology*. Sinauer Associates, Sunderland, MA.

Guillot, G., et al. (2009) Statistical methods in spatial genetics. *Molecular Ecology* **18**, 4734–56.

Guy, T.J., Gresswell, R.E., & Banks, M.A. (2008) Landscape-scale evaluation of genetic structure among barrier-isolated populations of coastal cutthroat trout, *Oncorhynchus clarkii clarkii*. *Canadian Journal of Fisheries and Aquatic Sciences* **65**, 1749–62.

Halpern, B.S., et al. (2008) A global map of human impact on marine ecosystems. *Science* **319**, 948–52.

Hänfling, B. & Weetman, D. (2006) Concordant genetic estimators of migration reveal anthropogenically enhanced source-sink population structure in the river sculpin, *Cottus gobio*. *Genetics* **173**, 1487–501.

Harley, C.D.G., et al. (2006) The impacts of climate change in coastal marine systems. *Ecology Letters* **9**, 228–41.

Hart, M.W. & Marko, P.B. (2010) It's about time: divergence, demography, and the evolution of developmental modes in marine invertebrates. *Integrative and Comparative Biology* **50**, 643–61.

Hauser, L. & Carvalho, G.R. (2008) Paradigm shifts in marine fisheries genetics: ugly hypotheses slain by beautiful facts. *Fish and Fisheries* **9**, 333–62.

Hedgecock, D. & Pudovkin, A.I. (2011) Sweepstakes reproductive success in highly fecund marine fish and shellfish: a review and commentary. *Bulletin of Marine Science* **87**, 971–1002.

Hedgecock, D., Choti, V., & Waple, R.S. (1992) Effective population numbers of shellfish broodstocks estimated from temporal variance in allelic frequencies. *Aquaculture* **108**, 215–32.

Hellberg, M.E., Balch, D.P., & Roy, K. (2001) Climate-driven range expansion and morphological evolution in a marine gastropod. *Science (New York, NY)* **292**, 1707–10.

Hemmer-Hansen, J., et al. (2007) Evolutionary mechanisms shaping the genetic population structure of marine fishes; lessons from the European flounder (*Platichthys flesus* L.). *Molecular Ecology*, **16**, 3104–18.

Hendry, A.P. & Day, T. (2005) Population structure attributable to reproductive time: isolation by time and adaptation by time. *Molecular Ecology*, **14**, 901–16.

Hendry, A.P. & Stearns, S. (eds.) (2004) *Evolution Illuminated: Salmon and Their Relatives*, Oxford University Press, Oxford.

Hendry, A.P., Farrugia, T.J., & Kinnison, M.T. (2008) Human influences on rates of phenotypic change in wild animal populations. *Molecular Ecology* **17**, 20–9.

Hey, J. (2010) Isolation with migration models for more than two populations. *Molecular Biology and Evolution* **27**, 905–20.

Hice, L.A., et al. (2012) Spatial scale and divergent patterns of variation in adapted traits in the ocean. *Ecology Letters* **15**, 568–75.

Hickerson, M.J. & Cunningham, C.W. (2005) Contrasting quaternary histories in an ecologically divergent sister pair of low-dispersing intertidal fish (*Xiphister*) revealed by multilocus DNA analysis. *Evolution; International Journal of Organic Evolution* **59**, 344–60.

Hilborn, R., et al. (2003) Biocomplexity and fisheries sustainability. *Proceedings of the National Academy of Sciences of the United States of America* **100**, 6564–8.

Hitt, N.P. & Angermeier, P.L. (2008) Evidence for fish dispersal from spatial analysis of stream network topology. *Journal of the North American Benthological Society* **27**, 304–20.

Hoegh-Guldberg, O., et al. (2007) Coral reefs under rapid climate change and ocean acidification. *Science* **318**, 1737–42.

Homola, J.J., et al. (2012) Genetically derived estimates of contemporary natural straying rates and historical gene flow among Lake Michigan lake sturgeon populations. *Transactions of the American Fisheries Society* **141**, 1374–88.

Horne, J., et al. (2012) Searching for common threads in threadfins: phylogeography of Australian polynemids in space and time. *Marine Ecology Progress Series* **449**, 263–76.

Host, G.E., et al. (2011) High-resolution assessment and visualization of environmental stressors in the Lake Superior basin. *Aquatic Ecosystem Health and Management* **14**, 376–85.

Hubbs, C.L. & Lagler, K.F. (1964) *Fishes of the Great Lakes Region*. University of Michigan Press, Ann Arbor, MI.

Huelsenbeck, J.P. & Andolfatto, P. (2007) Inference of population structure under a Dirichlet process model. *Genetics* **175**, 1787–802.

Huey, J.A., et al. (2011) High gene flow and metapopulation dynamics detected for three species in a dryland river system. *Freshwater Biology* **56**, 2378–90.

Hughes, J.M., Schmidt, D.J., & Finn, D.S. (2009) Genes in streams: using DNA to understand the movement of freshwater fauna and their riverine habitat. *BioScience* **59**, 573–83.

Humphrey, S., et al. (2012) The effects of water currents on walleye (*Sander vitreus*) eggs and larvae and implications for the early survival of walleye in Lake Erie. *Canadian Journal of Fisheries and Aquatic Sciences* **69**, 1959–67.

Jakobsson, M. & Rosenberg, N.A. (2007) CLUMPP: a cluster matching and permutation program for dealing with label switching and multimodality in analysis of population structure. *Bioinformatics (Oxford, England)* **23**, 1801–6.

Jombart, T., Pontier, D., & Dufour, A.-B. (2009) Genetic markers in the playground of multivariate analysis. *Heredity*, **102**, 330–41.

Joost, S., Kalbermatten, M., & Bonin, A. (2008) Spatial analysis method (SAM): a software tool combining molecular and environmental data to identify candidate loci for selection. *Molecular Ecology Resources* **8**, 957–60.

Jørgensen, H.B.H., et al. (2005) Marine landscapes and population genetic structure of herring (*Clupea harengus* L.) in the Baltic Sea. *Molecular Ecology* **14**, 3219–34.

Kanno, Y., Vokoun, J.C., & Letcher, B.H. (2011) Fine-scale population structure and riverscape genetics of brook trout (*Salvelinus fontinalis*) distributed continuously along headwater channel networks. *Molecular Ecology* **20**, 3711–29.

Kennett, R.M. & Georges, A. (1990) Habitat utilization and its relationship to growth and reproduction of the Eastern long-necked turtle, *Chelodina longicollis* (Testudinata: Chelidae), from Australia. *Herpetologica* **46**, 22–33.

Kershaw, F. & Rosenbaum, H.C. (2014) Ten years lost at sea: response to Manel and Holderegger. *Trends in Ecology and Evolution* **29**, 69–70.

Kininmonth, S., Van Oppen, M.J.H., & Possingham, H.P. (2010) Determining the community structure of the coral *Seriatopora hystrix* from hydrodynamic and genetic networks. *Ecological Modelling* **221**, 2870–2880.

Knutsen, H., et al. (2012) Population genetic structure in a deepwater fish *Coryphaenoides rupestris*: patterns and processes. *Marine Ecology Progress Series* **460**, 233–46.

Koizumi, I., Yamamoto, S., & Maekawa, K. (2006) Decomposed pairwise regression analysis of genetic and geographic distances reveals a metapopulation structure of stream-dwelling Dolly Varden charr. *Molecular Ecology* **15**, 3175–89.

Kool, J.T., et al. (2010) Complex migration and the development of genetic structure in subdivided populations: an example from Caribbean coral reef ecosystems. *Ecography* **33**, 597–606.

Kuparinen, A. & Merilä, J. (2007) Detecting and managing fisheries-induced evolution. *Trends in Ecology and Evolution* **22**, 652–9.

Lafontaine, P. & Dodson, J.J. (1997) Intraspecific genetic structure of white sucker (*Catostomus commersoni*) in northeastern North America as revealed by mitochondrial DNA polymorphism. *Canadian Journal of Fisheries and Aquatic Sciences* **54**, 555–65.

Lande, R. (1988) Genetics and biological demography in conservation. *Science* **241**, 1455–60.

Landguth, E.L., et al. (2010) Quantifying the lag time to detect barriers in landscape genetics. *Molecular Ecology* **19**, 4179–91.

Landguth, E.L., Muhlfeld, C.C., & Luikart, G. (2011) CDFISH: an individual-based, spatially-explicit, landscape genetics simulator for aquatic species in complex riverscapes. *Conservation Genetics Resources* **4**, 133–6.

Leblois, R., Estoup, A., & Rousset, F. (2009) IBDSim: a computer program to simulate genotypic data under isolation by distance. *Molecular Ecology Resources* **9**, 107–9.

Leibold, M.A., et al. (2004) The metacommunity concept: a framework for multi-scale community ecology. *Ecology Letters* **7**, 601–13.

Leibold, M.A., Economo, E.P., & Peres-Neto, P. (2010) Metacommunity phylogenetics: separating the roles of environmental filters and historical biogeography. *Ecology Letters* **13**, 1290–9.

Leis, J.M., Siebeck, U., & Dixson, D.L. (2011) How Nemo finds home: the neuroecology of dispersal and of population connectivity in larvae of marine fishes. *Integrative and Comparative Biology*, **51**, 826–43.

Lichstein, J.W. (2006) Multiple regression on distance matrices: a multivariate spatial analysis tool. *Plant Ecology* **188**, 117–31.

Liggins, L., Treml, E.A., & Riginos, C. (2013) Taking the plunge: an introduction to undertaking seascape genetic studies and using biophysical models. *Geography Compass* **7**, 173–96.

Limborg, M.T., et al. (2012) Imprints from genetic drift and mutation imply relative divergence times across marine transition zones in a pan-European small pelagic fish (*Sprattus sprattus*). *Heredity* **109**, 96–107.

Lowe, W.H. & Allendorf, F.W. (2010) What can genetics tell us about population connectivity? *Molecular Ecology* **19**, 3038–51.

Loxterman, J.L. & Keeley, E.R. (2012) Watershed boundaries and geographic isolation: patterns of diversification in cutthroat trout from western North America. *BMC Evolutionary Biology*, **12**, 38.

Luque, S., Saura, S., & Fortin, M.-J. (2012) Landscape connectivity analysis for conservation: insights from combining new methods with ecological and genetic data. *Landscape Ecology*, **27**, 153–7.

Maggs, C.A., et al. (2008) Evaluating signatures of glacial refugia for North Atlantic benthic marine taxa. *Ecology* **89**, S108–22.

Manel, S., et al. (2003) Landscape genetics: combining landscape ecology and population genetics. *Trends in Ecology & Evolution* **18**, 189–97.

Mann, K. & Lazier, J. (2006) *Dynamics of Marine Ecosystems: Biological–Physical Interactions in the Oceans*, 3rd edition. Wiley-Blackwell, MALDEN, MA.

Marko, P.B. (2004) "What's larvae got to do with it?" Disparate patterns of post-glacial population structure in two benthic marine gastropods with identical dispersal potential. *Molecular Ecology* **13**, 597–611.

Marko, P.B. & Hart, M.W. (2011) The complex analytical landscape of gene flow inference. *Trends in Ecology and Evolution* **26**, 448–56.

Marshall, D.J., et al. (2010) Phenotype-environment mismatches reduce connectivity in the sea. *Ecology Letters* **13**, 128–40.

McDole, T., et al. (2012) Assessing coral reefs on a Pacific-wide scale using the microbialization score. *PloS One* **7**, e43233.

McGovern, T.M., et al. (2010) Divergence genetics analysis reveals historical population genetic processes leading to contrasting phylogeographic patterns in co-distributed species. *Molecular Ecology* **19**, 5043–60.

Miller, T. (2007) Contribution of individual-based coupled physical–biological models to understanding recruitment in marine fish populations. *Marine Ecology Progress Series* **347**, 127–38.

Mitarai, S., Siegel, D., & Winters, K.B. (2008) A numerical study of stochastic larval settlement in the California current system. *Journal of Marine Systems* **69**, 295–309.

Mitarai, S., et al. (2009) Quantifying connectivity in the coastal ocean with application to the Southern California Bight. *Journal of Geophysical Research* **114**, C10026.

Moalic, Y., et al. (2011) Travelling in time with networks: revealing present day hybridization versus ancestral polymorphism between two species of brown algae, *Fucus vesiculosus* and *F. spiralis*. *BMC Evolutionary Biology* **11**, 33.

Moalic, Y., et al. (2012) Biogeography revisited with network theory: retracing the history of hydrothermal vent communities. *Systematic Biology* **61**, 127–37.

Mock, K.E., et al. (2010) Genetic structuring in the freshwater mussel *Anodonta* corresponds with major hydrologic basins in the western United States. *Molecular Ecology* **19**, 569–91.

Mokhtar-Jamaï, K., et al. (2011) From global to local genetic structuring in the red gorgonian *Paramuricea clavata*: the interplay between oceanographic conditions and limited larval dispersal. *Molecular Ecology*, **20**, 3291–305.

Morgan, S. & Fisher, J. (2010) Larval behavior regulates nearshore retention and offshore migration in an upwelling shadow and along the open coast. *Marine Ecology Progress Series* **404**, 109–26.

Moore, J.A., Tallmon, D.A., Nielsen, J., & Pyare, S. (2011) Effects of the landscape on boreal toad gene flow: does the pattern-process relationship hold true across distinct landscapes at the northern range margin? *Molecular Ecology* **20**, 4858–69.

Morreale, S.J., Gibbons, J.W., & Congdon, J.D. (1984) Significance of activity and movement in the yellow-bellied slider (*Pseudemys scripta*). *Canadian Journal of Zoology* **62**, 1038–42.

Morrissey, M.B. & de Kerckhove, D.T. (2009) The maintenance of genetic variation due to asymmetric gene flow in dendritic metapopulations. *The American Naturalist* **174**, 875–89.

Mouquet, N., et al. (2012) Ecophylogenetics: advances and perspectives. *Biological Reviews of the Cambridge Philosophical Society* **87**, 769–85.

Mullen, L.B., et al. (2010) Scale-dependent genetic structure of the Idaho giant salamander (*Dicamptodon aterrimus*) in stream networks. *Molecular Ecology*, **19**, 898–909.

Murphy, M.A., Evans, J.S., & Storfer, A. (2010) Quantifying *Bufo boreas* connectivity in Yellowstone National Park with landscape genetics. *Ecology*, **91**, 252–61.

Narum, S.R., et al. (2010) Adaptation of redband trout in desert and montane environments. *Molecular Ecology* **19**, 4622–37.

Narum, S.R., et al. (2013) Thermal adaptation and acclimation of ectotherms from differing aquatic climates. *Molecular Ecology* **22**, 3090–7.

Neuenschwander, S. (2006) AQUASPLATCHE: a program to simulate genetic diversity in populations living in linear habitats. *Molecular Ecology Notes* **6**, 583–5.

Nielsen, E.E., et al. (2009) Genomic signatures of local directional selection in a high gene flow marine organism; the Atlantic cod (*Gadus morhua*). *BMC Evolutionary Biology* **9**, 276.

Ozerov, M.Y., Veselov, A.E., Lumme, J., & Primmer, C.R. (2012) "Riverscape" genetics: river characteristics influence the genetic structure and diversity of anadromous and freshwater Atlantic salmon (*Salmo salar*) populations in northwest Russia. *Canadian Journal of Fisheries and Aquatic Sciences* **69**, 1947–58.

Palumbi, S.R., & Pinsky, M. (2014) Marine dispersal, ecology, and conservation. In: Bertness, M., et al. (eds.), *Marine Community Ecology and Conservation*. Sinauer Associates, Inc., Sunderland, MA.

Parmesan, C. (2006) Ecological and evolutionary responses to recent climate change. *Annual Review of Ecology, Evolution, and Systematics* **37**, 637–69.

Pechenik, J. (1999) On the advantages and disadvantages of larval stages in benthic marine invertebrate life cycles. *Marine Ecology Progress Series* **177**, 269–97.

Peres-Neto, P.R., Leibold, M.A., & Dray, S. (2012) Assessing the effects of spatial contingency and environmental filtering on metacommunity phylogenetics. *Ecology* **93**, S14–S30.

Perkin, J.S. & Gido, K.B. (2012) Fragmentation alters stream fish community structure in dendritic ecological networks. *Ecological Applications: A Publication of the Ecological Society of America* **22**, 2176–87.

Pineda, J., Hare, J., & Sponaugle, S. (2007) Larval transport and dispersal in the Coastal Ocean and consequences for population connectivity. *Oceanography* **20**, 22–39.

Pineda, J., et al. (2010) Causes of decoupling between larval supply and settlement and consequences for understanding recruitment and population connectivity. *Journal of Experimental Marine Biology and Ecology* **392**, 9–21.

Pinsky, M.L., Montes, H.R., & Palumbi, S.R. (2010) Using isolation by distance and effective density to estimate dispersal scales in anemonefish. *Evolution; International Journal of Organic Evolution* **64**, 2688–700.

Poff, N.L. (2009) Managing for variability to sustain freshwater ecosystems. *Journal of Water Resources Planning and Management* **135**, 1–4.

Poissant, J., Knight, T.W., & Ferguson, M.M. (2005) Nonequilibrium conditions following landscape rearrangement: the relative contribution of past and current hydrological landscapes on the genetic structure of a stream-dwelling fish. *Molecular Ecology* **14**, 1321–31.

Postel, S. & Carpenter, S. (1997) Freshwater ecosystem services. In: Daily, G. (ed.), *Nature's Services: Societal Dependence on Natural Ecosystems*. Island Press, Washington, DC.

Pringle, C. (2003) What is hydrologic connectivity and why is it ecologically important? *Hydrological Processes* **17**, 2685–9.

Pritchard, J.K., Stephens, M., & Donnelly, P. (2000) Inference of population structure using multilocus genotype data. *Genetics* **155**, 945–59.

Puebla, O., Bermingham, E., & Guichard, F. (2009) Estimating dispersal from genetic isolation by distance in a coral reef fish (*Hypoplectrus puella*). *Ecology* **90**, 3087–98.

Puebla, O., Bermingham, E., & McMillan, W.O. (2012) On the spatial scale of dispersal in coral reef fishes. *Molecular Ecology* **21**, 5675–88.

Purcell, J.F.H., et al. (2006) Weak genetic structure indicates strong dispersal limits: a tale of two coral reef fish. *Proceedings. Biological Sciences/The Royal Society* **273**, 1483–90.

Puritz, J.B. & Toonen, R.J. (2011) Coastal pollution limits pelagic larval dispersal. *Nature Communications* **2**, 226.

Reusch, T.B.H. & Wood, T.E. (2007) Molecular ecology of global change. *Molecular Ecology* **16**, 3973–92.

Ricciardi, A. & Rasmussen, J.B. (1999) Extinction rates of North American freshwater fauna. *Conservation Biology* **13**, 1220–2.

Riginos, C. & Liggins, L. (2013) Seascape genetics: populations, individuals, and genes marooned and adrift. *Geography Compass* **7**, 197–216.

Rivera, M.A.J., et al. (2011) Genetic analyses and simulations of larval dispersal reveal distinct populations and directional connectivity across the range of the Hawaiian grouper (*Epinephelus quernus*). *Journal of Marine Biology* **2011**, 1–11.

Robinson, B.W. & Schluter, D. (2000) Natural selection and the evolution of adaptive genetic variation in northern freshwater fishes. In: Mousseau, T., Sinervo, B., & Endler, J. (eds.), *Adaptive Genetic Variation in the Wild*. Oxford University Press, Oxford.

Rodriguez-Iturbe, I., et al. (2009) River networks as ecological corridors: A complex systems perspective for integrating hydrologic, geomorphologic, and ecologic dynamics. *Water Resources Research* **45**, 1–22.

Roe, J.H., Brinton, A.C., & Georges, A. (2009) Temporal and spatial variation in landscape connectivity for a freshwater turtle in a temporally dynamic wetland system. *Ecological Applications: A Publication of the Ecological Society of America* **19**, 1288–99.

Ronce, O. (2007) How does it feel to be like a rolling stone? Ten questions about dispersal evolution. *Annual Review of Ecology, Evolution, and Systematics* **38**, 231–53.

Rousset, F. (2004) *Genetic Structure and Selection in Subdivided Populations*. Princeton University Press, Princeton, NL.

Roy, D., Hurlbut, T.R., & Ruzzante, D.E. (2012) Biocomplexity in a demersal exploited fish, white hake (*Urophycis tenuis*): depth-related structure and inadequacy of current management approaches. *Canadian Journal of Fisheries and Aquatic Sciences* **69**, 415–29.

Ryman, N. & Palm, S. (2006) POWSIM: a computer program for assessing statistical power when testing for genetic differentiation. *Molecular Ecology* **6**, 600–2.

Sagarin, R.D., Gaines, S.D., & Gaylord, B. (2006) Moving beyond assumptions to understand abundance distributions across the ranges of species. *Trends in Ecology and Evolution* **21**, 524–30.

Saura, S. & Rubio, L. (2010) A common currency for the different ways in which patches and links can contribute to habitat availability and connectivity in the landscape. *Ecography* **33**, 523–37.

Scheltema, R.S. (1971) Larval dispersal as a means of genetic exchange between geographically separated populations of shallow-water benthic marine gastropods. *Biological Bulletin* **140**, 284–322.

Schindler, D.E., et al. (2010) Population diversity and the portfolio effect in an exploited species. *Nature* **465**, 609–12.

Schlosser, I.J. & Angermeier, P.L. (1995) Spatial variation in demographic processes of lotic fishes: conceptual models, empirical evidence, and implications for conservation. *American Fisheries Society Symposium* **17**, 392–401.

Schluter, D. (1996) Ecological causes of adaptive radiation. *The American Naturalist* **148**, S40–S64.

Schmitt, R.J. & Holbrook, S.J. (2007) The scale and cause of spatial heterogeneity in strength of temporal density dependence. *Ecology* **88**, 1241–9.

Schunter, C., et al. (2011) Matching genetics with oceanography: directional gene flow in a Mediterranean fish species. *Molecular Ecology* **20**, 5167–81.

Scribner, K.T., et al. (1986) Genetic divergence among Ppopulations of the yellow-bellied slider turtle (*Pseudemys scripta*) separated by aquatic and terrestrial habitats. *Copeia* **1986**, 691–700.

Scribner, K.T., et al. (1995) Factors contributing to temporal and age-specific genetic variation in the freshwater turtle *Trachemys scripta*. *Copeia* **1995**, 970–7.

Scribner, K.T., et al. (2000) Environmental correlates of toad abundance and inter-population variation in measures of genetic diversity: estimates derived from single locus minisatellites. *Biological Conservation* **98**, 201–10.

Scribner, K.T., Page, K.S., & Bartron, M.L. (2001) Hybridization in freshwater fishes: a review of case studies and cytonuclear methods of biological inference. *Reviews in Fish Biology and Fisheries* **10**, 293–323.

Selkoe, K.A. & Toonen, R.J. (2011) Marine connectivity: a new look at pelagic larval duration and genetic metrics of dispersal. *Marine Ecology Progress Series* **436**, 291–305.

Selkoe, K.A., Henzler, C.M., & Gaines, S.D. (2008) Seascape genetics and the spatial ecology of marine populations. *Fish and Fisheries* **9**, 363–77.

Selkoe, K.A., et al. (2010) Taking the chaos out of genetic patchiness: seascape genetics reveals ecological and oceanographic drivers of genetic patterns in three temperate reef species. *Molecular Ecology* **19**, 3708–26.

Selkoe, K.A., Gaggiotti, O., ToBo Laboratory, Bowen, B.W., & Toonen, R.J. (2014) Emergent patterns of population genetic structure for a coral reef community. *Molecular Ecology* **23**, 3064–79.

Semlitsch, R.D. (1998) Biological delineation of terrestrial buffer zones for pond-breeding salamanders. *Conservation Biology* **12**, 1113–19.

Semlitsch, R.D. & Bodie, J.R. (2003) Biological criteria for buffer zones around wetlands and Riparian habitats for amphibians and reptiles. *Conservation Biology* **17**, 1219–28.

Shen, C., et al. (2010) Estimating longitudinal dispersion in rivers using acoustic Doppler current profilers. *Advances in Water Resources* **33**, 615–23.

Short, A. & Caterino, M. (2009) On the validity of habitat as a predictor of genetic structure in aquatic systems: a comparative study using California water beetles. *Molecular Ecology* **18**, 403–14.

Shulman, M.J. & Bermingham, E. (1995) Early-life histories, ocean currents, and the population-genetics of Caribbean reef fishes. *Evolution* **49**, 897–910.

Siegel, D., et al. (2003) Lagrangian descriptions of marine larval dispersal. *Marine Ecology Progress Series* **260**, 83–96.

Slatkin, M. (1993) Isolation by distance in equilibrium and non-equilibrium populations. *Evolution* **47**, 264–79.

Slatkin, M. & Maddison, W.P. (1990) Detecting isolation by distance using phylogenies of genes. *Genetics* **126**, 249–60.

Soranno, P.A., et al. (2010) Using landscape limnology to classify freshwater ecosystems for multi-ecosystem management and conservation. *BioScience* **60**, 440–54.

Soria, G., et al. (2012) Linking bio-oceanography and population genetics to assess larval connectivity. *Marine Ecology Progress Series* **463**, 159–75.

Spear, S.F. & Storfer, A. (2008) Landscape genetic structure of coastal tailed frogs (*Ascaphus truei*) in protected vs. managed forests. *Molecular Ecology* **17**, 4642–56.

Stanford, J.A. (2006) Landscapes and riverscapes. In: Hauer, F.R. & Lamberti, G.A. (eds.), *Methods in Stream Ecology*, 2nd edition. Academic Press/Elsevier, San Diego, CA, USA.

Stanford, J.A., Lorang, M.S., & Hauer, F.R. (2005) The shifting habitat mosaic of river ecosystems (Plenary Lecture). *Proceedings of the International Society of Limnology* **29**, 123–36.

Stefan, H.G., Fang, X., & Eaton, J.G. (2001) Simulated fish habitat changes in North American Lakes in response to projected climate warming. *Transactions of the American Fisheries Society* **130**, 459–77.

Stockwell, C.A., Hendry, A.P., & Kinnison, M.T. (2003) Contemporary evolution meets conservation biology. *Trends in Ecology and Evolution* **18**, 94–101.

Storfer, A., et al. (2007) Putting the "landscape" in landscape genetics. *Heredity* **98**, 128–42.

Storfer, A., et al. (2010) Landscape genetics: where are we now? *Molecular Ecology*, **19**, 3496–514.

Strathmann, R.R. (1985) Feeding and nonfeeding larval development and life-history evolution in marine invertebrates. *Annual Review of Ecology and Systematics* **16**, 339–61.

Swearer, S.E., et al. (2002) Evidence of self-recruitment in demersal marine populations. *Bulletin of Marine Science* **70**, 251–71.

Sweeney, B.W., et al. (2004) Riparian deforestation, stream narrowing, and loss of stream ecosystem services. *Proceedings of the National Academy of Sciences of the United States of America* **101**, 14132–7.

Tamkee, P., Parkinson, E., & Taylor, E.B. (2010) The influence of Wisconsinan glaciation and contemporary stream hydrology on microsatellite DNA variation in rainbow trout (*Oncorhynchus mykiss*). *Canadian Journal of Fisheries and Aquatic Sciences* **67**, 919–935.

Taylor, E. & Bentzen, P. (1993) Evidence for multiple origins and sympatric divergence of trophic ecotypes of smelt (*Osmerus*) in northeastern North America. *Evolution* **47**, 813–32.

Taylor, B.L., et al. (2000) Evaluating dispersal estimates using mtDNA data: comparing analytical and simulation approaches. *Conservation Biology* **14**, 1287–97.

Teel, D.J., et al. (2011) Introduction to a Special Section: Genetic Adaptation of Natural Salmonid Populations. *Transactions of the American Fisheries Society*, **140**, 659–664.

Theobald, D.M., Norman, J.B., Peterson, E.E., & Ferraz, S.B. (2005) *FLoWS v1: Functional Linkage of Watersheds and Streams Tools for ArcGIS v9*. Natural Resource Ecology Lab, Colorado State University.

Treml, E.A., et al. (2007) Modeling population connectivity by ocean currents, a graph-theoretic approach for marine conservation. *Landscape Ecology* **23**, 19–36.

Treml, E.A., et al. (2012) Reproductive output and duration of the pelagic larval stage determine seascape-wide connectivity of marine populations. *Integrative and Comparative Biology* **52**, 525–37.

Turgeon, J., Estoup, A., & Bernatchez, L. (1999) Species flock in the North American Great Lakes: molecular ecology of Lake Nipigon Ciscoes (Teleostei: Coregonidae: Coregonus). *Evolution* **53**, 1857–71.

Underhill, J. (1986) The fish fauna of the Laurentian Great Lakes, the St. Lawrence Lowland, Newfoundland and Labrador. In: Hocutt, C.H. & Wiley, E. (eds.), *The Zoogeography of North American Freshwater Fishes*. John Wiley & Sons, Inc., New York.

Valdal, E.J. & Quinn, M.S. (2011) Spatial analysis of forestry related disturbance on westslope cutthroat trout (*Oncorhynchus clarkii lewisi*): implications for policy and management. *Applied Spatial Analysis and Policy* **4**, 95–111.

Vörösmarty, C.J., et al. (2010) Global threats to human water security and river biodiversity. *Nature* **467**, 555–61.

Wagner, H.H. & Fortin, M.-J. (2013) A conceptual framework for the spatial analysis of landscape genetic data. *Conservation Genetics* **14**, 253–61.

Wang, L.Z., Seelbach, P., & Hughes, R.M. (2006) Introduction to landscape influences on stream habitats and biological assemblages. *American Fisheries Society Symposium* **48**, 1–23.

Wang, L.Z., et al. (2008) Landscape based identification of human disturbance gradients and reference conditions for Michigan streams. *Environmental Monitoring and Assessment* **14**, 1–17.

Waples, R. (1990) Conservation genetics of Pacific salmon. II. Effective population size and the rate of loss of genetic variability. *Journal of Heredity* **81**, 267–76.

Waples, R.S. (1998) Separating the wheat from the chaff: patterns of genetic differentiation in high gene flow species. *The Journal of Heredity* **89**, 438–50.

Ward, R.D. (2006) The importance of identifying spatial population structure in restocking and stock enhancement programmes. *Fisheries Research* **80**, 9–18.

Ward, J.V. & Stanford, J.A. (1995) The serial discontinuity concept: extending the model to floodplain rivers. *Regulated Rivers Research and Management* **10**, 159–68.

Watson, J.R., et al. (2012) Changing seascapes, stochastic connectivity, and marine metapopulation dynamics. *The American Naturalist* **180**, 99–112.

Weersing, K. & Toonen, R.J. (2009) Population genetics, larval dispersal, and connectivity in marine systems. *Marine Ecology Progress Series* **393**, 1–12.

Werner, F.E., Cowen, R.K., & Paris, C.B. (2007) Coupled biological and physical models: present capabilities and necessary developments for future studies of population connectivity. *Oceanography* **20**, 54–69.

White, T.A., Stamford, J., & Rus Hoelzel, A. (2010) Local selection and population structure in a deep-sea fish, the roundnose grenadier (*Coryphaenoides rupestris*). *Molecular Ecology* **19**, 216–26.

Whitlock, M.C. & McCauley, D.E. (1999) Indirect measures of gene flow and migration: FST not equal to 1/(4Nm + 1). *Heredity* **82**, 117–25.

Wiens, J.A. (1999) Landscape ecology – scaling from mechanisms to management. In: Farina, A. (ed.), *Perspectives in Ecology*. The Netherlands: Backhuys Publishers, Leiden.

Wiley, M.J., et al. (2010) A multi-modeling approach to evaluating climate and land use change impacts in a Great Lakes River Basin. *Hydrobiologia* **657**, 243–62.

Williamson, C.E., et al. (2008) Lakes and streams as sentinels of environmental change in terrestrial and atmospheric processes. *Frontiers in Ecology and the Environment* **6**, 247–54.

Wilson, M.A. & Carpenter, S. (1999) Economic valuation of freshwater ecosystem services in the United States: 1971–1997. *Ecological Applications* **9**, 772–83.

Winemiller, K.O. & Rose, K.A. (1992) Patterns of life-history diversification in North American fishes: implications for population regulation. *Canadian Journal of Fisheries and Aquatic Sciences* **49**, 2196–218.

Woodson, C.B. & McManus, M.A. (2007) Foraging behavior can influence dispersal of marine organisms. *Limnology and Oceanography* **52**, 2701–9.

Zacherl, D., Gaines, S.D., & Lonhart, S.I. (2003) The limits to biogeographical distributions: insights from the northward range extension of the marine snail, *Kelletia kelletii* (Forbes, 1852). *Journal of Biogeography* **30**, 913–24.

Zedler, J.B. & Kercher, S. (2005) Wetland resources: status, trends, ecosystem services, and restorability. *Annual Review of Environment and Resources* **30**, 39–74.

Zhao, Y., et al. (2009) A biophysical model of Lake Erie walleye (*Sander vitreus*) explains interannual variations in recruitment. *Canadian Journal of Fisheries and Aquatic Sciences* **66**, 114–25.

# Chapter 14

# CURRENT STATUS, FUTURE OPPORTUNITIES, AND REMAINING CHALLENGES IN LANDSCAPE GENETICS

*Niko Balkenhol,[1] Samuel A. Cushman,[2] Lisette P. Waits,[3] and Andrew Storfer[4]*

[1]*Department of Wildlife Sciences, University of Göttingen, Germany*
[2]*Forest and Woodlands Ecosystems Program, Rocky Mountain Research Station, United States Forest Service, USA*
[3]*Fish and Wildlife Sciences, University of Idaho, USA*
[4]*School of Biological Sciences, Washington State University, USA*

## 14.1 INTRODUCTION

Just over a decade ago, advances in geographic information systems and genetic methodology helped usher in the field of landscape genetics, an amalgamation of landscape ecology, population genetics, and spatial statistics aimed at testing how landscape heterogeneity shapes patterns of spatial genetic variation. It is clear that the tools developed in landscape genetics have shaped a new paradigm for conducting empirical population genetics studies; although population genetics theory itself remains largely unchanged, the majority of empirical studies of spatial genetic structure now include spatially explicit tests of landscape influences on gene flow. As a result, landscape genetics has

essentially replaced the isolation-by-distance paradigm with spatially explicit tests of effects of landscape variables on genetic population structure. More generally, landscape genetics has advanced the field of evolutionary ecology by providing a direct focus on relationships between landscape patterns and population processes, such as gene flow, selection, and genetic drift.

Since the inception of landscape genetics, there have been many calls for: (1) better communication among spatial statisticians, landscape ecologists, and population geneticists (Storfer et al. 2007; Balkenhol et al. 2009a); (2) more rigorous hypothesis testing (Storfer et al. 2007, 2010; Cushman & Landguth 2010; Segelbacher et al. 2010; Manel & Holderegger 2013); (3) increased consideration of proper sampling design (Storfer et al. 2007;

Holderegger & Wagner 2008; Anderson et al. 2010; Spear et al. 2010; Landguth et al. 2012; see Chapter 4); (4) developing predictive models for conservation and management, particularly in the face of climate change (Landguth & Cushman 2010; Segelbacher et al. 2010; Manel et al. 2010; Bollinger et al. 2014); (5) selecting appropriate analytical methods and understanding their underlying assumptions (e.g., Balkenhol et al. 2009b; Wagner & Fortin 2013; Chapters 3 to 5): and (6) understanding the interactions of scale and landscape definition with the heterogeneity of the environment in affecting the strength and detectability of landscape effects on genetic variation (e.g., Cushman & Landguth 2010; Cushman et al. 2013a; Chapter 2).

In this book, it was our goal to summarize the current state of knowledge with respect to these topics and foreshadow a number of emerging opportunities and challenges the field will face in the coming decades. Landscape genetics is a field in its infancy and is characterized by rapid growth, multiple lines of parallel and sometimes contradictory work, and an apparent cultural and conceptual divide between practitioners coming from genetic versus landscape ecological backgrounds. Given this, the field appears to be producing a confusing thicket of ideas. We are confident that over time the competition among ideas and approaches will lead to increasing clarity and, we hope, a true synthesis of population and evolutionary genetic theory with spatial ecology. For now, we will offer our own view of some of the current and emerging challenges and opportunities, which we hope may focus and facilitate future progress in the field. Based on the previous book chapters and other published literature, we believe that at least ten conclusions can be drawn about the current state-of-the-art in landscape genetics.

## 14.2   CONCLUSION 1: ISSUES OF SCALE NEED TO BE CONSIDERED

Most landscape genetic studies have not carefully considered the effects of spatial scale and the definition of the landscape, even though these aspects can substantially affect our ability to detect relationships between population genetic structure and landscape features. As explained in Chapter 2, landscape definitions that differ in thematic content, thematic resolution, and spatial scale may dramatically alter the statistical relationships between genetic variation and landscape structure (e.g., Blair & Melnick

2012; Keller et al. 2013). While there have been a few empirical and simulation studies on the effects of scale and landscape definition in the field (e.g. Cushman & Landguth 2010; Galpern et al. 2012), the vast majority of studies published in landscape genetics have not addressed these issues at all. Indeed, many past studies have naively sought correlations between the genetic structure of a population and a putative barrier feature, such as a road or river, without considering that other landscape features might also be important. Similarly, too few studies quantitatively assess the influences of alternative scales or landscape definitions. While some researchers have begun to address these challenges by different optimization procedures (e.g. Shirk et al. 2012; Galpern & Manseau 2013; Castillo et al. 2014), this topic has not been thoroughly explored. Given the fundamental importance of scale optimization and correct landscape definition in quantifying any pattern–process relationship, we strongly feel that much more attention should be given to investigating the relationships between landscape definition, spatial scale, and the accurate quantification of landscape–genetic relationships.

## 14.3   CONCLUSION 2: SAMPLING NEEDS TO SPECIFICALLY TARGET LANDSCAPE GENETIC QUESTIONS

One of the largest limitations of most landscape genetics research conducted to date is that sampling is often done without *a priori* consideration of expected landscape effects on genetic variation and underlying processes. However, sampling for landscape genetics can only be effective if it is based on these expectations, which should be stated as testable hypotheses (see below and Chapter 4). To derive such hypotheses, we need to develop a theory that includes the multifaceted influences of landscape heterogeneity on genetic variation (see conclusion 10). Nevertheless, even if hypotheses are clearly formulated at the beginning of a study, the complexity of landscape genetic research will always be a challenge for optimal study design and sampling. Thus, we strongly advocate simulations as a means to test different sampling options before beginning a study, and to conduct a power analysis to determine whether certain landscape–genetic relationships can actually be detected within a given study (Cushman 2014).

## 14.4  CONCLUSION 3: CHOICE OF APPROPRIATE STATISTICAL METHODS REMAINS CHALLENGING

We have not yet reached a true consensus on which analytical methods to use for the three analytical steps of landscape genetics described in Chapter 1. Reflective of the rapid growth of a new field, there are a number of alternative methods currently in use to analyze landscape genetic data in node-, neighborhood-, and link-based frameworks (see Wagner & Fortin 2013; Chapter 5). Few of these methods have been rigorously compared, and there is no general agreement as to which methods are best for a specific question. This is an area of utmost importance to advance the field and should receive very high priority for future research. Given the complexity of the task, we will probably never have a single method that fits all research questions and data, but the conceptual framework suggested by Wagner and Fortin (2013; see Chapter 5) helps to guide future efforts for developing appropriate methods. To address some of the challenges associated with existing methods used in landscape genetics, increasingly complex statistical approaches are suggested. Unfortunately, some of the more complex analytical approaches seem to be viewed as all-embracing and impeccable, and are simply used "as is" in many landscape genetic studies. A good example is the use of genetic assignment and clustering methods, which are often simply used to infer a single "best" number of populations contained in the data. As shown in Chapter 7, this task is far from trivial, making the appropriate use of assignment-based methods much more challenging than is often realized. At the same time, assignment and clustering methods can do much more than infer the most likely number of populations (e.g., Murphy et al. 2008; Balkenhol et al. 2014), and their potential for landscape genetic analyses remains high. Overall, the overwhelming variety of different analytical options will continue to be one of the most challenging aspects in landscape genetics in the next years, especially for beginners. However, we are actually encouraged by the great diversity of alternative methods that are being developed and evaluated, as they are proof of the tremendous interest of the scientific community in the field. This "free-market" of ideas will eventually select the best approaches, provided that competing methods are thoroughly evaluated using empirical and simulated data.

## 14.5  CONCLUSION 4: SIMULATIONS PLAY A KEY ROLE IN LANDSCAPE GENETICS

Simulation modeling will play a vital role for the future development of landscape genetics. Analysis of data is the foundation of empirical science, and is critical to advance landscape genetics. However, when analyzing empirical data a researcher never knows the true process that governs the observed response. One can only *infer* the process from the pattern of response and its association with one or multiple hypotheses. The great power of simulation is that it allows this inferential pathway to be inverted. That is, in simulation modeling the researcher stipulates and controls the process being modeled and then can generate the patterns of genetic structure that would result from that process. This provides critical control over the pattern–process relationship that is essential to reliably evaluate such things as effect of different landscape definitions, spatial scale, effectiveness of alternative sampling schemes, and power of different statistical methods (Cushman 2014). While there is clearly much more work to be done in developing realistic and efficient simulation models for landscape genetics, much progress has already been made (Chapter 6). Importantly, simulations will also be an essential component for developing and testing general theories of landscape genetics (see conclusion 10). However, to realize the full potential of landscape genetic simulations, there is a need to incorporate the various influences of landscape heterogeneity on neutral and adaptive genetic variation (see conclusion 6), so that we can directly evaluate the interaction between landscapes, processes, and resulting patterns in genetic variation (e.g. Balkenhol & Landguth 2011; Cushman 2014, 2015).

## 14.6  CONCLUSION 5: MEASURES OF GENETIC VARIATION ARE RARELY DEVELOPED SPECIFICALLY FOR LANDSCAPE GENETICS

To accurately and reliably quantify landscape–genetic relationships, we also need to develop measures of genetic variation that explicitly account for the spatial complexity of reality. For the quantification of landscape heterogeneity (analytical step 2), multiple novel approaches have been suggested in recent years, and several of them are specifically intended for landscape genetics (e.g.,

McRae 2006, Van Strien et al. 2012). In contrast, very few measures of genetic diversity and structure (analytical step 1) have been designed specifically for landscape genetics (Chapter 3). Some of the existing measures are not ideal for the field, because populations that inhabit heterogeneous landscapes rarely conform to classical population genetic assumptions such as random mating and Hardy–Weinberg equilibrium, and more often are characterized by metapopulation or gradient structure (Chapter 2). Nevertheless, a few notable exceptions exist. For example, Shirk and Cushman (2011) developed a spatially explicit estimate of genetic diversity (*sGD*) based on grouping individuals into potentially overlapping genetic neighborhoods that match the population structure, whether discrete or clinal. Recently, Shirk and Cushman (2014) further explored spatial effects on genetic diversity by applying Wright's neighborhood concept (Wright 1943) to isolation-by-resistance to produce spatially explicit estimates of local population size from genetic data in continuous populations. Similarly, to estimate genetic structure within a network of potentially connected metapopulations, the ***conditional genetic distance*** (*cGD*) was developed by Dyer and Nason (2004) (see also Dyer et al. 2010; Chapter 10). This distance reflects gene flow among subpopulations in a much more realistic way than traditional measures of genetic differentiation, because it does not assume a simple stepping stone model of migration; nor does it assume gene flow to be homogeneous across the landscape. Thus, both *sGD* and *cDG* were designed specifically to avoid some of the problematic assumptions of classical measures of genetic variation, and improve our ability to accurately quantify genetic patterns in heterogeneous environments. These recent efforts illustrate that it is both possible and important to develop measures of genetic variation that more explicitly match the data and research questions typical of landscape genetics. We particularly encourage geneticists to help us advance towards measures that better reflect the spatial complexity of natural populations and the various processes that influence them.

## 14.7    CONCLUSION 6: LANDSCAPE RESISTANCE IS JUST ONE OF THE POSSIBLE LANDSCAPE–GENETIC RELATIONSHIPS

The majority of published landscape genetics studies has focused on quantifying the effects of landscape features *in between* sampling locations on gene flow and resulting genetic structure, for example, by incorporating barriers or estimates of landscape resistance (Chapters 2 and 8). However, landscape characteristics *at* or *around* sampling locations also affect the processes that shape genetic variation. These should find more consideration in future studies, as suggested by Wagner and Fortin (2013; Chapter 5), Murphy et al. (2010; Chapter 9), and Pflüger and Balkenhol (2014). For example, under isolation-by-environment (IBE; Wang & Bradburd 2014), genetic differentiation can be predicted based on the dissimilarity of local environmental variables. Such isolation-by-environment was shown to explain more variation in gene flow than isolation-by-resistance (IBR) in 17 species of Caribbean *Anolis* lizards (Wang et al. 2013). In order to evaluate the relative importance of IBE versus IBR more generally, we need to incorporate both local environmental variables and matrix resistance in our studies. For example, Cushman et al. (2013b), in a study comparing IBE and IBR models of genetic differentiation of a riparian tree found that both riverine network connectivity and climatic gradients contribute to genetic differentiation. Overall, landscape genetics should not limit itself to evaluating the effects of landscape resistance on genetic structure, but embrace the full complexity of interactions between spatial environmental heterogeneity, ecological and evolutionary processes, and the resulting genetic patterns.

## 14.8    CONCLUSION 7: GENOMICS PROVIDES NOVEL OPPORTUNITIES, BUT ALSO CREATES NEW CHALLENGES

Landscape genomics is one way to fully grasp the various influences that landscape can have on neutral and adaptive genetic variation. Genomic approaches are increasingly used in landscape genetics and offer revolutionary abilities to understand selection and adaptation in complex landscapes (Chapter 9). Next-generation sequencing has allowed collection of vast amounts of sequence data, even in non-model organisms. Development of numerous analytical methods has followed, but several issues still plague landscape genomics studies with complicating factors that make detecting true signatures of selection difficult. A major challenge in landscape genomics has been differentiating signals of selection as detected by changes in allele frequencies from demographic history. For example, nearby populations tend to have correlated allele frequencies, as well as shared environmental variables, which can result in high false-positive rates of correlations between allele frequencies and environmental variables (Coop et al. 2010; Joost et al. 2013). Moreover, recent range expansions can confound drift caused by serial founder effects (also known as "allele surfing"; Excoffier & Ray 2008) with selection, resulting in false positives. In general, the absence of demographic data can result in overestimation of rates and strengths of selection because not accounting for population structure can bias diversity and allele frequency estimates among populations (Joost et al. 2013; Lotterhos & Whitlock 2014). Some researchers have developed methods for joint estimation of demographic parameters and selection (Eyre-Walker & Keightley 2009). However, due to their stepwise nature, the resulting models tend to over-fit the data (Joost et al. 2013).

Overall, these new ways for deriving genetic data also provide new challenges, as summarized in Chapter 9. A major challenge, then, is to analyze data sets that integrate both the genomic landscape and the ecological landscape in order to understand the spatial distribution of adaptive genetic variation. Landscape genomics has an exceptionally high potential for increasing our understanding of the feedbacks between ecological and evolutionary dynamics, but only if we realize that correlation tests to detect selection are only a first step, and that local adaptation is not only influenced by local environmental pressures but also by the complex interactions between local selection, gene flow, and drift – and all of these aspects can be affected by landscape heterogeneity and change (e.g., Lowe & McPeek 2014; Fordham et al. 2014; Richardson et al. 2014). Thus, linking complex genotypes to phenotypes and local adaptation is one of evolutionary ecology's grand challenges in the coming decade(s) (Cushman 2014).

## 14.9 CONCLUSION 8: THE SCOPE OF LANDSCAPE GENETICS NEEDS TO EXPAND

In addition to considering neutral and adaptive processes, landscape genetics generally must broaden its scope, for example, with respect to study organisms. A large number of studies have used landscape genetics for various organisms in aquatic and terrestrial systems (Chapters 11 to 13; Storfer et al. 2010). However, the taxonomic range of landscape genetic research should be substantially expanded. Studying taxonomic groups, such as insects, which are small in size, highly mobile, and have short generation times, will facilitate controlled and replicated experiments, which are necessary to confirm putative relationships seen in empirical field studies or suggested in simulation modeling (e.g. Cushman 2014, 2015). Such experimental landscape genetics will help to clarify pattern–process relationships governing genetic patterns, especially if they are conducted in multiple study areas (e.g. Short Bull et al. 2011). In addition, it is essential for more studies to combine landscape genetic approaches with other research methods, so that the actual processes underlying observed genetic patterns can be identified and their relative importance evaluated (e.g. Cushman and Lewis 2010; Reding et al. 2013; Shafer et al. 2012; Weckworth et al. 2013). Another way that landscape genetics needs to broaden is in the geographic focus of studies and training of scientists. Currently landscape genetic studies are strongly biased toward temperate climates and developed countries (Storfer et al. 2010).

## 14.10 CONCLUSION 9: SPECIFIC HYPOTHESES ARE RARELY STATED IN CURRENT LANDSCAPE GENETIC STUDIES

In our opinion, one of the most important steps for increasing scientific rigor and broadening the scope of landscape genetics is the formulation of clear, testable hypotheses that describe how genetic variation and underlying processes are expected to be influenced by the landscape. Currently, a disturbingly large portion of the published landscape genetics literature does not define clear hypotheses and considers only a few of the possible landscape-genetic relationships (see conclusion 6). If we were to derive such hypotheses before beginning a study, we would probably often realize that factors other than landscape resistance can affect our genetic data, and it would help us to identify how other types of data and research approaches could help to answer specific research questions and disentangle the influence of multiple underlying processes. Indeed, we argue that we need to move away from single-species, single-landscape studies that focus on finding statistical significant correlations between spatial patterns in genetic and landscape data (statistical, pattern-focused;

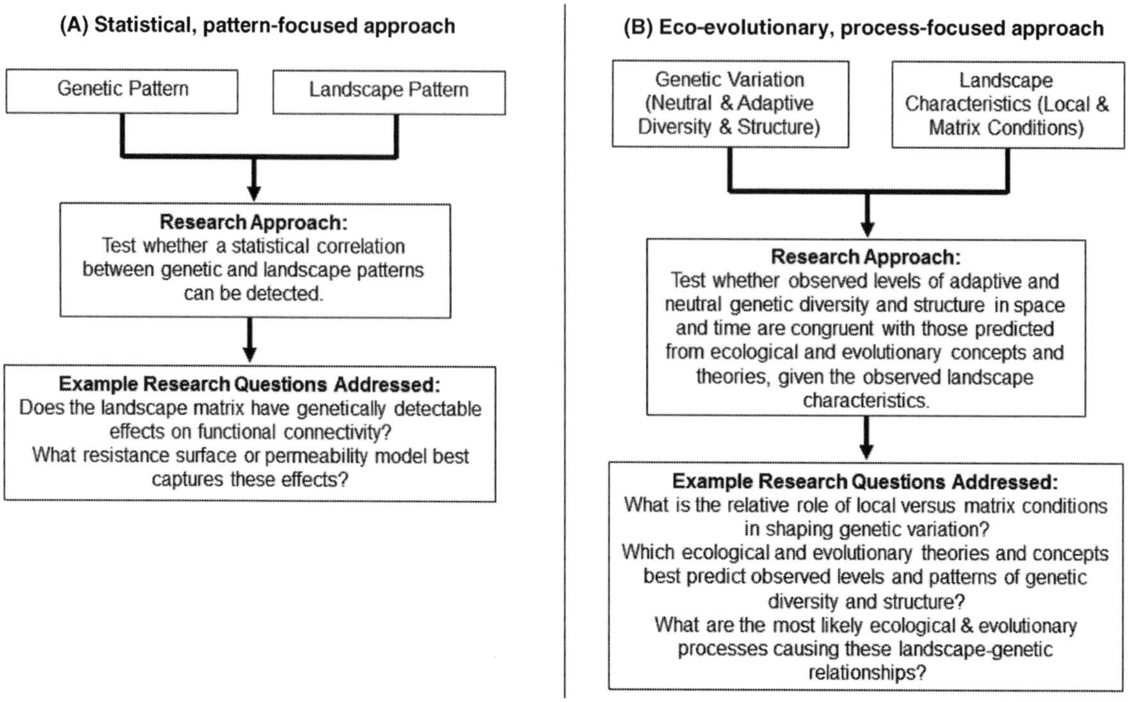

**(A) Statistical, pattern-focused approach**

| Genetic Pattern | Landscape Pattern |

**Research Approach:**
Test whether a statistical correlation between genetic and landscape patterns can be detected.

**Example Research Questions Addressed:**
Does the landscape matrix have genetically detectable effects on functional connectivity?
What resistance surface or permeability model best captures these effects?

**(B) Eco-evolutionary, process-focused approach**

| Genetic Variation (Neutral & Adaptive Diversity & Structure) | Landscape Characteristics (Local & Matrix Conditions) |

**Research Approach:**
Test whether observed levels of adaptive and neutral genetic diversity and structure in space and time are congruent with those predicted from ecological and evolutionary concepts and theories, given the observed landscape characteristics.

**Example Research Questions Addressed:**
What is the relative role of local versus matrix conditions in shaping genetic variation?
Which ecological and evolutionary theories and concepts best predict observed levels and patterns of genetic diversity and structure?
What are the most likely ecological & evolutionary processes causing these landscape-genetic relationships?

**Fig. 14.1** Schematic comparison of (A) current landscape genetic approaches that are often pattern-focused and correlation-based with (B) a suggested approach that focuses on the eco-evolutionary processes shaping landscape–genetic relationships.

Fig. 14.1A) and towards studies that focus on finding meaningful relationships between landscape characteristics and the processes that shape genetic variation across space and time (eco-evolutionary, process-focused; Fig. 14.1B).

## 14.11  CONCLUSION 10: A COMPREHENSIVE THEORY FOR LANDSCAPE GENETICS IS CURRENTLY MISSING

One aspect related to all of the above conclusions is the current lack of a comprehensive theory that links landscape heterogeneity in space and time to patterns in neutral and adaptive genetic variation, by considering the many different processes that affect gene flow, drift, and selection. One major goal of landscape genetics should be to develop a more mechanistic understanding of how genetic variation interacts with species distribution, population size and density, dispersal ability, differential resistance, selection gradients, and

individual fitness in heterogeneous environments. Trying to move towards a theoretical framework that accomplishes this would substantially improve our ability to derive testable hypotheses, to sample and analyze landscape and genetic data in a targeted way, and to combine it with other data and research approaches for more meaningful inferences. We urge researchers in landscape genetics to focus more strongly on developing conceptual and theoretical foundations for landscape genetics, and suggest that synthesizing results provided by existing studies is one way to achieve a more general understanding of landscape influences on genetics.

## 14.12  THE FUTURE OF LANDSCAPE GENETICS

As can be seen above, much remains to be done in landscape genetics. We are convinced that, despite the current challenges, the field has a bright and dynamic future and that landscape genetic approaches will continue to provide exciting new insights in various scientific disciplines.

This is a time of explosive growth in landscape genetics, with a panoply of different ideas, methods, and scopes of analysis. It is likely that this complexity is going to increase even more, for example, because whole genome sequencing and bioinformatics are driving a transformative paradigm shift that presents tremendous opportunities and challenges to our field. Similarly, increasingly sophisticated approaches for analyzing data related to landscape genetics, such as animal movement and population demography, are constantly being developed (e.g., Kranstauber et al. 2014; Merow et al. 2014). We believe that for our field to take full advantage of these opportunities, we must effectively combine the ongoing work in landscape genetics, which has focused on describing population structure and associating it with landscape features, with genomics, bioinformatics, and other research approaches in experimental and field settings (Cushman 2014). To accomplish this, landscape geneticists need to think even more carefully in advance about hypotheses, suitable data, study design, and data analysis so that results can be generalized and their implications can be explored across scales of biological organization, from nucleotides to ecosystems.

We are very interested to see what directions landscape genetics will take in the future. On the one hand, we foresee that the different ways landscape genetics can be used (see Chapter 1) will lead to more differentiated and more disparate developments in "landscape genetics for ecology" versus "landscape genetics for genetics". Ecologists will focus even more on evaluating the processes that can be inferred from analyzing landscape genetic data and will hopefully develop new ways for combining landscape genetic data with other types of research methods, including telemetry, stable isotopes, mark-recapture, or demographic population modeling. In contrast, geneticists will put more emphasis on developing novel ways for analyzing genomic data in a landscape context, on testing existing evolutionary theory, and on deriving new theory for adaptive processes in complex landscapes. Importantly, we also see much potential for merging ecological and evolutionary principles within a landscape genetic framework. Neutral and adaptive processes do not occur isolated from one another and need to be considered in combination to analyze eco-evolutionary dynamics (Cushman 2014; Landguth et al. 2011), which is a prerequisite for truly understanding the emergence and maintenance of biological diversity in spatially and temporally varying environments (Hand et al. 2015).

Currently, it seems that our technical abilities to gather genetic data in the field, to analyze it in the lab, and to obtain landscape data over very large spatial extents (e.g., via remote sensing) have surpassed our abilities to draw meaningful conclusions from these data. Landscape genetics is probably not the only scientific field where this is currently evident.

Given the complexities and current challenges in landscape genetics, it is not surprising that many researchers diving into the field are somewhat overwhelmed by the variety of research approaches, and feel that they urgently need help with analytical aspects of landscape genetics. This book tries to provide some help to get started, but we also hope that the book shows that landscape genetics is not just a set of tools, and thus requires much more than the combined use of landscape and genetic data. While we absolutely realize the importance of using appropriate methods for this task (see conclusion 3), we feel that too little emphasis is currently given to the development of landscape genetic theories, the derivation of explicit hypotheses, and the synthesis of results obtained from existing landscape genetic studies. We should keep in mind that the ultimate goal of landscape genetics is not just to provide new tools for the joint analysis of genetic and landscape data but also to enhance our understanding of how landscapes shape patterns in neutral and adaptive genetic diversity and structure. Thus, we hope that in the not too distant future, we can move away from the currently dominating question in the field of *"How do we do it?"* towards the more fundamental question of *"What do we learn from it?"* This book hopefully serves as another small step towards reaching this ambitious goal.

## REFERENCES

Anderson, C.D., Epperson, B.K., Fortin, M.-J., Holderegger, R., James, P.M.A, Rosenberg, M.S., Scribner, K.T., & Spear, S. (2010) Considering spatial and temporal scale in landscape-genetic studies of gene flow. *Molecular Ecology* **19**, 3565–75.

Balkenhol, N. & Landguth, E. L. (2011) Simulation modeling in landscape genetics: on the need to go further. *Molecular Ecology* **20**, 667–70.

Balkenhol, N., Gugerli, F., Cushman, S.A., Waits, L.P., Coulon, A., Arntzen, J.W., Holderegger, R., & Wagner, H.H. (2009a)

Identifying future research needs in landscape genetics: where to from here? *Landscape Ecology* **24**, 455–63.

Balkenhol, N., Waits, L.P., & Dezzani, R.J. (2009b) Statistical approaches in landscape genetics: an evaluation of methods for linking landscape and genetic data. *Ecography* **32**, 818–30.

Balkenhol, N., Holbrook, J.D., Onorato, D., Zager, P., White, C., & Waits, L.P. (2014) A multi-method approach for analyzing hierarchical genetic structures: a case study with cougars *Puma concolor*. *Ecography* **37**, 552–63.

Blair, M.E. & Melnick, D.J. (2012) Scale-dependent effects of a heterogeneous landscape on genetic differentiation in the Central American squirrel monkey (*Saimiri oerstedii*). *PloS One* **7**, e43027.

Bolliger, J., Lander, T., & Balkenhol, N. (2014) Landscape genetics since 2003: status, challenges and future directions. *Landscape Ecology* **29**, 361–6.

Castillo, J.A., Epps, C.W., Davis, A.R., & Cushman, S.A. (2014) Landscape effects on gene flow for a climate-sensitive montane species, the American pika. *Molecular Ecology* **23**, 843–56.

Coop, G., Witonsky, D., Di Rienzo, A., & Pritchard, J.K. (2010) Using environmental correlations to identify loci underlying local adaptation. *Genetics* **185**, 1411–23.

Cushman, S.A. (2014) 'Grand challenges in evolutionary and population genetics: the importance of integrating epigenetics, genomics, modeling, and experimentation.' *Frontiers in Genetics* **5**, 197.

Cushman, S.A. (2015) Pushing the envelope in genetic analysis of species invasion. *Molecular Ecology* **24**, 259–262.

Cushman, S.A. & Landguth, E.L. (2010) Spurious correlations and inference in landscape genetics. *Molecular Ecology*, vol. **19**, no. 17, pp. 3592–602.

Cushman, S.A. & Lewis, J.S. (2010) Movement behavior explains genetic differentiation in American black bears. *Landscape Ecology* **25**, 1613–25.

Cushman, S.A., Wasserman, T.N., Landguth, E.L., & Shirk, A.J. (2013a) Re-evaluating causal modeling with Mantel tests in landscape genetics. *Diversity* **5**, 51–72.

Cushman, S.A., Max, T., Meneses, N., Evans, L.M., Ferrier, S., Honchak, B., Whitham, T.G., & Allan, G.J. (2013b) Landscape genetic connectivity in a riparian foundation tree is jointly driven by climatic gradients and river networks. *Ecological Applications* **24**, 1000–14.

Dyer, R. J., & Nason, J. D. (2004) 'Population Graphs: the graph theoretic shape of genetic structure. *Molecular Ecology* **13**, 1713–27.

Dyer, R.J., Nason, J.D., & Garrick, R.C. (2010) Landscape modeling of gene flow: improved power using conditional genetic distance derived from the topology of population networks. *Molecular Ecology* **19**, 3746–59.

Excoffier, L. & Ray, N. (2008) Surfing during population expansions promotes genetic revolutions and structuration. *Trends in Ecology and Evolution* **23**, 347–51.

Eyre-Walker, A. & Keightley, P.D. (2009) Estimating the rate of adaptive molecular evolution in the presence of slightly deleterious mutation and population size change. *Molecular Biology and Evolution* **26**, 2097–108.

Fordham, D.A., Brook, B.W., Moritz, C., & Nogués-Bravo, D. (2014) Better forecasts of range dynamics using genetic data. *Trends in Ecology and Evolution* **29**, 436–43.

Galpern, P. & Manseau, M. (2013) Finding the functional grain: comparing methods for scaling resistance surfaces. *Landscape Ecology* **28**, 1269–81.

Galpern, P., Manseau, M., & Wilson, P. (2012) Grains of connectivity: analysis at multiple spatial scales in landscape genetics. *Molecular Ecology* **21**, 3996–4009.

Hand, B.K., Lowe, W.H., Kovach, R.P., Muhlfeld, C.C., & Luikart, G. (2015) Landscape community genomics: understanding eco-evolutionary processes in complex environments. *Trends in Ecology and Evolution* **20**, 161–8.

Holderegger, R. & Wagner, H.H. (2008) Landscape genetics. *BioScience* **58**, 199–207.

Joost, S., Vuilleumier, S., Jensen, J.D., Schoville, S., Leempoel, K., Stucki, S., Widmer, I., Melodelime, C., Rolland, J., & Manel, S. (2013) Uncovering the genetic basis of adaptive change: on the intersection of landscape genomics and theoretical population genetics. *Molecular Ecology* **22**, 3659–65.

Keller, D., Holderegger, R., & Van Strien, M.J. (2013) Spatial scale affects landscape genetic analysis of a wetland grasshopper. *Molecular Ecology* **22**, 2467–82.

Kranstauber, B., Safi, K., & Bartumeus, F. (2014) Bivariate Gaussian bridges: directional factorization of diffusion in Brownian bridge models. *Movement Ecology* **2**, 5.

Landguth, E.L., & Balkenhol, N. (2012) Relative sensitivity of neutral versus adaptive genetic data for assessing population differentiation. *Conservation Genetics* **13**, 1421–6.

Landguth, E.L. & Cushman, S.A. (2010) cdpop: a spatially explicit cost distance population genetics program. *Molecular Ecology Resources* **10**, 156–61.

Landguth, E.L., Johnson, N. & Cushman, S.A. (2011) 'Simulating selection in landscape genetics.' *Molecular Ecology Resources*, **12**, 363–86.

Landguth, E.L., Fedy, B.C., Oyler-McCance, S.J., Garey, A.L., Emel, S.L., Mumma, M., Wagner, H.H., Fortin, M.-J., & Cushman, S.A. (2012) Effects of sample size, number of markers, and allelic richness on the detection of spatial genetic pattern. *Molecular Ecology Resources* **12**, 276–84.

Lotterhos K.E. & Whitlock, M.C. (2014) Genome scans for FST outliers are unreliable without accurate demographic history. *Molecular Ecology* **23**, 2178–92.

Lowe, W.H. & McPeek, M.A. (2014) Is dispersal neutral? *Trends in Ecology and Evolution* **29**, 444–50.

Manel, S. & Holderegger, R. (2013) Ten years of landscape genetics. *Trends in Ecology and Evolution* **28**, 614–21.

Manel, S., Joost, S., Epperson, B.K., Holderegger, R., Storfer, A., Rosenberg, M.S., Scribner, T.M., Bonin, A., & Fortin, M.-J. (2010) Perspectives on the use of landscape genetics to detect genetic adaptive variation in the field. *Molecular Ecology* **19**, 3760–72.

McRae, B.H. (2006) Isolation by resistance. *Evolution* **60**, 1551–61.

Merow, C., Dahlgren, J.P., Metcalf, C.J. E., Childs, D.Z., Evans, M.E.K., Jongejans, E., Record, S., Rees, M., Saluero-Gómez, R., & McMahon, S.M. (2014) Advancing population ecology with integral projection models: a practical guide. *Methods in Ecology and Evolution* **5**, 99–110.

Murphy, M.A., Evans, J.S., Cushman, S.A., & Storfer, A. (2008) Representing genetic variation as continuous surfaces: an approach for identifying spatial dependency in landscape genetic studies. *Ecography* **31**, 685–97.

Murphy, M.A., Evans, J.S., & Storfer, A. (2010) Quantifying *Bufo boreas* connectivity in Yellowstone National Park with landscape genetics. *Ecology* **91**, 252–61.

Pflüger, F. & Balkenhol, N. (2014) A plea for simultaneously considering matrix quality and local environmental conditions when analyzing landscape impacts on effective dispersal. *Molecular Ecology* **23**, 2146–56.

Reding, D.M., Cushman, S.A., Gosselink, T.E., & Clark, W.R. (2013) Linking movement behavior and fine-scale genetic structure to model landscape connectivity for bobcats (*Lynx rufus*). *Landscape Ecology* **28**, 471–86.

Richardson, J.L., Urban, M.C., Bolnick, D.I., & Skelly, D.K. (2014) Microgeographic adaptation and the spatial scale of evolution. *Trends in Ecology and Evolution* **29**, 165–76.

Segelbacher, G., Cushman, S.A., Epperson, B.K., Fortin, M.-J., Francois, O., Hardy, O. J., Holderegger, R., Taberlet, P., Waits, L.P., & Manel, S. (2010) Applications of landscape genetics in conservation biology: concepts and challenges. *Conservation Genetics* **11**, 375–85.

Shafer, A.B.A., Northrup, J.M., White, K.S., Boyce, M.S., Côté, S.D., & Coltman, D.W. (2012) Habitat selection predicts genetic relatedness in an alpine ungulate. *Ecology* **93**, 1317–29.

Shirk, A.J. & Cushman, S.A. (2011) sGD: software for estimating spatially explicit indices of genetic diversity. *Molecular Ecology Resources* **11**, 922–34.

Shirk, A.J. & Cushman, S.A. (2014) Spatially-explicit estimation of Wright's neighborhood size in continuous populations. *Frontiers in Ecology and Evolution* **2**, 62.

Shirk, A.J., Cushman, S.A., & Landguth, E.L. (2012) Simulating pattern-process relationships to validate landscape genetic models. *International Journal of Ecology* **2012**, 539109.

Short Bull, R.A., Cushman, S.A., Mace, R., Chilton, T., Kendall, K.C., Landguth, E.L., Schartz, M.K., McKelvey, K., Allendorf, F.W., & Luikart, G. (2011) Why replication is important in landscape genetics: American black bear in the Rocky Mountains. *Molecular Ecology* **20**, 1092–107.

Spear, S.F., Balkenhol, N., Fortin, M.-J., McRae, B.H., & Scribner, K.T. (2010) Use of resistance surfaces for landscape genetic studies: considerations for parameterization and analysis. *Molecular Ecology* **19**, 3576–91.

Storfer, A., Murphy, M.A., Evans, J.S., Goldberg, C.S., Robinson, S., Spear, S.F., Dezzani, R., Delmelle, E., Vierling, L., & Waits, L.P. (2007) Putting the "landscape" in landscape genetics. *Heredity* **98**, 128–42.

Storfer, A., Murphy, M.A., Spear, S.F., Holderegger, R., & Waits, L.P. (2010) Landscape genetics: Where are we now? *Molecular Ecology* **19**, 3496–514.

Van Strien, M.J., Keller, D., & Holderegger, R. (2012) A new analytical approach to landscape genetic modeling: least-cost transect analysis and linear mixed models. *Molecular Ecology* **21**, 4010–23.

Wagner, H.H. & Fortin, M.-J. (2013) A conceptual framework for the spatial analysis of landscape genetic data. *Conservation Genetics* **14**, 253–61.

Wang, I.J. & Bradburd, G.S. (2014) Isolation by environment. *Molecular Ecology* **23**, 5649–62.

Wang, I.J., Glor, R.E., & Losos, J.B. (2013) Quantifying the roles of ecology and geography in spatial genetic divergence. *Ecology Letters* **16**, 175–82.

Weckworth, B.V., Musiani, M., Decesare, N.J., Mcdevitt, A.D., Hebblewhite, M., & Mariani, S. (2013) Preferred habitat and effective population size drive landscape genetic patterns in an endangered species. *Proceedings of the Royal Society B* **280**, 20131756.

Wright, S. (1943) Isolation-by-distance. *Genetics* **28**, 114–38.

# INDEX

Note: Page references in *italics* refer to Figures; those in **bold** refer to Tables and Boxes

*Landscape Genetics: Concepts, Methods, Applications*, First Edition. Edited by Niko Balkenhol, Samuel A. Cushman,
Andrew T. Storfer, and Lisette P. Waits.
© 2016 John Wiley & Sons, Ltd. Published 2016 by John Wiley & Sons, Ltd.